Preparing for the
PROVEN INEVITABLE

An Urban Operations Training Strategy for America's Joint Force

Russell W. Glenn, Jody Jacobs, Brian Nichiporuk, Christopher Paul,
Barbara Raymond, Randall Steeb, Harry J. Thie

Prepared for the
Office of the Secretary of Defense
and U.S. Joint Forces Command

Approved for public release;
distribution unlimited

NATIONAL DEFENSE
RESEARCH INSTITUTE

The research described in this report was prepared for the Office of the Secretary of Defense (OSD) and U.S. Joint Forces Command. The research was conducted in the the RAND National Defense Research Institute, a federally funded research and development center sponsored by the OSD, the Joint Staff, the Unified Combatant Commands, the Department of the Navy, the Marine Corps, the defense agencies, and the defense Intelligence Community under Contract DASW01-01-C-0004.

Library of Congress Cataloging-in-Publication Data

Preparing for the proven inevitable : an urban operations training strategy for America's joint force / Russell W. Glenn ... [et al.].
    p. cm.
    "MG-439."
    Includes bibliographical references.
    ISBN 0-8330-3871-0 (pbk. : alk. paper)
    1. Urban warfare—United States. 2. Unified operations (Military science)—United States. 3. Soldiers—Training of—United States. 4. Military education—United States. I. Glenn, Russell W. II. Title.

U167.5.S7P84 2006
355.5'2—dc22

                                                                    2005030031

The RAND Corporation is a nonprofit research organization providing objective analysis and effective solutions that address the challenges facing the public and private sectors around the world. RAND's publications do not necessarily reflect the opinions of its research clients and sponsors.

RAND® is a registered trademark.

Cover design by Stephen Bloodsworth
Cover photo images courtesy Captain R. J. Bodisch. The photograph is of First Lieutenant Aaron C. Smithley's "Comanche-5" tank, Kilo Company, Third Battalion, First Marines during a security mission conducted under the command of Captain Timothy J. Jent. The photograph was taken on November 9, 2004, in the Jolan District of Al Fallujah.

Published 2006 by the RAND Corporation
1776 Main Street, P.O. Box 2138, Santa Monica, CA 90407-2138
1200 South Hayes Street, Arlington, VA 22202-5050
201 North Craig Street, Suite 202, Pittsburgh, PA 15213-1516
RAND URL: http://www.rand.org/
To order RAND documents or to obtain additional information, contact
Distribution Services: Telephone: (310) 451-7002;
Fax: (310) 451-6915; Email: order@rand.org

Preface

It is evident to virtually everyone that future military operations will include urban operations far more often than not. In fact, operations in densely populated, built-up areas already frequently dominate U.S. armed forces deployments. Over the past decade, Service training initiatives have reflected a renewed interest in preparing for such contingencies. However, members of Congress have expressed concern that these efforts were insufficiently coordinated. Therefore, Congress requested that a study be conducted of how the military community might better orchestrate its resources to improve readiness for force-wide urban operations. Three sponsors—the Office of the Secretary of Defense Readiness; J7 U.S. Joint Forces Command; and Joint Urban Operations Office, J9, U.S. Joint Forces Command—asked the RAND Corporation to undertake the task of developing a joint urban training strategy for the period 2005–2011 to assist in meeting this objective.

This monograph presents that strategy. It will be of interest to individuals in the government, nongovernmental organizations, private volunteer organizations, and the commercial sector whose responsibilities include the planning, policy, doctrine, training, funding, and conduct of actions undertaken in or near urban areas in both the immediate future and the longer term.

This research was conducted for the Department of Defense within the International Security and Defense Policy Center and the Forces and Resources Policy Center of the RAND National Defense Research Institute, a federally funded research and development center sponsored by the Office of the Secretary of Defense, the Joint Staff, the Unified Combatant Commands, the Department of the Navy, the Marine Corps, the defense agencies, and the defense Intelligence Community.

For more information on RAND's International Security and Defense Policy Center, contact the Director, James Dobbins. He can be reached by email at James_Dobbins@rand.org; by phone at 703-413-1100, extension 5134; or by mail at RAND, 1200 South Hayes Street, Arlington, VA 22202-5050.

Contents

CHAPTER EIGHT

APPENDIX

Figures

Tables

Summary

Overview

Urban operations have challenged and continue to challenge the world's most sophisticated militaries. Still reliant on technologies, doctrines, and training at times overly influenced by the Cold War—a period during which neither major adversary wished to fight in large metropolitan areas—operations in built-up areas have subsequently often proven unpleasantly difficult for U.S. forces. Despite the passage of more than a decade since the end of the Cold War and the momentous change in the strategic environment, the U.S. armed forces have thus far been unable to adequately reproduce the challenges their soldiers, sailors, marines, and airmen meet in the towns and cities of Afghanistan and Iraq.

That is not to imply that the Services of the U.S. military have ignored this challenge. The desperate October 1993 fighting on the streets of Mogadishu triggered U.S. Army development of a new type of urban training facility, one designed to be less like the pristine villages of northwest Europe and more akin to the chaotic environments found in densely populated areas of the developing world. The Marine Corps built "Yodaville," an innovative training site in Arizona that vividly replicates the difficulties of engaging urban targets from aircraft. Service and joint simulation initiatives have likewise focused on efforts to better represent urban scenarios.

Such training initiatives influenced and were influenced by the simultaneous development of new Service and joint urban doctrine. Yet while both Service and joint doctrine received attention, improvements in urban training were almost exclusively limited to efforts within the four Services. Requests to Congress for urban training-facility construction reflected this Service centrism. As a result, the Senate Armed Services Committee requested a review of "the desired distribution and total number of [urban training] facilities, the extent to which MOUT [military operations on urbanized terrain] facilities can be shared among the military departments and active and reserve components, and whether such facilities are required at installations, such as Lackland Air Force Base, conducting basic and advanced training in addition to operational units."[1] These issues are addressed as follows in the present analysis:

- "The desired distribution and total number of [urban training] facilities." We explicitly recommend that urban training facilities capable of supporting a platoon (facilities we define as approximately 25 structures in size) be located at each home station permanently hosting a brigade or larger maneuver element. We further recommend development of four CONUS sites sufficient to train a battalion task force or larger (approximately 300 structures) and that each of the sites include a nearby air-ground urban training capability. The closeness of home-station installations and training demand are among the factors that influence our recommendations regarding the locations of these four facilities. We suggest that the facilities be located in the Kentucky–North Carolina–Georgia region; at Ft. Polk, LA; at Ft. Hood, TX; and in the U.S. southwest. These points are addressed on pages 230–240.
- "The extent to which MOUT facilities can be shared among the military departments and active and reserve components." Our research further suggests that Service retention of urban

---

[1] "National Defense Authorization Act for Fiscal Year 2003 Report [to Accompany S. 2514] on Authorizing Appropriations for Fiscal Year 2003 for Military Activities of the Department of Defense, for Military Construction, and for Defense Activities of the Department of Energy, to Prescribe Personnel Strengths for Such Fiscal Year for the Armed Forces, and for Other Purposes Together with Additional and Minority Views," Senate Committee on Armed Services Report 107-151, May 9, 2002, p. 428.

training-site ownership is desirable given that the preponderance of such training will take place within the Services. However, that should by no means preclude joint use or inter-Service sharing of these facilities. Joint training either (1) does not require use of urban-specific facilities (e.g., upper-echelon headquarters training exercises), (2) can fulfill joint requirements via occasional use of Service capabilities (e.g., Joint National Training Center (JNTC) events conducted in 2004 and 2005), or (3) can be an organic part of training sponsored by a single Service. Joint usage, to include that by both active and Reserve components, is both feasible and desirable.[2] We further recommend that the joint community be assigned responsibility for the oversight and supervisory management of major urban training-facility scheduling; requests for funding to develop live, virtual, or constructive training capabilities; and allocation of funds provided for that development. A fuller discussion of these points appears on pages 251–256.

- "Whether such facilities are required at installations, such as Lackland Air Force Base, conducting basic and advanced training in addition to operational units." This study deliberately maintains a focus on the establishment and maintenance of *joint* urban training capabilities. However, it also heartily endorses the traditional building-block approach to training, in which individual and smaller-unit readiness provides the foundation

---

[2] Because the audiences for this study include both civilian and military at all echelons, the terms *requirement, capability*, and others such as *shortfall* are used throughout this study in accordance with their commonly understood definitions. This usage does not contradict but does at times expand word meanings beyond the specific usages noted in Joint Publication JP 1-02, *Department of Defense Dictionary of Military and Associated Terms*, April 12, 2001, as amended through May 23, 2003. For example, JP 1-02 defines a military requirement as "an established need justifying the timely allocation of resources to achieve a capability to accomplish approved military objectives, missions, or tasks." Usage here at times includes this understanding but also appears in the sense of "something wanted or needed" (*Merriam Webster's Collegiate Dictionary, Tenth Edition*, Springfield, MA: Merriam-Webster, 1997). The meaning of these and other terms should be apparent when taken in context. Our choice for use in a broader sense of meaning avoids reliance on other terms that have specific implications when employed in a military context but might lead to misunderstanding when read by a wider audience.

for developing preparedness at higher echelons. Recent increased emphasis on urban operations preparedness by joint and active and Reserve components is encouraging; initiatives focused on tactical-level preparation are notably so. Services should retain the responsibility and authority to determine the extent to which urban training is necessary at entry level and during advanced individual training. Such training would involve only the lowest echelons, e.g., squad clearing of rooms and air-ground controller instruction. Courses including such preparation would require limited urban-specific infrastructure. Underutilized portions of training bases or low-cost, purpose-built facilities should be sufficient to meet the majority of requirements. (The Dutch and British armies, for example, use very simple, partially open structures for room- and building-clearing instruction. The approximate 2005 cost of each such "building" was less than $15,000 equivalent.) Some advanced individual and other school training (e.g., that supporting WMD-related instruction) will require more-substantial capabilities. As noted in the bullet immediately above, we recommend that requests for training facilities be forwarded to the joint entity assigned responsibility for reviewing such proposals and allocating funds for their construction (or their development, in the case of virtual and constructive training). The need for specialized training (in WMD) is addressed on pages 42, 111, 221–222, 227, 268.

This study identifies areas in need of redress and proposes ways in which the Services—Army, Navy, Marine Corps, and Air Force—and other critical components of national capability can better ready themselves cooperatively for future operations in cities around the world. The result is a joint urban training (JUT) strategy for the period 2005–2011. The foundation for this strategy is the current *Doctrine for Joint Urban Operations* presented in the joint publication of that name (JP 3-06). The guidance in JP 3-06 includes the valuable understand, shape, engage, consolidate, and transition (USECT) concept for joint urban operations (JUO). These five phases of an urban operation are interdependent and overlapping. Together, they effectively articulate the nature of urban contingencies and the functions that Service and joint leaders must take into account.

We took a modular approach toward constructing the JUT strategy. A "module," as used in this context, is a collection of resources normally associated with a type of facility, simulation, or other capability used in the design or execution of training. The modules ultimately selected collectively serve as the components of the JUT strategy developed in this study, meet all JUT requirements identified in the study to the extent feasible, and provide a means of comparing costs associated with very different capabilities. Requirement attainment, rather than dollar cost, becomes the primary metric for determining the value of a module and its suitability as a component of a comprehensive JUT strategy. Further, the modules are internally flexible. They can be adapted to allow for comparison of similar but not perfectly matched capabilities.

Centering the JUT strategy on modules led to a five-step analytical approach: (1) identify JUT requirements; (2) identify current and pending JUT capabilities; (3) identify the short-term (2005–2007) and longer-term (2008–2011) gaps between JUT requirements and capabilities; (4) complete initial steps toward a JUT strategy, including defining modules and assessing how well the modules address the final set of JUT requirements; and (5) complete the final steps toward a JUT strategy, including considering the costs of the modules used in developing the strategy in terms of their ability to meet JUT requirements, and address the short- and longer-term training shortfalls identified.

Identifying JUT Requirements

Figure S.1 shows the three-step process by which we arrived at the final set of requirements used in the analysis.

The first step was a comprehensive review of Service and joint doctrine, various official and unofficial source materials, and input

Figure S.1
Process of Identifying Joint Urban Operations Training Requirements

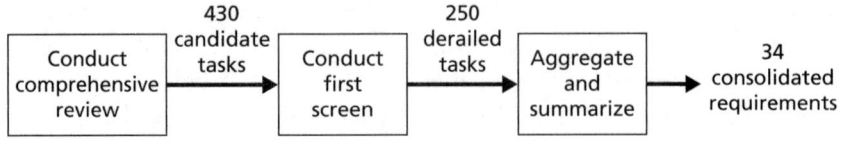

from interview subjects, combatant commands, and Service representatives. This initial review produced 430 candidate tasks. The next
step, conducting a first screen, eliminated tasks that were redundant
or neither essentially joint nor urban; this reduced the list to 250 detailed JUT tasks. The third step consisted of further synthesis and
aggregation of the 250 tasks into 34 consolidated tasks that are comprehensive (i.e., leave no pertinent tasks uncovered), of manageable
scope, and appropriate to the assessment of capabilities.

It should be noted that these requirements overlap; it is infeasible to designate them in such a manner that they do not. Such is the
complexity of military operational environments, a complexity increased multifold in cases where the environment involves a significant urban component. "Conduct stability operations in the urban
environment" and "conduct support operations in the urban environment" are inseparable from "govern in the urban environment";
many subtasks are shared. Indeed, the same is true of "conduct stability operations" and "conduct support operations" when instability
is an issue; without provision of employment, life's necessities, and
other forms of support, achievement of stability is virtually impossible. Definition of mutually exclusive requirements would be rife with
artificiality—to fail in listing any of the requirements discussed below
would risk leaving unidentified a critical element necessary in preparing the U.S. joint force for future contingencies.

Table S.1 shows the final set of 34 consolidated JUT requirements we derived and used in conjunction with JUT capabilities.

Table S.1
Consolidated Joint Urban Training Requirements

Avoid fratricide

Communicate in the urban environment

Conduct airspace coordination

Synchronize joint rules of engagement

Conduct stability operations in the urban environment

Conduct support operations in the urban environment

Conduct urban human intelligence (HUMINT) operations

Conduct urban signal intelligence (SIGINT), imagery intelligence (IMINT), measurement and signatures intelligence (MASINT), communications intelligence (COMINT), electronic intelligence (ELINT), and other intelligence efforts

Conduct urban operations exercises

Integrate urban operations with other relevant environments

Coordinate maneuvers in the urban environment

Coordinate multinational and interagency resources

Govern in the urban environment

Identify critical infrastructure nodes and system relations

Navigate in the urban environment

Plan urban operations

Provide common situational awareness

Provide fire support

Provide security during urban transition operations

Rehearse/war-game urban operations

Conduct urban noncombatant evacuation operations (NEOs)

Conduct U.S. domestic urban operations

Conduct urban combat search and rescue (CSAR)

Conduct urban operations during and after a WMD event

Consolidate success in the urban environment

Disembark, base, protect, and move in urban environments

Engage in the urban environment

Orchestrate resources during urban operations

Shape the urban environment

Sustain urban operations

Transition to civilian control

Understand the urban environment

Achieve simultaneity in meeting requirements

Conduct training across multiple levels of war

Their ordering does not imply primacy or any other form of prioritization. Every task is essential to the development of a comprehensive JUT strategy for the period 2005–2011. All address at least one of the five USECT elements; many span several, if not all, of the demands inherent in understanding, shaping, engaging, consolidating, and transitioning during urban operations. The lack of prioritization, however, does not imply that some tasks will not be more significant than others for given JUT aspects. The tasks that are most important to a given combatant commander, subordinate joint commander, operation, or mission will vary. That variation will be reflected in the appropriate commander's joint mission-essential task list or other written guidance, including his personal prioritization of requirements to prepare for particular contingencies.

Identifying JUT Capabilities

In identifying JUT capabilities, we focused on three capability groups that will play primary roles in the development of a JUT strategy: (1) purpose-built urban training sites (i.e., current and planned U.S. urban training sites and the capabilities found at such facilities), commonly called MOUT complexes; (2) the current and projected state of simulations, simulators, and training involving synthetic environments (hereafter collectively referred to as *simulations*); and (3) innovative or novel urban training sites. Such less-traditional approaches to urban training as those in this third category may offer benefits either in the generic sense or in cases of specific instructional needs. The approaches include the use of ships, factory complexes, abandoned urban areas, closed military installations, commercially available sites or those leased by public institutions, amusement parks, and other innovative complexes.

We relied on a wide range of sources in compiling the comprehensive list of facilities. Our ten-plus years of work in the urban operations field helped in expanding initial lists provided by the Office of the Secretary of Defense–Readiness. Searches of the U.S. armed forces Non-Classified Internet Protocol Network (NIPRNET), as well as the Internet, expanded the roster and enhanced the information available on individual sites. These sources frequently contained references to other pertinent materials, allowing an inductive expansion of the initial source base. Additional lists provided by representatives from headquarters within the several Services and studies conducted prior to this effort further increased the number of facilities identified.

Starting with this comprehensive list of urban training sites, we then selected a set of those deemed to have the greatest potential to support joint preparation for urban operations. More specifically, sites were selected for their uniqueness or because they possessed characteristics thought to be of value in determining what JUT resources a site should possess. For these sites, we decided to gather more information through site visits or (when site visits were not feasible) off-location interviews. To facilitate this approach, we designed a site survey instrument which we used as a tool to guide the data collection effort, either sending it in advance to sites we visited or using it as part of the interview process for off-location interviews.

The selected urban training sites (both purpose-built and novel) are listed below, along with the way the information was collected (by site visit or by off-location interview). This list is not exhaustive of major urban training sites within the United States. Rather, it includes a significant sampling of urban training capabilities as well as others representative of the functions and approaches currently being employed in the preparation of American armed forces personnel for future urban undertakings:

- Camp Pendleton, CA (site visit)
- Twentynine Palms, CA (interview)
- Yodaville Training Range, Yuma, AZ (site visit)
- Yuma Proving Grounds "little Baghdad" test range, Yuma, AZ (interview)
- Nellis AFB, NV (site visit)
- Ft. Irwin, CA (site visit)
- Muscatatuck, IN (site visit)
- Joint Readiness Training Center, Ft. Polk, LA (site visit)

- Blackwater Inc., Moyock, NC, training facility (site visit)
- Ft. Knox, KY (site visit)
- 2nd Special Naval Warfare Group training facility, Norfolk, VA (site visit)
- Marine Corps Security Force training facility, Chesapeake, VA (site visit)
- Hurlburt Field, FL (site visit)
- Playas, NM (site visit)
- Dutch Army Oostdorp and Marnehuizen urban training facilities (site visits)
- British Army Copehill Down Village training facility (site visit) and Operational Training and Advisory Group (OPTAG) installation (interview)
- Bagram AFB, Afghanistan, urban training site (site visit)
- Ft. Benning, GA, McKenna MOUT site (site visit)

All the purpose-built sites identified, in both the comprehensive list and the screened list, contain some combination of five types of sites: (1) MOUT complexes; (2) urban target ranges; (3) shoot houses; (4) aerial ranges; and (5) temporary or façade ranges.

We also assessed simulation and simulated capabilities. In particular, we assessed many of the individual simulation and modeling systems—JANUS, JCATS, IUSS, OneSAF, Full Spectrum Warrior, Full Spectrum Command, Diamond, and MANA—available to the JUT community. Each of these was examined in terms of its near- and long-term application to urban operations training. We also explored enhanced versions of these systems, along with large-scale training systems that incorporate multiple simulations and can link to live exercises.

What Are the Shortfalls Between Requirements and Capabilities?

Having enumerated JUT requirements and existing and planned JUT capabilities, we next examined the shortfalls between what is needed to prepare the U.S. armed forces for urban operations and what exists in that regard. Eliminating, or at a minimum mitigating, the effects of these shortfalls is essential if America's joint force is to properly prepare for near- and longer-term challenges.

There are seven primary reasons why an organization might fail to meet a training requirement:

1. **Lack of capability.** Current capability cannot satisfy the requirement.
2. **Inadequate throughput capacity.** While capabilities are adequate to train for a requirement, there is an insufficient quantity of those capabilities available to accommodate joint training demand.
3. **Accessibility.** While there is sufficient capability and capacity, the capability is not available within the bounds of reasonable financial cost and travel time.
4. **Inadequate linkage or synchronization of capabilities.** Capabilities exist in sufficient capacity and accessibility, but they are geographically, functionally, or technologically separated to the extent that collective training requirements cannot be met. For example, live and virtual training capabilities exist that enable a pilot to engage targets in urban areas through a simulator while ground-based fire support coordinators (FSCs) occupy the area replicated on the pilot's screen. However, there is no effective link to allow the FSC and the pilot to communicate in real time and credibly appraise the effects of their respective actions or measure the utility of the interaction itself (though Voice over Internet Protocol (VoIP) technologies show considerable promise in this regard).
5. **Legal, regulatory, and policy constraints.** Environmental issues preclude using the full potential of otherwise effective JUT capabilities.
6. **Recognition of need.** If a requirement has only recently been identified, as may well be the case during periods of intense force commitment, no one may have previously recognized the need to train for it, regardless of whether a capability to do so exists.

7. **Training prioritization**. Unit commanders may choose to spend available training time or other resources on things other than JUT requirements.

The bulk of our study's JUT strategy development focused on the first four of these causes—lack of capabilities, inadequate throughput capacity, accessibility, and inadequate linkage or synchronization of capabilities, all of which relate to resource adequacy. These collectively address "what" is needed and "how much" of those capabilities will be sufficient to meet requirements.

Whether any one facility has the capability to actually address a requirement depends on a number of urban training-site characteristics that fundamentally impact a facility's potential as a joint training venue. These include the size/scope of the facility, how much urban complexity is represented at the site, types of forces accommodated, instrumentation in support of urban training, the existence of opposing force (OPFOR) and noncombatant role players, and the range of live-fire activities allowed at the site.

It is not enough to merely have a particular resource on hand for use by U.S. joint force elements. It is also essential to have a sufficient number of the required capabilities available. Adequate availability means that all personnel and organizations requiring training can obtain that training with the frequency necessary. Therefore, the problem is not only the number of capabilities, but also resource throughput capacity: How many organizations can cycle through the capability in a given unit of time?

Factors affecting throughput for a given facility include:

- Number of days needed for a unit to complete training at a facility;
- Standard of training required;
- Quality of instruction provided (related to number of days needed, as training quality will influence the time required to achieve task proficiency at a given standard);
- Potential for simultaneous use (personnel or unit training is complementary or the training resource is designed to allow for

independent but simultaneous use, e.g., separate Situation Tactical Exercises (STX) training);

- Initial level of student expertise;
- Perishability of skill(s) being taught;
- Availability of essential training augmentation (e.g., OPFOR, joint headquarters elements);
- Time necessary to maintain, adapt, or "reset" training capability between rotations;
- Amount of downtime required for trainees (e.g., leave, attendance at courses, deployments to active theaters).

Finally, environmental, safety, and other constraints limit the bounds of what can and will be accomplished through urban live training in the 2005–2011 period; thus, we assessed capabilities in terms of these constraints.

On the basis of the issues raised above and historical study, interviews with serving officers of all Services, and recent reports from active operations, we identified the shortfalls most critical to adequately preparing the U.S. joint force for urban undertakings.

As a result of these combined analyses, we also determined that the U.S. armed forces are thus far unable to adequately reproduce the challenges their soldiers, sailors, marines, and airmen meet in the towns and cities of Afghanistan and Iraq.

Several of the reasons for this shortcoming are immediately evident when one reviews the gaps between identified JUT requirements and existing live, virtual, and constructive training capabilities. The most evident is lack of size. Training in complexes of 25, 50, or even 150 buildings is inadequate preparation for actual tactical actions in which structures number in the hundreds, if not thousands or tens of thousands. That quantity of buildings implies correspondingly greater numbers of people, vehicles, infrastructures, and other elements that imbue actual cities with a complexity that is altogether lacking in current live exercises. Simulations supporting virtual and constructive training are unfortunately similarly overly simplistic. Regardless of how many buildings they might replicate, the notional behaviors of opposing forces and noncombatants fall far short of reproducing the range of actual interactions and the scope of higher-order effects potentially precipitated by each action and decision.

Analogous oversimplification likewise inhibits the effectiveness of urban exercises that attempt to replicate the operational and strategic levels of war.[3] These exercises too greatly ease the burden on participants by focusing almost exclusively on combat operations, marginalizing the influence of agencies other than the Department of Defense (DoD), effectively ignoring noncombatants' support requirements or their attitudes toward the friendly force, and glossing over governing responsibilities. While much improvement is also necessary in both Service and joint tactical-level training, preparation at this stratum by and large employs the accepted building-block process of first schooling the components and then educating the larger units of which they are a part. The same cannot be said for readying those who participate in higher-level training events. Service and joint schools rarely address governing responsibilities, interfacing with indigenous populations, or urban concerns in general.

In urban operations, it is no longer enough to "train as you fight." Winning battles is but one element of success, and often not the dominant one. Joint urban training must prepare the U.S. armed forces for the entirety of conflict's spectrum, the complete hierarchy of tactical to strategic; and it must integrate these many parts into a single whole, for that is what awaits its trainees overseas and, potentially, at home.

---

[3] Although the levels of war are not formally delineated by echelon, a rough guide for determining what type of organization would tend to receive a given type of urban training is as follows (the overlap is deliberate):

- Strategic: joint staff, specified and unified commands, Service staffs.
- Operational level: combatant command, component, and large unit (e.g., corps, army, joint task force).
- Tactical: component organizations of corps and smaller size, smaller joint and Special Operations Forces (SOF) elements.

Deriving JUO Training Modules

Given the identified shortfalls, we derived training modules that simultaneously include identified existing capabilities (including those pending in future years out to 2011) and those needed to close the shortfalls. We developed an original list of candidate modules and then assessed them in terms of their capability to close shortfalls. On the basis of that assessment, we eliminated some of the modules that did not adequately apply to the development of a JUO training strategy to produce the final set of modules to be used in constructing the strategy. All those removed have pertinence to Service or very limited joint applications, but their loss does not reopen any shortfalls closed in the original development of the modules.

We developed a modular approach because modules provide essential flexibility and adaptability. Instead of each individual training site or simulation being a module in and of itself, a training module consists of *categories* of facilities or simulations. This limits the number of modules to a manageable size. Defining modules in terms of categories also permits adaptation over time. Periodically editing module definitions will account for evolutions in field conditions, which means that a strategy that relies on a set of modules will not be rendered invalid as time progresses. Users can also adapt modules to account for change as capabilities change—as the joint community develops new training technologies, software, methods, or doctrine. Any financial impact of module modification can likewise be incorporated into an updated version. Thus, a training strategy that incorporates a given module can be adjusted, and the new costs associated with the strategy can be readily determined.

Table S.2 presents our first cut at training modules. The 39 modules listed are divided into five broad categories: (1) purpose-built facilities; (2) use of populated urban areas; (3) alternative/other training concepts; (4) simulation capabilities; and (5) training support elements. We then screened the 39 modules with respect to additional considerations. Only those that passed through all gradations

Table S.2
Initial List of 39 Modules

No.	Module
Purpose-Built Facilities	
1	Battalion and larger purpose-built facility
2	Company purpose-built facility
3	Platoon purpose-built facility
4	Modular purpose-built facility
5	Façade-based facility
6	Commercially manufactured portable training facility
7	Hybrid facility
8	Air-ground facility
9	Shoot house
Use of Populated Urban Areas	
10	Terrain walks
11	Urban navigation
12	Urban simulated engagement
13	Urban live fire in populated areas
14	Use of vacant buildings in populated areas
15	Use of buildings scheduled for demolition
16	Use of public facilities during hours of closure
Alternative/Other Training Concepts	
17	Use of abandoned domestic urban areas
18	BRAC'd installations[a]
19	Ships as permanent urban training facilities
20	Mothballed ships temporarily used for urban training
21	Abandoned factories and surrounding urban infrastructure
22	Abandoned/constructed overseas urban areas
23	Use of existing other-agency and commercially available urban training facilities
24	Classroom instruction
25	Conduct of combatant command or joint task force (JTF) headquarters, large-scale schools, or multi-echelon/interagency exercises
Simulation Capabilities	
26	Tactical behaviors in and around structures
27	Higher-echelon planning and coordination
28	Joint, multinational, and interagency operations
29	Specialized-technology simulation
30	Scenario-variant generation
31	Physiological and other stress simulation
32	Geographically distributed joint simulation
33	Environmental degradation and urban biorhythm

Table S.2 (continued)

No.	Module
	Training Support Elements
34	Infrastructure trappings
35	OPFOR
36	Noncombatant role players
37	Targets to support urban training
38	Instrumentation/connectivity
39	Joint force headquarter(s)

[a]Installations subject to Base Realignment and Closure (BRAC).

of this sieving process merit possible inclusion in the ultimate training strategy design. The sieves, or categories of filters, through which the initial set of modules were viewed are:

- Does the module meet a sufficient number of JUT requirements? If so, does it provide the force with a sufficient level of proficiency?
- Are there environmental, ergonomic, or other considerations that make use of the module impractical?
- Is the module cost-effective in terms of dollars and time spent in its application to training?

In short, does a module provide sufficient joint training effectiveness to merit continued consideration as a component of a U.S. joint urban training strategy? The cost-effectiveness sieve was applied to the modules that made it through the first two sieves (cost-effectiveness is discussed in the following section).

In assessing how well a module filled each of our previously identified 34 JUT requirements, we assigned the module one of four ratings (the definitions of these ratings are our own):

- C. Permits achievement of a "crawl" standard of readiness, defined as attainment of foundation skills necessary as precursors to developing more-advanced skills or combinations of skills. A module supporting a "crawl" measure of ability would have to

support development of base-level skills translatable to application under actual operational conditions in the field.

- W. Permits achievement of a "walk" standard of readiness, defined as achievement of greater sophistication in task accomplishment and the ability to coordinate several "crawl"-level or other "walk"-level skills in servicing mission accomplishment. A module supporting attainment of a "walk" measure would require managing several skills under realistic field conditions sequentially or simultaneously as demanded by the situation.

- R. Permits achievement of a "run" standard of readiness, defined as accomplishment of complete operational preparedness (combat readiness, for missions involving combat action). A "run" status implies proficiency in all supporting tasks and the orchestration of those tasks to accomplish assigned missions. A module supporting attainment of a "run" status would have to provide sufficient challenge to replicate the most adverse operational conditions.

- S. "Supports" meeting a training requirement. A support module cannot fulfill the needs of the requirement under consideration by itself, but the use of such a module adds realism, provides additional challenges, or otherwise enhances another module in the attainment of a C, W, or R rating in servicing a requirement.

Table S.3 synthesizes the results. Leaving S entries unchanged, we assigned numerical values of 1, 2, and 3, to C, W, and R modules, respectively. Given that there are 34 requirements, the maximum score a module could achieve would be 102 (i.e., 34 times 3). As an example, the Module 1 score of 84 out of 102 is the result of 16 "run" evaluations (3s), 18 "walks" (2s), and 0 "crawls" (1s). If the module does not meet the "crawl," "walk," or "run" criteria for a particular requirement, it can either support other modules (receive an S rating) or not meet a requirement (be assigned a score of 0). The numerical effect is a score of 0 in either case.

Table S.3
Final List of Modules Retained

No.	Module	Score
	Purpose-Built Facilities	
1	Battalion and larger purpose-built facility	84
2	Company purpose-built facility	55
3	Platoon purpose-built facility	44
4	~~Modular purpose-built facility~~	~~32~~
5	~~Facade-based facility~~	~~30~~
6	~~Commercially manufactured portable training facility~~	~~31~~
7	Hybrid facility	81
8	Air-ground facility	31
9	~~Shoot house~~	~~16~~
	Use of Populated Urban Areas	
10	Terrain walks	39
~~11~~	~~Urban navigation~~	~~26~~
~~12~~	~~Urban simulated engagement~~	~~29~~
~~13~~	~~Urban live fire in populated area~~	~~18~~
~~14~~	~~Use of vacant buildings in populated area~~	~~32~~
15	Use of buildings scheduled for demolition	41
16	Use of public facilities during hours of closure	52
	Alternative/Other Training Concepts	
17	Use of abandoned domestic urban areas	90
18	BRAC'd installations	91
~~19~~	~~Ships as permanent urban training facilities~~	~~34~~
~~20~~	~~Mothballed ships temporarily used for urban training~~	~~33~~
21	Abandoned factories and surrounding urban infrastructure	40
22	Abandoned/constructed overseas urban areas	84
~~23~~	~~Use of existing other-agency and commercially available urban training facilities~~	~~34~~
24	Classroom instruction	45
25	Conduct of combatant command or JTF headquarters, large-scale schools, or multi-echelon/interagency exercises	73
	Simulation Capabilities	
26	Tactical behaviors in and around structures	38
27	Higher-echelon planning and coordination	43
28	Joint, multinational, and interagency operations	41
29	Specialized-technology simulation	18
30	Scenario-variant generation	1
~~31~~	~~Physiological and other stress simulation~~	~~1~~
32	Geographically distributed joint simulation	4
33	Environmental degradation and urban biorhythm	1

Table S.3 (continued)

No.	Module	Score
	Training Support Elements	
34	Infrastructure trappings	2
35	OPFOR	10
36	Noncombatant role players	22
37	Targets to support urban training	1
38	Instrumentation/connectivity	3
39	Joint-force headquarter(s)	27

The crossed-out entries represent the modules that were deleted from the initial list because of their low scores during this ranking process. Several modules with very low numerical values were retained, however, because they contain many S ratings and therefore have value in conjunction with other modules that they support.

It is notable that many of these modules have application to operational challenges beyond those of urban missions. This is a sometimes less-than-obvious benefit of analyses involving urban environments: Much of the training and other preparation for urban contingencies applies to portions of the conflict spectrum well beyond operations in villages, towns, and cities. The greater densities and increased complexities found in urban areas mean that more often than not, a force prepared for action in built-up areas can readily adapt to other environments. The reverse is less often the case: Preparing for missions in deserts, jungles, or mountains leaves significant gaps in Service and joint readiness to conduct urban undertakings.

Conducting Cost Analysis

As noted above, the modules must ultimately go through a third sieve to determine whether they are cost-effective in terms of dollars and time spent in application to training. Regardless of how effective a module is in addressing requirements, it will lead to its own demise if it does so at prohibitive cost.

To derive the costs of the modules, we followed standard DoD procedures, using a combination of engineering data, parametric analysis, analogy, and interviews with subject-matter experts. For analytical purposes, we did not include certain costs that are generally common to all modules (e.g., local transportation), nor did we include minor costs that would not be germane to the conclusions derived from the assessments, e.g., those associated with coordinating use of a facility. Operational training costs for such items as controllers and role players are provided separately, while other operational training costs for such things as range safety and scheduling are not included because they are generally encompassed in base operation budgets, regardless of the range used or the type of training conducted. Finally, the joint training tasks are not ammunition- or equipment-intensive, so costs of these items are not included.

As a starting point, we constructed a comprehensive cost-breakdown structure and then modified that structure as needed to accommodate the specific characteristics defining each of the modules. Ultimately, the assessment focuses on the life-cycle cost categories of investment (nonrecurring) and sustainment (recurring). Because detailed costs were not available for many of the modules, we used aggregate recurring and nonrecurring costs. When more than one source of costs was available for a module, we used blended costs in developing our estimate. Each module was assessed on a life-cycle basis, using standard factors for discount rates and inflation derived from the Army's FORCES cost-model website. The *DoD Facilities Costs Factors Handbook* was used as a source of data and methodology. To the extent possible, all costs were computed on a constant FY2004 dollar basis and then discounted to their net present value.

In summary, many of the training modules (primarily those eliminated from final consideration) are of marginal value for training large numbers of people because of capacity limits. Thus, they should be considered as niche training opportunities that could be part of an annual training budget or Service initiative rather than part of a long-term JUO training investment strategy.

Examining costs of the individual modules leads to the observation that the investment strategy should be based on the approaches adopted in answering the following three questions:

1. Is joint training a separate entity or an augmentation of Service preparation?
2. Should the training capabilities be built, adapted, rented, or acquired in other ways?
3. Should virtual and constructive training be alternatives or supplements?

Question 1. Joint Training: A Separate Entity or an Augmentation of Service Preparation?

The structures, facilities, and simulations in which joint training takes place are almost exclusively Service structures, facilities, and simulations. The trainees might be Service individuals or units or people staffing joint headquarters. These considerations bear on how to cost the different modules for joint training. For example, if a new purpose-built facility is needed purely to satisfy a joint training requirement, its associated costs could be determined as exclusively joint. However, the cost is incremental and possibly minimal if the joint training requirement could be satisfied by adding it to an existing training regimen at a Service facility or occasionally using that facility for a joint-headquarters-controlled urban exercise. Another possibility might be that the joint requirement adds a day or more to an existing urban training regimen at an existing facility; this could ultimately require more facilities—or possibly not. It depends on throughput needs. These approaches tend to imply that the primary training audience is in most cases a Service unit or individual and that the joint training requirement is contextual to the training. However, the training audience might also be an inherently joint organization, such as a joint force headquarters. Because much training at this level will involve primarily higher-echelon staffs rather than maneuver units, deployment to a live urban training facility might not be needed. Simulation or conduct of a joint headquarters exercise at some generic location could well be sufficient. Ultimately, the investment strategy must account for either the full cost of new JUT means or the incremental cost to existing training means.

Question 2. Build, Adapt, Rent, or Otherwise Acquire Training Capabilities?

There are two primary tradeoffs for an investment strategy. The first is between building training facilities and structures at installations where soldiers, sailors, airmen, and marines are located and moving them to existing facilities and structures that can be bought or leased. In essence, is it more effective to build at facilities heavily populated with user units or to move those units to fewer sites used by organizations from multiple installations? The second tradeoff is between building battalion-sized facilities and building smaller ones. Both of these options depend on troop density at installations, throughput requirements, availability of non–purpose-built facilities, and the distances to such field training capabilities.

To address this question, we analyzed selected training modules and compared their costs on an annual cost-per-person basis. Specifically, we examined three modules in which facilities are built at installations where a substantial number of tactical units are home-based. For costing purposes, no transportation expenses are associated with them. A fourth module involves movement of half the personnel that use it to its location from remote sites (i.e., installations not in the immediate vicinity of the training capability). Such travel is not an unrealistic demand given that the facility offers the opportunity for an entire battalion to train simultaneously for urban operations. Three other modules require movement of all trainees to the sites from remote home stations. The last module involves a hybrid facility, one that also hosts half of its trainees from remote locations. Figure S.2 shows these modules and their related per-person costs.

The first four modules have high initial (first-year) construction costs and substantial sustainment costs thereafter relative to the size of the unit they can host. The other four modules represent facilities capable of supporting training for up to a battalion-size unit.

Figure S.2
Average Annual Cost per Person (FY2005–FY2011) Based on a 30-Year
Life Cycle

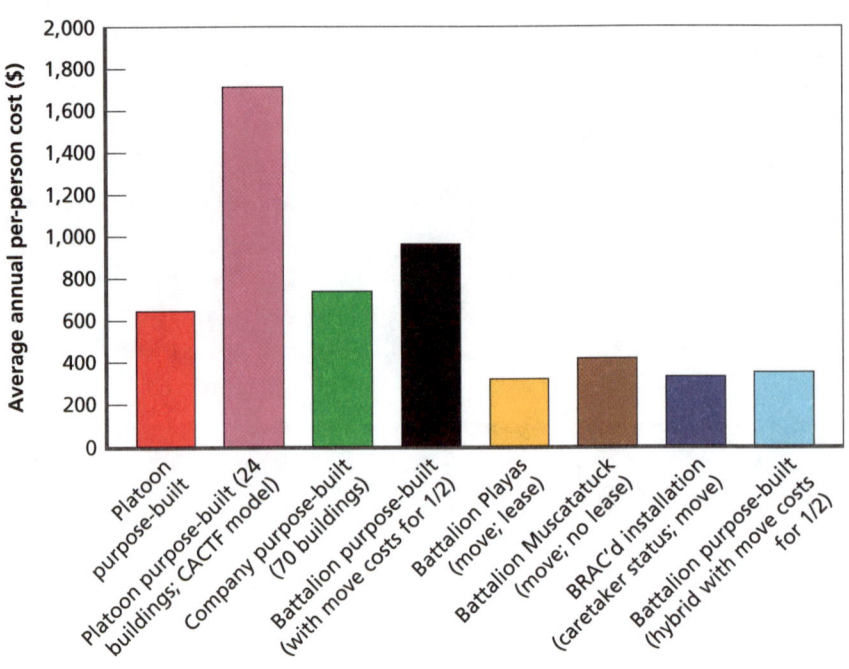

Several points stand out. First, the use of an abandoned commercially owned urban area, a BRAC'd facility, or other less-traditional sites tends to be notably less costly on a per-person-trained basis than more-familiar purpose-built alternatives. (It is notable that the U.S. Army is already investigating the suitability of Cannon AFB, which appeared on the spring 2005 BRAC list, as a possible large-scale urban training facility.) Second, the all-movement modules (those requiring all users to travel significant distances to train) are economical for round-trip travel of up to about 2,500 miles; they become more costly after that. Third, non-hybrid purpose-built facilities are costly, as noted, and combined arms collective training facility (CACTF) designs (given that they meet a standard of training only up to a platoon at a time) are extraordinarily so. Fourth, the calculations shown here are based on live urban training usage rates of about 210 days per year. If more days of training take place, the efficiency of the purpose-built modules increases faster than that of modules requiring movement. Fifth, the comparison is not complete until costs associated with the *total* number of these types of facilities is factored into the calculations (i.e., so that total Service and/or joint force throughput can be calculated). Increasing the total number of purpose-built facilities does not impact *average* costs as long as the facilities are used to the capacity used for computations here. However, total costs do increase as facilities increase in number. For example, if five separate, 70-building, company-size purpose-built facilities are needed to train the force, the average cost per person for this option is about $750, the same as that for a single such facility, but the total costs quintuple.

Sensitivity analysis on movement distance and even on training-rotation duration raises a sixth and final point. Increasing movement distance increases the per-person costs of those options involving movement. Options with the most other costs change least with movement distance. Increasing movement distance from 1,000 to 3,000 miles round trip increases the per-person costs of most of the move modules to higher than that for some of the platoon and company purpose-built modules. The hybrid facility is the least expensive on a per-person basis. At a round-trip distance of 5,000 miles, the all-move modules approach the per-person costs of the battalion purpose-built module. This analysis suggests that regional movement options are more cost-effective than national ones. If movement distances tripled, (non-CACTF) platoon and company purpose-built facilities would be competitive on a cost-per-person basis.

Decreasing training-rotation duration to less than seven days has the effect of decreasing per-person cost proportionally for the purpose-built facilities (halving duration halves per-person cost) and not changing per-person cost as significantly for the options requiring movement. In contrast to the effect of movement distance above, here the recurring and nonrecurring costs are the costs that go down on a per-person basis. The platoon purpose-built facility and the company purpose-built facility become competitive with the move options. The hybrid option is the least expensive over the broadest range of durations. Also, there is interplay between reduction in rotation duration and increases in movement distance that is not captured in this analysis. Decreases in event duration would lead to more than half of the trainees moving to fixed facilities, because facility capacity would now be higher, compared with local population density.

Question 3. Virtual and Constructive Training: Alternatives or Supplements?

The potential capabilities, and the resultant value, of virtual and constructive training will continue to increase during the 2005–2011 period. However, the inherent complexity of urban areas and the difficulty of determining likely second-, third-, and higher-order effects means that computing requirements will tend to demand greater capabilities than can be provided by software advances or PC hardware in this time frame. Simulations will abet the quality of urban training, but advances in the virtual and constructive realms will be insufficient to allow them to serve in other than a complementary role to live training in most cases.

Developing and Implementing a DoD-Wide JUO Training Strategy

Developing the Strategy

Several guidelines and conditions influence the development of a JUO training strategy:

- The training strategy must be comprehensive.
- The training strategy must be dynamic.
- Much improvement is needed in lower tactical-level JUO training, but the greatest shortfalls are at the highest echelons.
- U.S. trainers must remain in "receive mode."
- Joint training modules are only some of a training strategy's building blocks.

- Systems of effective capabilities underpin successful training.
- Even the best training and the most effective training strategy can sometimes not fully prepare a force.
- Size has a quality all its own. Corollary 1: Size can be cheated. Corollary 2: If the consequences of an action in an urban area are not reflected during training, the instruction is flawed.
- Bigger is better. Bigger and denser is better yet.
- If a capability exists in the field, find a way to replicate it for training.
- The total training audience in, around, or over an urban training site may not equate to the number of personnel actually receiving substantive urban training on relevant requirements.
- Simulations, virtual and constructive training, and synthetic environments will not be capable of fully replacing live training during the 2005–2011 period.
- Promote innovation; reconsider proven methods.

Furthermore, development of JUT standards is essential. Standards are fundamental to the design and development of joint doctrine, exercises, experiments, simulations, and live training facilities. They establish uniformity across the Services and guide the development of supporting Service standards, doctrine, and training. They provide similar, if less prescriptive, guidance to the many non-U.S. militaries that look to our nation as they seek to develop their capabilities in the urban arena.

With these guidelines in mind, we developed a short-term and a longer-term strategy to meet the requirements at the R capability level. Table S.4 summarizes the results of the strategy. Five of the 34 requirements (shaded in the table) cannot be met by any of the modules developed: (1) conduct urban human intelligence (HUMINT) operations; (2) provide fire support; (3) consolidate success in the urban environment; (4) shape the urban environment; and (5) transition to civilian control. Of the 29 remaining requirements, 13 can be met in the short term by a combination of Module 25—conduct of combatant command or JTF headquarters, large-scale school, or multi-echelon/interagency exercises (12 modules)—and Module 10—

Table S.4
Summary of Short-Term and Longer-Term Strategies

Requirements	Module Can Meet Requirement at R Level	
	Short Term	Longer Term
Avoid fratricide		Module 18
Communicate in the urban environment		Module 7
Conduct airspace coordination		Module 18
Synchronize joint rules of engagement		Module 18
Conduct stability operations in the urban environment	Module 25	
Conduct support operations in the urban environment	Module 25	
Conduct urban HUMINT operations	No module	No module
Conduct urban SIGINT, IMINT, MASINT, COMINT, ELINT, and other intelligence efforts	Neither selected module	Neither selected module
Conduct urban operations exercises	Module 25	
Integrate urban operations with other relevant environments	Module 25	
Coordinate maneuver in the urban environment		Module 18
Coordinate multinational and interagency resources	Module 25	
Govern in the urban environment	Module 25	
Identify critical infrastructure nodes and system relations	Module 25	
Navigate in the urban environment	Module 10	
Plan urban operations	Module 25	
Provide common situational awareness		Module 18
Provide fire support	No module	No module
Provide security during urban transition operations		Module 18
Rehearse/war-game urban operations		Module 18
Conduct urban noncombatant evacuation operations (NEOs)		Module 18
Conduct U.S. domestic urban operations		Module 18
Conduct urban combat search and rescue (CSAR)		Module 18
Conduct urban operations during and after a WMD event	Neither selected module	Neither selected module
Consolidate success in the urban environment	No module	No module
Disembark, base, protect, and move in urban environments		Module 18
Engage in the urban environment		Module 18
Orchestrate resources during urban operations	Module 25	
Shape the urban environment	No module	No module
Sustain urban operations		Module 18

Table S.4 (continued)

Requirements	Module Can Meet Requirement at R Level	
	Short Term	Longer Term
Transition to civilian control	No module	No module
Understand the urban environment	Module 25	
Achieve simultaneity in meeting requirements	Module 25	
Conduct training across multiple levels of war	Module 25	
Total Requirements Met	**13**	**14**

terrain walks (1 module). (Additionally, but not shown in Table S.4, Module 24—classroom instruction—training can lend a "walk" level of preparedness in 8 of these 12 instances and a "crawl" level in another three, an excellent example of how building-block training can provide preparation for both more-complex training events and real-world undertakings, since the classroom instruction could provide the training needed for maximizing the benefit derived from higher-echelon exercises.) There is further good news: High-level headquarters and similar exercises are fairly economical, costing an estimated $2.9 million to $7.1 million per event if the headquarters deploys and the event includes links to units in the field. Such deployment and outside-organization participation are often not necessary, allowing for the conduct of such events at lower cost. That these events could address more than one-third of the requirements of concern means that the return for dollar invested is excellent. Further, classroom JUO training outlays are negligible, as they are properly measured more in terms of opportunity costs than of dollar expenditures (the loss of course instruction on topics that would be covered is the time not being spent on urban material).

The thirteenth requirement that is readily within "run" training status for units in the immediate term is "navigate in an urban environment." Module 10, terrain walks in actual urban areas, provides this opportunity, again at little cost. Nearby towns or cities offer the training environment necessary, although commanders wanting to challenge their personnel with unfamiliar terrain may choose to go farther afield.

With those 13 requirements addressed in the short term (and five that cannot be met by any module), 16 requirements remain to be met in the longer term (34 − 13 − 5 = 16). Five modules address varying amounts of these requirements:

- Module 1: Battalion-size and larger purpose-built facility (addresses 10 of the 16);
- Module 7: Hybrid facility (addresses 9 of the 16);
- Module 17: Use of abandoned domestic urban areas (addresses 11 of the 16);
- Module 18: BRAC'd installations (addresses 13 of the 16);
- Module 22: Abandoned/constructed overseas urban areas (addresses 9 of the 16).

Of these five modules, we found Module 22 to be of limited value to this analysis. Abandoned or constructed overseas facilities should not be relied on as primary urban training capabilities (except for the occasional instance in which U.S. units are located in close proximity to such assets). Module 1 is very attractive from an availability perspective in that DoD will own such battalion-size and larger purpose-built facilities, but the high cost of this option, especially when movement is necessary for use, is a serious drawback (see Figure S.2 above). A drawback to Module 17 is that users may have to move considerable distances to use abandoned domestic urban areas. Further, availability is questionable unless DoD leases civilian-owned facilities on a long-term basis, a key concern during unpredictable times when surge training is essential, as was the case for preparation in support of operations in Iraq and Afghanistan throughout 2004 and into 2005. Restrictions on live fire and environmentally related issues are also likely at these facilities. Finally, the realistic life expectancy of these facilities is estimated to be only five years, barring considerable upkeep, meaning that DoD would inherently be relying on civilian entities to find and develop such sites repeatedly, and to do so in a manner conducive to sophisticated military exercises. The risk in that regard seems a significant one.

Given these concerns, we chose the remaining two modules, Module 18 and Module 7, for further analysis. Module 18 is immediately attractive because it addresses all but three of the outstanding requirements (13 out of 16), more than any other option. A reconsideration of how such facilities might be managed potentially enhances this attractiveness. Original BRAC'd facility cost estimates assumed use on a lease basis, with user payments being a fairly economical $10,000 per day (average) and annual expenses being a rather low $3 million to $3.6 million. However, this option would be potentially even more economical if DoD were to retain ownership of one or more closed military installations, and the cost estimates would share many of the characteristics of the armed forces using an abandoned civilian area, but with the added benefit of retaining complete control. This would comprise less a BRAC than transition of an installation from one set of functions to another (e.g., from housing a headquarters to support of urban training). Benefits would be numerous, including less negative impact on the local civilian community (e.g., retention of jobs). This social/political benefit is potentially quite significant. The residents in the vicinity of Muscatatuck, IN (previously a state educational facility now employed by the National Guard for urban training), are in many ways very supportive of the military assuming responsibility for the site, because of the economic benefits continued use promises for the local community.

An added attraction of the BRAC-transition/BRAC'd facility lease option is the potential to select future and already BRAC'd facilities from locations that minimize travel times for potential user units. An offshoot of this option is the use of parts of active installations that have been abandoned or are underutilized, as was done in selecting the location for the relatively recently opened urban training site at Ft. Lewis, WA.

BRAC review procedures and other considerations would need to be revised for those facilities thought to have potential use as urban training sites if the BRAC-transition option is adopted. Moreover, as attractive as this alternative is, it suffers from the major shortfall that establishing facility availability is difficult. Base closures—in truth, any significant alteration of a military facility's status—involve lengthy and multiple processes over which DoD has only partial control. Practical implications and military necessity will have only limited impact on decisions. This should not be an argument for abandoning initiatives to develop such resources, especially in cases of already closed locations, but common sense dictates consideration of other options.

We therefore propose Module 7—purpose-built hybrid facilities—as a backup. Such facilities support only nine of the outstanding 16 JUT requirements at a "run" level of readiness, but they include some of the same ones as Module 18 does. Module 7 also includes one—communicating in the urban environment—that is not at a "run" level in Module 18.

Implementing the Strategy

We also examine some of the key considerations in developing a JUO training investment strategy, including what to build; how many facilities to build; the best locations for battalion- and larger-capable BRAC, hybrid, or other types of urban training facilities; and what should be upgraded among current capabilities.

Regarding what, how many, and where to build, we advise construction of an urban training facility capable of supporting at least a platoon at every installation that is a home station for a ground maneuver element of brigade or larger size. Such a facility should consist of 25 or more structures (depending on the underlying terrain, building type and density, and other factors); it may be purpose-built, created from existing underutilized portions of the installation, or a combination of these and other alternatives.

In additional, a minimum of four brigade-size or larger urban training complexes are needed, each with an adjacent or nearby air-ground training capability. The brigade-size sites should include some 300 structures (with the same qualifications noted above) and both OPFOR and noncombatant role-player support for all units conducting training. Based on factors such as proximity of Service home stations and major training rotations, we determined that one of these four facilities should be included in each of the following locations or regions:

- The Kentucky–North Carolina–Georgia area
- Ft. Polk, LA
- Ft. Hood, TX
- The U.S. southwest, likely in the vicinity of Twentynine Palms and Ft. Irwin, CA

Improving existing facilities (e.g., Ft. Polk's Shugart-Gordon) is one way of creating such capabilities. Transitioning installations selected for BRAC is the most desirable option when such an alternative is available in the desired locations, and in fact we recommend that suitability for use as urban training sites be an element of consideration during BRAC review. Hybrid facilities constructed from combinations of existing building complexes, purpose-built facilities, movable façades, and the like are an alternative to BRAC facilities for creation of brigade-size training sites. Regardless of their makeup or origin, the number of supporting noncombatant role players should better replicate real-world conditions, where civilians almost inevitably outnumber combatants. (This is but one issue that needs to be addressed during the development of heretofore largely nonexistent OPFOR and noncombatant role-player doctrine.)

A building-block approach to both Service and joint urban training is crucial: Individual and lower-echelon proficiency is key to successful training at higher echelons. The recommendations for platoon- and brigade-size facilities reflect this dictate. There is a very significant need to do likewise for higher-echelon staffs and leaders. Current Service and joint classroom training fails to adequately prepare U.S. armed forces personnel for urban contingencies. Exercises for military and civilian leaders at these higher echelons are similarly too rare and, when conducted, generally fail to adequately challenge participants. The result is that the nation's military and other leaders at the upper tactical, operational, and strategic levels require training involving the full spectrum of urban operations, including related instruction, such as that pertaining to governing, to address notable shortfalls.

Hard decisions need to be made regarding which urban-related simulations merit retention and which do not, who should have authority for urban facility scheduling, and responsibility for both creating JUT standards and overseeing the funding of urban training initiatives. We recommend that these be assigned to a joint entity, one given responsibility for the oversight of armed-forces-wide urban training in the interest of efficiency, joint cooperation, and better training.

Four challenges should be examined in implementing the training strategy: (1) whether to build "born joint" training facilities or to employ Service capabilities in support of JUO training; (2) whether a joint entity should assume authority for coordinating range usage; (3) whether the joint community should have oversight of range funding; and (4) the need for joint urban live, virtual, and constructive training standards.

Figure S.3 provides a concise overview of how primary actions in the recommended JUT strategy should be undertaken during the 2005–2011 period. The modules associated with each action appear in parentheses after the descriptions in the first column.

Finally, Table S.5 provides an illustrative cost estimate for the proposed strategy. A number of factors will impact the actual cost of implementation, including the number of BRAC'd (versus hybrid) facilities used for urban training, the ultimate disposition of units returning to the United States from Germany and elsewhere, combatant commander decisions about whether to conduct urban-related exercises, and choices about the types and numbers of OPFOR and noncombatant role-player capabilities.

Concluding Thoughts

Having laid the foundations for a JUO training strategy, we present some concluding thoughts on the process of developing that strategy. First, there is **a need to improve linkage of lessons from the field and joint force urban training**. Taking lessons from operations and training and passing them to those who can benefit is key to success-

Figure S.3
Overview of How Primary Actions of the JUO Training Strategy Should Be Undertaken During the 2007–2011 Period

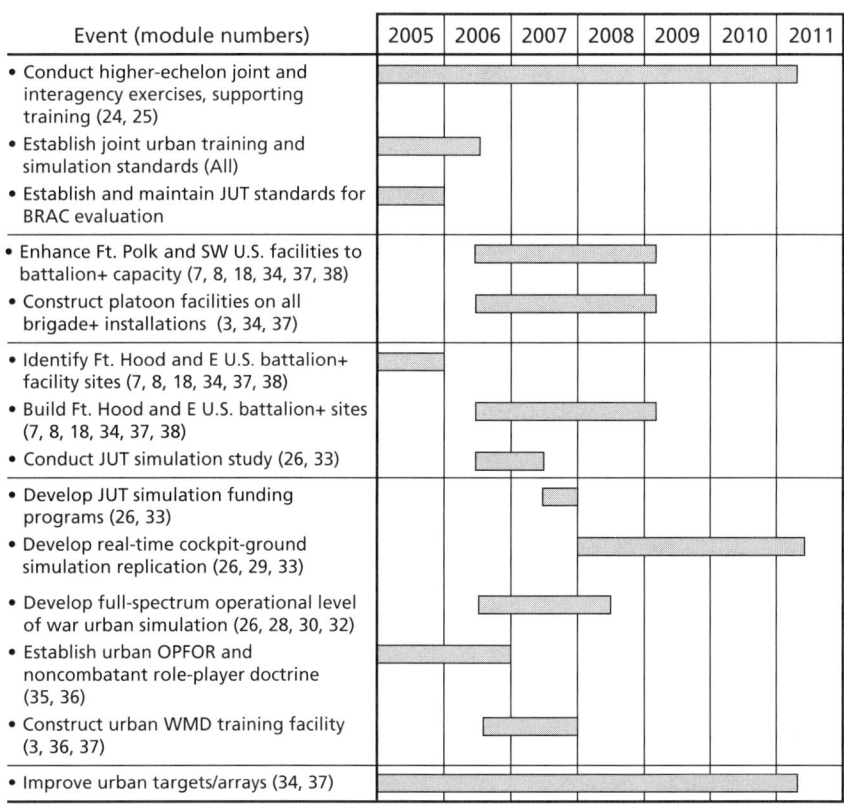

Event (module numbers)	2005	2006	2007	2008	2009	2010	2011

RAND *MG439-S.3*

fully implementing a JUO training strategy. It is both an area of demonstrated success and one requiring further enhancement. The measure of success or failure is, as always, men's and women's lives.

Second, it is critical to **avoid conducting JUO training in isolation**. Both the joint community and the Service community will be well served by supporting specific individual training in preparation for as-yet-unidentified contingencies and to complement pre-deployment instruction after arrival in a theater of operations. The

Table S.5
Estimated Costs for Sample JUT Strategy Implementation ($ millions)

	2005	2006	2007	2008	2009	2010	2011	Total
Headquarters/JTF exercises[a] (7/year)	7.6	7.5	7.3	7.1	6.9	6.8	6.6	49.8
Classroom training	0	0	0	0	0	0	0	0
Navigation training	0	0	0	0	0	0	0	0
OPFOR support for headquarters/JTF exercises[b]	27.0	27.0	27.0	27.0	27.0	27.0	27.0	189.0
Noncombatant role-player support for headquarters/JTF exercises[c]	5.3	5.3	5.3	5.3	5.3	5.3	5.3	37.1
BRAC (realigned/transition)[d]	0	1.5	1.4	1.3	1.3	5.4	1.2	12.1
Hybrid facility[d]	189.0	2.1	2.0	1.9	1.9	30.1	1.8	228.8
Platoon facility[e]	227.9	2.4	2.4	2.3	2.2	40.8	21.6	299.6
Air-ground facilities[f]	133.0	1.6	1.6	1.5	1.5	1.4	1.4	142.0
Joint headquarters support[e]	7.5	7.5	7.5	7.5	7.5	7.5	7.5	52.5
Fire-team simulations[e]	8.3	7.4	7.3	7.8	6.9	6.6	7.2	51.5
OPFOR support for BRAC/hybrid facilities[g]	35.4	35.4	35.4	35.4	35.4	35.4	35.4	247.8
Noncombatant role-player support for BRAC/hybrid facilities[h]	28.8	28.8	28.8	28.8	28.8	28.8	28.8	201.6
Totals	669.8	126.5	126.0	125.9	124.7	195.1	143.8	1,511.8

[a] Cost of headquarters exercises is (Annual cost for JTF Foxtrot alternative)(7 exercises/year).

[b] OPFOR support for headquarters/JTF exercises is assumed at 50 personnel (one-tenth of an OPFOR battalion) (7 exercises/year).

[c] Noncombatant role-player support is assumed at (25 personnel)(7 exercises/year).

[d] This estimate assumes one of the four battalion or larger and air-ground facilities is a BRAC (transition to urban training use) site. The remaining three are assumed to be hybrid facilities. Costs here are for the ground facility only.

[e] See discussion in Chapter Seven.

[f] One per battalion or larger ground facility. For estimate purposes, two of the four are assumed to be close to parts (thus reducing shipping costs of materials).

[g] It is assumed that units conducting training at platoon-size facilities provide their own OPFOR. It is also assumed that each battalion or larger training facility has a permanent active duty OPFOR company of 120 personnel. The costs are therefore ($42M/site)(120/570)(4 sites) = $35.4M.

[h] It is assumed that units conducting training at platoon-size facilities provide their own noncombatant role players. It is also assumed that each battalion or larger training facility has a permanent noncombatant role player cadre of 10 individuals at $30,000/year, 20 specialty individuals (e.g., representing specific cultures or with particular language skills), and 300 others paid $100/day for 210 days of training per annum. Given four sites, the annual cost is [$0.300M + $0.600M + (300)($100)(210)](4) = $28.8M.

need to govern and the need to meet the demands for better inter-agency cooperation are but two examples of instruction that overlaps urban and other instruction. While villages, towns, and cities do impose greater and sometimes unique difficulties, those designing and conducting urban training should recognize and capitalize on instruction involving other aspects of military operations as well. Further, the many demands of urban operations have to be incorporated in higher-echelon instruction. Exercises need to replicate the difficulties inherent in coordinating air, maritime, ground, and SOF component theater fires; intelligence activities; information operations; and logistics, including passage of personnel and supplies through urban aerial and sea points of debarkation (APODs and SPODs). Service and joint headquarters at multiple echelons should practice the command and control linkages and simultaneous use of urban areas of operation, control made more difficult by the fact that such towns and cities also house the daily residences and workplaces of thousands or millions of members of the indigenous population. JTF and other headquarters will similarly need to synchronize their activities with, support, or coordinate information campaigns with Special Operations foreign internal defense (FID), civil affairs (CA), psychological operations (PSYOP), and other missions in and around urban areas.

Third, there is a need to **train for both the generic and the specific.** Every urban area is unique, but all have common characteristics. Structures have more in common than do the people who inhabit them; training in buildings that look somewhat different from those in an operational theater can still be very effective. In short, from the standpoint of building design (whether in live, virtual, or constructive capabilities), sites that possess a variety of construction types, building materials, traveled ways, infrastructure (e.g., open sewerage versus enclosed), and other elements, either within a given site or between various sites, will serve trainees well. Trainers should seek to design capabilities that can be tailored to specific environments at minimal cost in time and funds (e.g., changing signs into regional languages and altering the nature of refuse, animals, furniture, stairwell and door locations and design, rooftop profiles, and the like). Designing "generic" training sites and adapting existing sites so that they can better reproduce conditions similar to those of current and near-term likely threat conditions will provide frames on which regionally specific details can be draped. This is true of synthetic terrain as well. (However, advances in speed and reductions in the cost of designing synthetic terrain based on actual theaters are such that calls for generic designs of synthetic terrain may diminish in the later years addressed by this strategy.)

Fourth, **there is a need to decide on the issue of instrumenting facilities to monitor performance and provide feedback during after-action reviews (AARs).** While we were uncommitted at the beginning of this study, we now tend to support greater reliance on a human-in-the-loop to monitor training. Cameras and supporting equipment are expensive both at the time of purchase and during ongoing maintenance and replacement because of upgrade demands or wear-and-tear. Instrumentation for the larger facilities called for in this study would likely be extremely expensive. Funding to support increased realism seems to provide a greater return on the training investment.

Fifth, **there is a need to decide on the value/importance of urban live-fire training.** Given the extraordinary safety precautions necessary for live-fire training, the impact on other training at an urban training site, and the too often exceptional preparation times taken to fire a low number of rounds, it is important to carefully consider the benefit gained from such training before forgoing similar training using less-than-lethal rounds. While we recognize the need for such exercises in some instances, a live-fire capability is not considered a necessary characteristic of large urban facilities designated for use in joint training.

Sixth, **there is a need to continue innovating in targetry.** Targetry innovation should be encouraged, as should the more formal developments in targetry that proceed apace with it. Targets and target arrays for pilots, in particular, could benefit from further development. To better prepare pilots for actual urban operations, moving targets must be intermixed with arrays of innocent civilians and private vehicles; dust, light, electronic, and other interference during engagements must be reproduced; and there must be a general increase in the complexity of the targeting process.

A JUT strategy is a starting point for more work. It provides guidance. It suggests a framework for understanding. But most of all, it imparts a responsibility to develop programs, plans, and guidance that address the many details needed to implement the strategy. It advises how those implementing should write doctrine (itself another form of guidance), spend funds, design instruction, and modify organizations in support of the objectives that initially motivated the strategy's development. In short, this study is an opportunity for many to participate in the refinement, augmentation, and constant maintenance of the JUO training strategy.

Acknowledgments

The list of individuals who lent their assistance and support to the authors during this study is a long and notable one, long because so many shared our concerns regarding preparing U.S. armed forces for ongoing and future urban contingencies, notable because of the unflagging dedication of the many who participated. The interviewees listed in the Bibliography were invaluable in developing what follows, and the authors sincerely appreciate their cooperation. Others whose names do not appear were no less helpful, but we have respected their requests for anonymity.

Another group deserving of recognition are those not interviewed but whose assistance was crucial in introducing their colleagues to the authors. We thank Majors James Cushnir, St. John Coughlan, and John Robson and long-time friend of RAND's urban operations efforts, Major General Jonathan Bailey, all members of the British Army. Dutch Army Lieutenant Colonels Henk Oerlemans and Johan van Houten were extraordinarily gracious in granting extensive time not only for interviews, but also for visits to training sites throughout their beautiful country. A significant number of the interviews with U.S. Service members could never have taken place without the characteristically selfless help of Johnny Brooks; he is one of the unsung heroes among those with many years in the urban operations community.

It is essential to recognize the gentlemen who provided in-progress subject-matter expert reviews. We thank Lt Gen (USMC, retired); G. R. Christmas; Captain Bob Harward, U.S. Navy; and

LTC Greg McMillan (U.S. Army, retired) for insights that were both valuable and timely. The formal RAND reviews by Maren Leed and LTG L. G. Holder (USA, retired) were exceptional in their detail and thoroughness, for which we heartily thank them.

Having three sponsors for a single project would normally constitute a daunting prospect. That it was otherwise is directly attributable to a dedication to the common cause of success unfailingly demonstrated by Colonel Frank DiGiovanni (OSD, Readiness), LtCol Preston Maclaughlin and Jay Reist (J7 JFCOM), and Duane Schattle and Scott Bamonte in JFCOM's J9 Joint Urban Operations Office. Many others in those organizations also provided valuable assistance throughout the year of research. It is important to note in particular Colonel DiGiovanni's continual and energetic guidance throughout the research effort. We very much appreciate his ceaseless dedication to meeting the combined needs of all parties interested in the vital topic under consideration.

As ever, there are individuals behind the scenes at RAND who are key to the report being on time and presented in a form that serves its readers well. Gayle Stephenson, Mary DeBold, and Maria Falvo have spent long hours and demonstrated saintly patience during the months of research and data compilation. RAND's Pittsburgh office librarian, Judy Lesso, was on several occasions key to obtaining critical information in a timely manner. We would also like to thank RAND colleagues Morgan Kisselburg, Phyllis Kantar, Brad Wilson, Phi Vu, Tom Herbert, and Yuna Huh Wong for their insights and contributions regarding constructive, virtual, and agent-based simulations. Paul Steinberg's artistry in shaping the final document through his notable editing skills added considerably to this report's accessibility, an intense effort for which we are most appreciative. Finally, Janet DeLand's ultimate review and edit removed points of potential confusion and otherwise polished the end product; the authors thank her for the invaluable service.

Acronyms and Abbreviations

AAR	After-action review
ABGS	Automated building generation system
ACR	Armored Cavalry Regiment
AFB	Air Force Base
AFSOC	Air Force Special Operations Command
AFTL	Air Force Task List
AOR	Area of responsibility
APOD	Aerial port of debarkation
ARL	Army Research Laboratory
ARNG	Army National Guard
ATEC	Army Test and Evaluation Center
AUTOCAD	A commercial, three-dimensional computer-aided design capability
AVCATT	Aviation Combined-Arms Tactical Trainer
BCTP	Battle Command Training Program
BDA	Battle damage assessment
BLUEFOR	Blue Force
BRAC	Base Realignment and Closure
C2	Command and control
C4ISR	Command, control, communications, computers, intelligence, surveillance, and reconnaissance

CA	Civil affairs
CACTF	Combined arms collective training facility
CALL	Center for Army Lessons Learned
CAMTF	Combined-arms MOUT task force
CAS	Close air support
CASEVAC	Casualty evacuation
Cave	Cavalry
CCTT	Close Combat Tactical Trainer
CDV	Copehill Down Village
CENTCOM	Central Command
CINC	Commander-in-chief (now "combatant commander" when used in reference to joint four-star command positions)
CJSOTF	Combined and Joint Special Operations Task Force
CMTC	Combat Maneuver Training Center
CNA	Center for Naval Analysis
COMINT	Communications intelligence
CONEX	Container express
CONUS	Continental United States
CPX	Command post exercise
CQC	Close-quarters combat
CS	Combat support
CSAR	Combat search and rescue
CSS	Combat services support
CTC	Combat training center
CTF	Collective training facility
DA/HR	Department of the Army/Human Resources
DARPA	Defense Advanced Research Projects Agency
DIS	Distribution interaction simulation
DISAF	Dismounted Infantry Semi-Automated Forces

DIS/HLA	Distributed interactive simulation/high-level architecture
DM	Dominant maneuver
DMSO	Defense Modeling and Simulation Office
DoD	Department of Defense
DREN	Defense Research and Engineering Network
DSTL	Defence Science and Technology Laboratory
ELINT	Electronic intelligence
EMRTL	Energetic Materials Research & Testing Laboratory
EW	Electronic warfare
FAARS	Future After Action Review System
FAC	Forward air controller
FBCB2	Force XXI Battle Command Brigade and below
FBI	Federal Bureau of Investigation
FCS	Future Combat System
FEMA	Federal Emergency Management Agency
FID	Foreign internal defense
FLIR	Forward-looking infrared radar
FO	Forward observer
FoF	Force on force
FORCES	Force and Organization Cost Estimation System
FORSCOM	Forces command
FoT	Force on targetry
FSC	Full Spectrum Command or fire support coordinator
FSW	Full Spectrum Warrior
GAO	General Accounting Office
GBU	Guided bomb unit
GPS	Global Positioning System
GPU	Graphic processing unit

HA	Humanitarian assistance
HLA	High-level architecture
HPMF	Hasty protective minefield
HRT	Hostage rescue team
HUMINT	Human intelligence
ICT	Institute for Creative Technology
IDA	Institute for Defense Analyses
IED	Improvised explosive device
I MEF	First Marine Expeditionary Force
IMINT	Imagery intelligence
IO	Information operation
IR	Infrared
IUSS	Integrated unit soldier simulation
JAWP	Joint Advanced Warfighting Program
JCATS	Joint Conflict and Tactical Simulation
JEC	Joint effects cell
JFACC	Joint force air component commander
JFCOM	Joint Forces Command
JFWC	Joint Forces Warfighting Center
JMET	Joint mission-essential task
JMETL	Joint mission-essential task list
JNTC	Joint National Training Center
JOA	Joint operations area
JOUST	Joint operations on urban synthetic terrain
JRTC	Joint Readiness Training Center
JSAF	Joint semi-automated forces
JSEAD	Joint suppression of enemy air defense
JSIMS	Joint simulation system
JTEP	Joint Training Experimentation Program

JTF	Joint task force
JTLS	Joint Theater-Level Simulation
JTT	Joint training task
JUO	Joint urban operations
JUT	Joint urban training
JWC	Joint Warfare Center
LOGFED	A training simulation focused on logistics
LOS	Line of sight
LVC	Live, virtual constructive
LZ	Landing zone
MAC	MOUT assault course
MAGTF	Marine air-ground task force
MANA	Map-Aware Non-uniform Automata
MAPEX	Map exercise
MASINT	Measurement and signatures intelligence
MAWTS-1	Marine Aviation Weapons and Tactics Squadron 1
MCA	Major CONUS activity
MCABWA	Marine Corps Air Bases, Western Area
MCANG	Military Construction, Army National Guard
MCAR	Military Construction, Army Reserve
MCAS	Marine Corps Air Station
MCWL	Marine Corps Warfighting Laboratory
MEDCOM	Medical command
MEDSAF	Medical semi-automated forces
MEF	Marine expeditionary forces
METL	Mission-essential task list
MEU	Marine expeditionary unit
MEU(SOC)	Marine Expeditionary Unit (Special Operations Capable)

MILCON	Military construction
MILES	Multiple integrated laser engagement system
MILPERS	Military personnel
MMR	Man marker round
MODSAF	Modular semi-automated forces
MOOTW	Military operation other than war
MOUT	Military operations on urbanized terrain
MP	Military Police
MRX	Mission readiness exercise
MSTP	MAGTF Staff Training Program
NBC	Nuclear, biological, or chemical
NCO	Non-commissioned officer
NDRF	National Defense Reserve Fleet
NEO	Noncombatant evacuation operation
NERTC	Northeast Regional Training Center
NGB	National Guard Bureau
NGO	Non-governmental organization
NIPRNET	Non-Classified Internet Protocol Network
NLOS	Not in line of sight
NTC	National Training Center
NTI	Nevada Training Initiative
NVESD	Night Vision and Electronic Sensors Directorate
O/C	Observer/controller
O&M	Operation and maintenance
OEF	Operation Enduring Freedom
OIF	Operation Iraqi Freedom
OJT	On-the-job training
OneSAF	One Semi-Automated Forces objective system, a computer simulation

OOS	OneSAF objective system
OOTW	Operations other than war
OPA	Officer Personnel Act
OPFOR	Opposing force
OPNET	A commercial network simulation originally developed at the Massachusetts Institute of Technology
OPTAG	Operational Training and Advisory Group
OSD P&R	Office of the Secretary of Defense, Personnel and Readiness
OTB	Objective test bed
PC	Personal computer
PGM	Precision guided munition
PM TRADE	Project Manager Training Devices (U.S. Army)
POL	Petroleum, oil, and lubricants
PPBES	Planning, programming, budget, and execution system
P&R	Personnel and Readiness
PRC	Populace and resource control
PSYOP	Psychological operations
PVO	Private volunteer organization
QUALNET	Qualitative network
RDT&E	Research, development, test, and evaluation
REFORGER	Return of Forces to Germany
ROE	Rules of engagement
RSO	Range safety officer
RSTA	Reconnaissance, surveillance, targeting, and acquisition
SAF	Semi-automated forces
SAFOR	Semi-automated forces

SAM	Surface-to-air missile
SCLA	Southern California Logistics Airport
SDZ	Surface danger zone
SEAD	Suppression of enemy air defenses
SEALs	Sea-Air Land (Navy special warfare units)
SIGINT	Signal intelligence
SIMCFS	Simulated close fire support
SIMNET	Simulator Networking
SJFHQ(CE)	Standing joint force headquarters (core element)
SOAR	Special Operations aviation regiment
SOC	Special Operations capable
SOCOM	Special Operations Command
SOF	Special Operations Forces
SOSO	Stability operations and support operations
SOUTHCOM	Southern Command
SPOD	Sea port of debarkation
SRTA	Short-range training ammunition
STX	Situation tactical exercises
TACSIM	Tactical Simulation
TAFC	Tactical area forward controller
TIREM	Terrain Integrated Rough Earth Model
TRADOC	Training and Doctrine Command (U.S. Army)
TTP	Tactics, techniques and procedures
UAC	Urban assault course
UAV	Unmanned aerial vehicle
UEx	Unit of employment focused primarily on tactical operations
UEy	Unit of employment focused primarily on theater-level support and land component forces macro-management

UGV	Unmanned ground vehicle
UHRB	Ultra-high-resolution building
UJTL	Universal Joint Task List
UKROE	British rules of engagement
UNTL	Universal Navy Task List
USAF	United States Air Force
USAR	United States Army Reserve
USARAK	United States Army Alaska
USARC	United States Army Reserve Command
USAREUR	United States Army Europe
USARPAC	United States Army Pacific
USARPAC NGB	United States Army Pacific National Guard Bureau
USASOC	United States Army Special Operations Command
USCG	United States Coast Guard
USD	Under Secretary of Defense
USECT	Understand, shape, engage, consolidate, transition
USMC	United States Marine Corps
USN	United States Navy
UTM	Ultimate training munition
VCP	Vehicle check point
VoIP	Voice over Internet Protocol
WARSIM	War fighter's simulation
WMD	Weapons of mass destruction

CHAPTER ONE

Introduction

He . . . shewed me that great city, the holy Jerusalem . . . and the
street of the city was pure gold, as it were transparent glass.

Revelation 21:10 and 21

The Romans are sure of victory . . . for their exercises are battles
without bloodshed, and their battles bloody exercises.

Flavius Josephus, 37–100 A.D.
The Jewish War

Background

Two millennia ago, the Western world's greatest empire faced an im-
passioned revolt by several insurgent factions in Jerusalem. The members of each faction were as interested in destroying their counterparts
as they were in defeating the Romans. Ill-trained, if at all, and relatively poorly equipped, these irregulars nonetheless drove the original
garrison from the city. However, Rome would not be denied. Its
army began a siege. Innocent civilians suffered the most during the
weeks of tribulation; it was they who would suffer further yet if the
city's streets and structures were to become a battleground. The empire's newly arrived Titus Caesar "pitied the common people, who
were helpless against the partisans, and over and over again he de-
layed the capture of the city and prolonged the siege in the hope that

the ringleaders would submit."[1] Only when a much larger Roman force finally threatened did the insurgents abandon Jerusalem to the besiegers.

Yet the population's difficulties had only begun. The city's citizens chafed at what they considered foreign occupation. Rome's military commander monitored events and kept watch on those locations key to maintaining peace. Places of worship and religious gatherings were particularly important in this regard. Titus ensured that a "Roman cohort stood on guard over the Temple colonnade [and that armed men were always] on duty at the feasts to forestall any rioting by the vast crowds."[2] The precautions were well considered, since even seemingly insignificant events triggered outbursts. Unfortunately, at times the Romans were their own worst enemies. During the feast celebration, for example:

> One of the soldiers pulled up his garment and bent over indecently, turning his backside towards the Jews and making a noise as indecent as his attitude. This infuriated the whole crowd. . . . The less restrained of the young men and the naturally tumultuous section of the people rushed into battle, and snatching up stones hurled them at the soldiers. [The Roman commander,] fearing that the whole population would rush to him, sent for more heavy infantry. When these poured into the colonnades the Jews were seized with uncontrollable panic, turned tail and fled from the Temple into the City. So violently did the dense mass struggle to escape that they trod on each other, and more than 30,000 were crushed to death. Thus the Feast ended in distress to the whole nation and bereavement to every household.[3]

The insurgency was renewed. Partisans, finding their weapons ineffective against those of their adversaries, adapted their tactics. Giving way before the longer-range Roman armaments, "once inside the minimum range of the far-flung missiles they assailed the Romans

---

[1] Flavius Josephus, *The Jewish War,* London: Penguin Classics, 1959, p. 28.

[2] Ibid., p. 144.

[3] Ibid.

furiously without a thought for life or limb, exhausted units being constantly replaced by fresh waves of attackers."[4] Rules of decency were cast aside in the brutal struggle. Insurgents carried off Roman equipment and despoiled the bodies of slain soldiers.[5] "One [partisan] who with a great many others had taken refuge in the caves begged [a soldier] to give him his hand as a pledge of protection and to help him climb out. The Roman incautiously gave him his hand, and with a swift upward thrust the man stabbed him in the groin and killed him instantly."[6]

Jerusalem's unrest spread from the city throughout the country:

> In all districts of Judaea there was a similar upsurge of terrorism, dormant hitherto; and as in the body if the chief member is inflamed all the others are infected, so when strife and disorder broke out in the capital the scoundrels in the country could plunder with impunity, and each group after plundering their own village vanished into the wilderness. There they joined forces and organized themselves in companies, smaller than an army but bigger than a gang of bandits, which swooped on sanctuaries and cities. Those they attacked suffered as severely as if they had lost a war, and were unable to retaliate as the raiders, like all bandits, made off as soon as they had got what they wanted. In fact, every corner of Judaea was going the way of the capital.[7]

Within Jerusalem itself, Roman restraint was rewarded with violence, as opposing insurgent factions united against the outsiders. Faction leaders "took humanity for weakness and imagined that it was through inability to take the rest of the City that"[8] the Roman army restrained itself. Partisans took advantage of their superior

---

[4] Ibid., p. 206.

[5] Ibid., p. 178.

[6] Ibid., p. 216.

[7] Ibid., p. 267.

[8] Ibid., p. 314.

knowledge of the capital's streets.[9] Both sides sought to undermine
the other's advantages through public declarations and other acts of
propaganda, recognizing that "very often the sword is less effective
than the tongue."[10] And once again it was the innocent who suffered
most. Hunger and other deprivations took their toll. The insurgents
forced noncombatants to support their respective causes. To resist
seemed hopeless. Irregulars dragged the families of "deserters onto the
wall with those members of the public who were ready to accept
Roman assurances, and showed them what happened to men who
deserted to the enemy, declaring that the victims were suppliants, not
prisoners. This caused many who were eager to desert to remain in
the City . . . but some crossed over without further delay, knowing
the fate that awaited them but regarding death at enemy hands as a
deliverance, compared with starvation."[11] Eventually the Romans
prevailed, despite their mistakes and the intensity of the resistance.
They crushed insurgent resistance largely through the application of
military might alone and at tremendous expense in innocents' lives.
"*Atque ubi colitudinum faciunt pacem appellant.*" ("They create a deso-
lation and call it a peace.")[12] So spoke the British king Calgacus, one
who had experienced the empire's might firsthand.

The Roman manner of achieving success is not one the civilized
nations of today would choose to emulate in urban operations.
Choosing instead to minimize unnecessary loss of both their soldiers'
and noncombatants' lives severely tasks today's militaries in those operations. That modern coalitions seek rather to ready nations to govern and defend themselves means that the peace sought must be one
of restoration, not destruction. Military men and women will have to
be trained not only to fight, but also to govern and to train others to

---

[9] Ibid., p. 314.

[10] Ibid., p. 317.

[11] Ibid., p. 326.

[12] The British king Calgacus as quoted in Tacitus, online at http://www. channel4.com/
history/microsites/H/history/guide03/part04.html (accessed November 23, 2004).

do likewise. The challenges for those who train and those who apply that training are far greater today than ever before.

The world's population is increasingly urban, perpetuating a long-standing trend of movement from rural environments to towns and cities. Those towns and cities are growing not only in population, but also in influence. They are local, regional, national, and even international hubs of power. Recent American military operations, not surprisingly, reflect this stature. Combat in Saigon and Hue was a critical factor in the outcome of the Vietnam War. Operations in Panama, Somalia, the Balkans, Afghanistan, and Iraq during the past 20 years have all been urban-centric or characterized by significant undertakings in cities and towns. Air operations over Baghdad and concerns about Kuwait City were vital parts of even Operation Desert Storm (1991), named for the most open of land features.

Such urban operations have challenged and continue to challenge the world's most sophisticated militaries. Still reliant on technologies, doctrines, and training at times overly influenced by the Cold War, a period during which neither major adversary wished to fight in large metropolitan areas, operations in built-up areas have subsequently often proven unpleasantly difficult for U.S. forces. Arguably the greatest challenges are those that confront American military leaders at the highest echelons. Urban operations are only in the rarest of circumstances a matter of defeating an adversary in force-on-force combat alone. Stability and support operations are ongoing simultaneously with offensive and defensive operations. Governing an urban area means dealing with myriad interest groups and leaders, some legitimate, others not; providing vital services to a population that is likely to be suffering the deprivations of war; and establishing financial, political, and social stability in addition to that relying on removal of threat forces. Such challenges face leaders at every echelon to some extent, but those at the operational and strategic levels of war meet them in greater variety, complexity, and sheer numbers. Current U.S. military instruction and exercises do little to prepare senior officers for such urban challenges. It is therefore not surprising that these echelons are currently the most in need of joint urban operations (JUO) training.

The Services of the U.S. military have not ignored this challenge as they seek to ready their forces for future conflicts. The desperate October 1993 fighting on the streets of Mogadishu triggered U.S. Army development of a new type of urban training facility, one designed to be less like the pristine villages of northwest Europe and more akin to the chaotic environments found in densely populated areas of the developing world. The Army constructed a system of urban training areas at Ft. Polk, LA, a system centered on Shughart-Gordon Village, named after two soldiers who were awarded the Medal of Honor for their bravery while fighting in Mogadishu, as well as other similar facilities at installations throughout the world.

The Marine Corps built "Yodaville," an innovative training site in Arizona that vividly replicates the difficulties of engaging urban targets from aircraft. Just as Mogadishu might be considered the clarion call for the nation's ground forces to improve urban training, the difficulties of engaging targets from the air during actions in the Balkans and elsewhere during the late 1990s motivated improvements in training for pilots and others in the nation's air arms. Service and joint simulation initiatives likewise focused on efforts to better represent urban scenarios.

Such training initiatives influenced and were influenced by the simultaneous development of new Service and joint urban doctrine. Yet while both Service and joint doctrine received attention, improvements in urban training were largely limited to efforts within the four Services. In the view of the Senate Armed Services Committee, requests to Congress for urban training-facility construction reflected this Service centrism.

The analysis and related recommendations that follow maintain an overarching perspective. Individual combatant commanders, leaders of joint task forces, and those heading Service components worldwide will have their own training gaps, influenced by regional variables, command missions, personnel turnover, time constraints, and the many other factors that leaders confront daily. It is important to keep in mind that any training strategy has to prepare the force so that it is ready to meet both universal and regionally specific challenges. This study seeks to provide solutions that can be readily tailored to meet these particular needs in addition to those with more universal application.

Objective and Scope

In response to that need for a review, this study identifies those areas in need of redress and proposes how the Services—Army, Navy, Marine Corps, and Air Force—and other critical components of national capability can better ready themselves cooperatively for future operations in cities around the world. The result is a joint urban training (JUT) strategy for the period 2005–2011.

The foundation for this strategy is the current *Doctrine for Joint Urban Operations* as presented in the joint publication of that name (JP 3-06). The guidance in JP 3-06 includes the valuable understand, shape, engage, consolidate, and transition (USECT) concept for joint urban operations. These five phases are interdependent and overlapping.[13] Together, they effectively articulate the nature of urban contingencies and the functions that Service and joint leaders must take into account. As such, they guide development of the JUT strategy and should similarly influence its implementation in the field.

That strategy must avoid emphasizing any point on the spectrum of conflict too greatly at the expense of others. For example, future urban operations will inevitably demand that the U.S. military be skilled in bringing its collective talents to bear to provide city residents support after natural or manmade disasters, to help restore the stability essential for residents to function, to defeat an armed foe bent on causing harm, and/or to preclude these and other misfortunes from befalling urban residents domestically and abroad. The U.S. military cannot afford to address the many tasks along this spectrum sequentially; it must be able to handle these demands simultaneously in time and space.

---

[13] Joint Publication 3-06, *Doctrine for Joint Urban Operations,* Washington, DC: Joint Chiefs of Staff, September 16, 2002, pp. II-6 to II-14.

Moreover, there will be allies in such a demanding undertaking—allies from other nations, from other U.S. governmental organizations, from nongovernmental organizations (NGOs), and from private volunteer organizations (PVOs), as well as the residents living in the villages, towns, or cities that are the focus of attention. All these participants should be part of any training strategy. While the focus in this report is primarily on inter-Service urban training, there is an equally dramatic call for cooperative interagency preparation.

The focus of this study is joint urban training: the preparation by multiple Services for the cooperative conduct of operations in an urban environment. While there is no doctrinal definition for joint urban training, its character is apparent in reading the definitions of "joint training" and "joint urban operations."[14] By combining the two, we have derived a working "joint urban operations training" definition:

> **Joint urban operations training:** Training for joint operations that are planned and conducted across the range of military operations on or against objectives on a topographical complex and its adjacent natural terrain where manmade construction or the density of noncombatants is the dominant feature. This training includes mission rehearsals; relevant instruction for joint-trained individuals, units, and staffs; and use of joint doctrine or joint tactics, techniques, and procedures to prepare joint forces or joint staffs to respond to strategic, operational, or tactical requirements considered necessary by the combatant commanders to execute their assigned or anticipated missions.

---

[14] Joint training is defined as "training, including mission rehearsals, of individuals, units, and staffs using joint doctrine or joint tactics, techniques, and procedures to prepare joint forces or joint staffs to respond to strategic, operational, or tactical requirements considered necessary by the Combatant Commanders to execute their assigned or anticipated missions" (Department of Defense Directive Number 1322.18, Subject: Military Training, September 3, 2004, p. 10).

Joint urban operations are defined as "all joint operations planned and conducted across the range of military operations on or against objectives on a topographical complex and its adjacent natural terrain where manmade construction or the density of noncombatants are the dominant features" (JP 1-02, *Department of Defense Dictionary of Military and Associated Terms,* April 12, 2001, as amended through May 23, 2003, p. 291).

It has already been noted that the Services have in recent years taken considerable strides toward training their forces for operations in built-up and densely populated areas. Continued improvement and maintenance of proficiencies are required for successful joint preparation, since they are the parts that form the joint whole. A constant assumption underlying the analysis in the pages that follow is that the Services will come to the joint urban training table well prepared, just as they have come to war prepared to support the greater good of U.S. and coalition success so many times in the past.

Further, while the focus on JUO training was a given, we decided that we would not limit deliberations to the construction of training sites alone, but that we would also investigate the potential inherent in computer simulations, simulators, synthetic environments, the use of actual U.S. urban areas, and other initiatives that might complement more-traditional ways of preparing the nation's armed forces. In other words, the research incorporates live, virtual, and constructive training.

Approach

The decision to give the investigation a broad scope—one involving the full spectrum of military operations and all training approaches rather than military construction alone—introduced analytical complications. Any thorough analysis would have to provide a description not only of the training means found to be best in terms of readying personnel for urban operations, but also of those that were most cost-effective. Comparing the effectiveness of training involving computer simulations with training wherein a unit deploys to an urban training complex would be challenging enough. To attempt to compare simulation development costs with military construction funding or with the rental of an abandoned small city adds another level of complexity. Moreover, the fact that some training alternatives are new or involve concepts still under development increases the difficulty. Further, contemporary operations have demonstrated that training requirements are changing very rapidly, as opposing sides in active theaters adapt and counteradapt at an extraordinary rate.

The diverse nature of urban operations themselves and the seemingly ever-expanding character of their demands means that, first, the JUO training strategy developed had to be flexible enough and adaptable enough to be pertinent to any type of urban scenario, while also able to absorb changes in field conditions. In short, a strategy perfectly suited to a moment in time would be far less valuable than one that future users could adapt to evolving scenarios. Second, the costs associated with any proposed strategy could not simply be stated in terms of a single dollar value. The nature of U.S. defense funding means that a dollar spent on building an urban facility (military construction, or MILCON, funds) is inherently of different cloth than one committed to maintaining that structure once it is built (operation and maintenance, or O&M, funds). Accurately estimating the expenditures that would be necessary to develop the capabilities outlined in various alternatives meant having to seek out these varied costs.

The result of these several challenges is that we took a modular approach toward constructing a joint urban training strategy. A "module," as used in this context, is a collection of resources normally associated with a type of facility, simulation, or other capability of value in the design or execution of training. The modules ultimately selected collectively serve as the components of the JUT strategy developed in this study; meet all JUT requirements identified in the study, albeit to different degrees; and provide a means of comparing costs associated with very different capabilities. Requirement attainment, rather than dollar cost, becomes the primary metric for determining the value of a module and its suitability as a component of a comprehensive JUT strategy. Further, the modules are internally flexible. They can be adapted to permit comparisons of similar but not perfectly matched capabilities.

Centering the JUT strategy on modules led to a five-step analytical approach, discussed briefly below. (The sources reviewed and the individuals interviewed in support of this research are listed in the Bibliography.)

Step 1: Identify Joint Urban Training Requirements

Accurately identifying requirements is the fundamental first step in training strategy development, because the requirements are the means of comparing very different JUT approaches. We derived training requirements through studies of the Universal Joint Task List (UJTL), various current and proposed joint training tasks (JTTs), combatant command and service component needs, more than 800 observations and insights in JUO studies conducted by RAND,[15] and other documents from the field. The analysis involved both compilation and considerable synthesis to avoid redundancy and inclusion of requirements with only tangential value.

Step 2: Identify Current and Pending Joint Urban Training Capabilities

The second step is determining what JUT capabilities already exist, to avoid redundant expenditures. Our compilation of in-place capabilities involved developing as comprehensive a list as possible of existing sites on U.S. soil (and selected others).[16] In addition, we looked at existing urban simulations capabilities and other live, virtual, and constructive resources. This initial search also sought to identify what capabilities in these several areas were under development or likely to be fielded in the 2005–2011 period. Further, joint and Service mili-

---

[15] The reports from the first two phases of the Operation Enduring Freedom (OEF) and Operation Iraqi Freedom (OIF) joint urban operations observations and insights study are Russell W. Glenn, Christopher Paul, and Todd C. Helmus, *Men Make the City: Joint Urban Operations Observations and Insights from Afghanistan and Iraq*, Santa Monica, CA: RAND Corporation, 2005; and Russell W. Glenn and Todd C. Helmus, *Men Make the City 2: More Joint Urban Operations Observations and Insights from Afghanistan and Iraq*, Santa Monica, CA: RAND Corporation, 2005.

[16] At times, we shared the same frustrations that confront others who attempt to compile a comprehensive list of the four Services' urban training facilities. While this was in part attributable to the currency of some compilations, it was more frequently a matter of interpretation (i.e., what qualifies as an urban training site?). Facilities that consisted of a significant number of urban structures or a single structure with obvious urban-specific training value (e.g., a live-fire shoot house) appeared on virtually every Service list. However, some installations might include a grenade range with a building façade, allowing trainees to throw grenades through a cutout representing a window. Other range managers would rightly conclude that so minimal an urban element did not qualify a site as "urban." The list in Appendix E does not include such marginal entries that could be identified.

tary school courses on urban operations and personal professional-education programs with a similar emphasis were included. These collectively comprise the American military's JUO training capabilities. An underlying study assumption is that there will be no substantial change in existing capabilities. Any actions that counter this assumption will obviously influence the final JUT strategy proposals and would require adjustments in the design of training and funding.

Step 3: Identify the Gap Between Requirements and Capabilities

Determining requirements and capabilities allows us to determine outstanding and future JUT shortfalls. Identification of those shortfalls comprises the third step in the JUT strategy development process. We relied on a simple requirements-minus-capabilities construct. The present shortfall between what is needed and what exists offers the starting point for developing a strategy to address immediate needs. The future shortfall, based on projected requirements and capabilities in the 2008–2011 time frame, similarly acts as a foundation for the longer-term component of a JUT strategy.

Step 4: Complete Initial Steps Toward a JUT Strategy

The fourth step focuses on two initial tasks in the creation of a JUT strategy. The first task is module definition. Each module represents a capability that either itself addresses the short- and longer-term shortfalls or supports other modules in closing the requirements-versus-capabilities gap. Initial module definition was completed without prejudice (i.e., feasibility, effectiveness, and cost were generally not considered in developing the initial module list). Only after each module was analyzed with respect to how well it addressed the final set of training requirements was an initial cut made. Those modules failing to substantially address outstanding needs were deemed unworthy of further analysis. We performed a series of additional reviews of the remaining modules with respect to more-rigorous feasibility (e.g., effects on the environment, safety concerns), effectiveness (the number of and extent to which identified requirements were met), and cost considerations; the latter analysis—cost analysis—is the second task in Step 4.

Step 5: Complete Final Steps Toward a JUT Strategy

In this final step, we considered the costs of the modules in association with their ability to meet JUT requirements. This allowed us to identify near-term (2005–2007) and longer-term (2008–2011) actions that address the training shortfalls. The modular approach to strategy design makes the proposals flexible enough to withstand likely changes in the operational environment, changes that will directly impact the preparation needed to meet the challenges of actual operations.

Experience teaches the value of a building-block approach to training, i.e., starting with simple tasks and moving on to more-complex ones, beginning with skills developed by individuals and melding them to construct collective talents. Readying the whole is best accomplished by first readying its parts. Only when the pieces have achieved a requisite level of expertise should they be joined to practice what they will ultimately execute in reality. This hierarchical approach to training underlies virtually every component of this study. Talents, individual and corporate, small group and large, build to the point of entire Services, agencies, and coalitions being able to together confront successfully even the most difficult challenges.

The building-block concept is not without risk. Effective training at each echelon, as well as the ultimate readying of the collective whole, depends on every component meeting a reasonable level of capability. Every block depends on the others. Each component has broad and multifaceted responsibilities. Service preparation will vastly outweigh training designed and managed by joint headquarters in terms of hours and dollars spent. Service schools and unit commanders will have to ready personnel and organizations, then assess the level of readiness to participate as part of the joint team in either exercises or actual operations. This is not to imply that combatant commanders, the Joint Chiefs of Staff, and the Office of the Secretary of Defense (OSD) should not also bear heavy burdens. Theirs is the collective responsibility to identify evolving requirements, set overarching standards, write the doctrine on which the Services depend for consistency throughout the force, educate those on or supporting joint headquarters, and gauge whether the force is capable of doing what must be done during international deployments or when there is call for military support of domestic contingencies. The components of this strategy are designed to provide tools of value across the levels of war and throughout joint and Service components. Leaders at every echelon have to put those implements to good use in preparing and employing the military element of U.S. national power.

Organization of This Document

Chapter Two provides a full explanation of the above-noted five-step process and presents the resultant list of requirements used in developing the JUT strategy, while Chapter Three provides the process for deriving the capabilities to address those requirements. Chapter Four identifies near- and long-term shortfalls that emerge from comparing the requirements identified in Chapter Two with the capabilities identified in Chapter Three. Chapter Five more fully explains the modular concept and defines modules to be used in developing the JUT strategy, while Chapter Six presents the results of the cost analysis conducted on the ultimate set of modules identified in Chapter Five. Chapter Seven covers development of the strategy itself, using the results from Chapters Five and Six to design a JUT strategy. It also further discusses how users can employ various combinations of modules to meet specific training requirements and combinations of requirements during the process of designing alternative training strategies. The final chapter provides a project synopsis and also presents recommendations to help in applying the JUT strategy in the field. For example, in addition to being the pieces from which strategies are made, the modules offer joint commanders a tool that can help them design urban training for their own or Service component organizations. Aligning modules with the requirements to be addressed during a training event gives these commanders a starting point for determining the type and number of joint urban capabilities they will require to meet training objectives.

A series of appendices support the results presented in the main document. Appendix A presents selected joint training definitions, some doctrinal and others developed by us to assist in strategy construction. Appendix B presents a roster of 250 JUT requirements and their sources, and Appendix C demonstrates the relationship between each requirement and one or more UJTL or JTT entry. Appendices D and E present the RAND urban training facility survey instrument that was sent to identified training facilities and a summary of the results of that survey. Appendix F provides a discussion of training retention. Appendix G is a detailed mapping of modules versus requirements, including the complete matrix and the matrix culled to include only the modules that become part of the JUT strategy.

CHAPTER TWO

Identifying Joint Urban Training Requirements

Build me a decent MOUT [military operations on urbanized terrain] facility that I can do the three block war in . . . that will hold an infantry battalion!

Anonymous USMC major

What we do not do very well is set the conditions . . . at the battalion and brigade level, for these guys to be successful not only during the fight, but after the fight.

LTG William Wallace
Commander, U.S. Army Combined Arms Center

Members of the Department of Defense shall receive, to the maximum extent possible, timely and effective individual, collective, unit, and staff training necessary to perform to standard during operations.

Department of Defense Directive Number 1322.18
Subject: Military Training,
September 3, 2004

Introduction

To determine shortfalls in current JUT needs, we must first identify the requirements that need to be trained for. This chapter describes the process we used to derive those requirements.

The Process

Development of a JUT strategy began with the aforementioned investigation of existing JUT doctrine. This provided the doctrinal construct that guided the strategy's design and assisted in identifying relevant training requirements. The three-step process employed to arrive at the final set of requirements used in the remainder of the analysis is shown in Figure 2.1.

The first step was a comprehensive review of Service and joint doctrine, various official and unofficial source materials, and input from interview subjects, combatant commands, and Service representatives. This initial review produced 430 candidate tasks. The next step—conducting a first screen—eliminated tasks that were redundant or neither essentially joint nor urban, which reduced the list to 250 detailed JUT tasks. The third step was a group process of expert judgment that further synthesized, aggregated, and summarized the 250 tasks into 34 consolidated tasks that are comprehensive (i.e., leave no pertinent tasks uncovered), of manageable scope, and appropriate to the assessment of capabilities.

Step 1: Conducting the Comprehensive Review

The intent of the first step was to cast as large a net as possible in identifying candidate JUT requirements. We started by working through the relevant documents that would contain such requirements. This document-based review involved three primary source types: doctrinal, nondoctrinal official (doctrine under development or nondoctrinal handbooks), and unofficial, including interview transcripts, lessons-learned documents, and reports. Relevant documents are listed in the Bibliography.

Individual documents were selected for either their doctrinal relevance, general applicability to joint urban operations, or the established expertise of their authors. Each source was examined thoroughly for both stated and implied requirements. Any task or requirement that seemed even peripherally related to JUO was included in the initial requirements list and joined the initial master list. It

Figure 2.1
Process of Identifying JUO Training Requirements

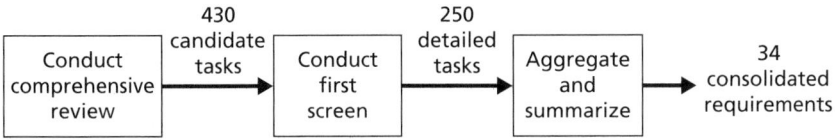

should be noted that this requirements process sought to identify both existing and future requirements through the year 2011, the time period of interest in this study.

We supplemented these documents with interviews conducted during visits to U.S., Dutch, British, Canadian, and Australian urban training facilities both during the project and prior to its formal initiation. Further input from combatant commander and Service component headquarters representatives about their JUT requirements complemented these written and interview sources. As noted in Chapter One, the Bibliography provides a list of sources accessed and individuals interviewed during the study.

Step 2: Conduct a First Screen

Compiling stated and implied requirements from the written sources, interviews, and combatant and component command inputs produced an initial set of some 430 candidate requirements, as shown in Figure 2.1. This set included redundant tasks listed in multiple sources, higher-level Service tasks captured from non–joint-specific sources, and tasks that are at best only tangentially related to urban operations.[1]

This initial task list was not exhaustive with respect to urban training needs. We recognized at the outset that the study focus was *joint* urban training. The initial set of 430 tasks generated from the

---

[1] Although this study articulates necessary JUT strategy in terms of requirements, such needs are often stated in task form when they appear in doctrinal or joint training publications. The terms *requirements* and *tasks* are used interchangeably in this chapter.

comprehensive review included some combined arms tasks that were later determined to be too Service-specific and others whose relevance to urban operations was considered too marginal to be of significant import to joint urban training. As a result, any task that was obviously neither urban nor joint was eliminated from further consideration during this first screen. Rejections included all requirements related to Service-specific tasks (room-clearing, placing a breaching charge on a door, assuming shooting stances for various wall apertures, and others for which training would be primarily the Services' responsibility) and any task that clearly was not affected by the urban environment.

The ultimate set of JUO training requirements can be envisioned as the intersection between all joint and all urban training needs (see Figure 2.2). The size and contents of that intersection vary depending on how one defines *joint* or *urban*. One interview respondent went so far as to suggest that there is no task that is inherently both joint and urban; his view was that joint tasks are strictly interoperability and communications issues and that these issues are no different in urban environments than they are in any other terrain or environment. Clearly, this is an extreme view that few experts would agree with; it does, however, point to a need to identify definitions used and any assumptions made.

"Urban" is treated as a "condition" in joint doctrine and is therefore minimally defined.[2] JP 1-02 (*DoD Dictionary of Military and Associated Terms*) lacks a separate entry for *urban*, but its essential characteristics are evident in viewing the definition of *joint urban operations*.[3] We used definitions presented in doctrine, using additional definitions only when they were necessary to provide a common understanding.

---

[2] See Chairman of the Joint Chiefs of Staff, *Universal Joint Task List (UJTL)*, Chairman of the Joint Chiefs of Staff Manual CJCSM 3500.04C, Washington, DC, July 1, 2002, for an example.

[3] JP 1-02, *DoD Dictionary of Military and Associated Terms*, April 12, 2001, as amended through May 23, 2003.

Figure 2.2
JUO, the Intersection Between Joint and Urban

Broad, inclusive definitions cause JUO requirements to be quite extensive

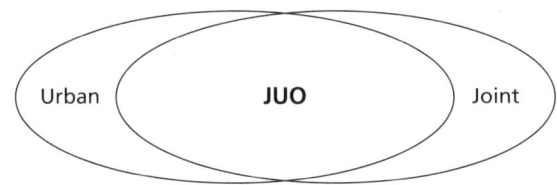

Narrow, exclusive definitions cause JUO requirements to be quite limited

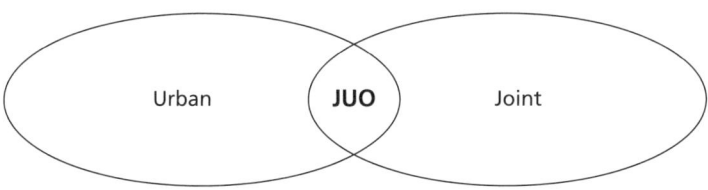

RAND *MG439-2.2*

Doctrinal definitions included the following:

- **Joint**: Connotes activities, operations, organizations, etc., in which elements of two or more military departments participate (JP 1-02, *DoD Dictionary of Military and Associated Terms*, April 12, 2001, as amended through May 23, 2003, p. 275).
- **Joint force**: A general term applied to a force composed of significant elements, assigned or attached, of two or more military departments operating under a single joint force commander (JP 1-02, *DoD Dictionary of Military and Associated Terms*, April 12, 2001, as amended through May 23, 2003, p. 279).
- **Joint urban operations**: All joint operations planned and conducted across the range of military operations on or against objectives on a topographical complex and its adjacent natural terrain where manmade construction or density of noncombatants is the dominant feature (JP 1-02, *DoD Dictionary of Military*

and Associated Terms, April 12, 2001, as amended through May 23, 2003, p. 291).

- **Joint training:** Training, including mission rehearsals, of individuals, units, and staffs using joint doctrine or joint tactics, techniques, and procedures to prepare joint forces or joint staffs to respond to strategic, operational, or tactical requirements considered necessary by the combatant commanders to execute their assigned or anticipated missions (Department of Defense Directive Number 1322.18, Subject: Military Training, September 3, 2004, p. 10).

Simplicity has a clarity and quality of its own. We applied the definitions above in answering the following straightforward question: If you did that in an urban environment, would you have to do it differently? If the answer was, "Not really," then the task or requirement in question was not included in the final enumeration. If the urban environment or one of its obvious correlates (e.g., built-up terrain, density and pattern of light and electronic signatures, high density of noncombatants) significantly impacted the execution of a given task, that task made the cut.

A similar screening method eliminated requirements that were insufficiently joint in character. It was this investigation that led us to reconsider our approach to the concept of joint training. Rather than considering training as exclusively joint or not, we concluded that the joint character of training is better viewed as akin to a continuum articulated by five levels, as shown in Table 2.1. We found this greater resolution helpful as we considered the design of a JUT strategy, but we also recognize that it has broader potential. Additional benefits include its use in support of a building-block strategy for JUT; e.g., a combatant or joint task force (JTF) commander can tailor exercises to improve preparedness via working from simpler joint coordination such as that characterizing Levels 1 or 2 to more-complex and demanding Level 4 events. These commanders can also use it for guidance as they develop specific training strategies to support their

Table 2.1
Levels of Joint Urban Training

Level	Description
0	• Single-Service urban training event with no participation by other Services
1	• Urban training event in which two or more separate Service components orchestrate their activities to achieve common objectives at one or more levels of war (strategic, operational, or tactical)
	• Single Service's actions dominate the event
	• Interoperability plays a minor or superficial role in accomplishing assigned missions or attaining specified objectives in the training scenario
	• Limited interaction occurs between command echelons other than within Services; significant vertical coordination may take place within Service components
	• No substantial joint headquarters or other joint synchronization element participation
2	• Urban training event in which two or more separate Service components orchestrate their activities to achieve common objectives at one or more levels of war (strategic, operational, or tactical)
	• While event involves greater reliance on one Service throughout or during some phases of training, frequent and substantive interoperability between at least two Services significantly influences accomplishing assigned missions or attaining specified objectives in training scenario
	• Limited interoperability occurs between command echelons other than within Services, though vertical integration may take place within Service components
	• May or may not involve participation by joint headquarters or other joint synchronization element
3	• Urban training event in which two or more separate Service components orchestrate their activities to achieve common objectives at one or more levels of war (strategic, operational, or tactical)
	• Event might involve greater reliance on one or more Services than others during some phases; virtually continuous orchestration of multiple Services' capabilities in a single primary environment (i.e., land, air, sea, or space) within and between phases dominates training and is essential to accomplishing assigned missions or attaining specified objectives in a training scenario
	• Significant interoperability occurs horizontally between command echelons across Service boundaries and vertically between Services and joint echelons
	• A joint headquarters, joint effects cell (JEC), or other joint synchronization element orchestrates joint organization and inter-Service capabilities
4	• Urban training event in which two or more separate Service components orchestrate their activities to achieve common objectives at one or more levels of war (strategic, operational, or tactical)

Table 2.1 (continued)

Level	Description
4 (cont.)	• Event might involve greater reliance on one or more Services than others during some phases; virtually continuous interoperability of multiple Services' capabilities in multiple primary environments (i.e., land, air, sea, or space) within and between phases dominates training and is essential to accomplishing assigned missions or attaining specified objectives in a training scenario • Significant interoperability occurs horizontally between command echelons across Service boundaries and vertically between Services and joint echelons • A joint headquarters, JEC, or other joint synchronization element orchestrates joint organization and inter-Service capabilities

individual command requirements (including which modules to incorporate in those strategies, as discussed in Chapter Five). Those responsible for facility and simulation designs could similarly turn to the levels for assistance when determining what characteristics these capabilities should possess.

We referred to the UJTL and JTTs to help delineate joint from Service urban training requirements.[4] Though both are fairly general in their articulation, they nonetheless provided another sieve to help in separating joint from Service-specific tasks. Removing Service-specific tasks, redacting redundancies, and eliminating tasks not significantly impacted by the urban environment reduced the initial 430 JUT tasks to approximately 250. These 250 tasks are listed in Appendix C.

Step 3: Aggregate and Summarize

Though a notable improvement over the initial list of 430 tasks, 250 is still an unwieldy number of detailed tasks to work with in comparing requirements and capabilities to develop a JUT strategy. Could the list be further culled without loss of resolution? The first

---

[4] The designation and definition of JTT were modified in the closing stages of this study. *Joint training tasks* were effectively replaced by *joint task articles*, which encompass tasks other than tactical ones, as well as JTT. This redesignation has no significant impact on the analysis performed in support of the study, in considerable part because of our simultaneous investigation of JTT and UJTL.

step in answering this question was to determine whether further commonality existed that would allow similar or overlapping tasks to be combined. Such a process was found to be feasible; the results provided the following 16 broad requirements categories:

- Overarching/unifying observations
- Fight together
- Maneuver together
- Manage unit adjacencies and transitions
- Manage mission transitions
- Avoid fratricide
- Treat, evacuate, and transport casualties
- Base and protect the force
- Integrate command, control, communications, computers, intelligence, surveillance, and reconnaissance (C4ISR)
- Coordinate interactions with the civilian population
- Coordinate with nonmilitary government agencies, NGOs, PVOs, and other relevant governments, agencies, or organizations
- Logistics
- Training needs, not elsewhere classified
- Facility requirements
- Service-level requirements, either new or unsatisfied
- Environment

These categories varied in their value to the study. Some, such as "avoid fratricide," represented a comprehensive and appropriately clear requirement definition. Others, such as "fight together," required consolidation or separation to capture the detail and clarity sufficient for use in developing a training strategy. The process of employing these common categories to reduce the roughly 250 requirements to a final, manageable list had three components:

- Retain the comprehensiveness that was the hallmark of the initial detailed task listing
- Reduce the requirements list to a manageable size

- Produce a list of requirements that addressed all pertinent UJTL and JTT elements

The character of this process is evident in viewing Appendix C. The approximately 250 detailed JUT requirements were organized using the 16 categories listed above. The original source for each task is indicated (e.g., UJTL, JTT, a particular document), as is the ultimate requirement to which they contributed.

These requirements overlap; it is infeasible to designate them in such a manner that they do not. Such is the complexity of military operational environments, a complexity increased multifold in cases where the environment involves a significant urban component. "Conduct stability operations in the urban environment" and "conduct support operations in the urban environment" are inseparable from "govern in the urban environment"; many subtasks are shared. The same is true of "conduct stability operations" and "conduct support operations" when instability is an issue; without provision of employment, life's necessities, and other forms of support, achievement of stability is virtually impossible. Definition of mutually exclusive requirements would be rife with artificiality; to fail in listing any of the requirements in our final set would risk leaving unidentified a critical element necessary for preparing the U.S. joint force for future contingencies.

The final 34 joint urban training requirements employed throughout the remainder of this analysis are listed in Table 2.2. The ordering does not imply primacy or any other form of prioritization; every task is essential to the development of a comprehensive JUT strategy for 2005–2011. However, the lack of prioritization does not imply that some tasks will not be more significant than others for given JUT aspects. Which of the tasks are most important to a given combatant commander, subordinate joint commander, operation, or mission will vary. That variation will be reflected in the appropriate commander's joint mission-essential task list or other written guidance, including his prioritization of requirements to prepare for particular contingencies.

Table 2.2
Consolidated Joint Urban Training Requirements

Avoid fratricide
Communicate in the urban environment
Conduct airspace coordination
Synchronize joint rules of engagement
Conduct stability operations in the urban environment
Conduct support operations in the urban environment
Conduct urban human intelligence (HUMINT) operations
Conduct urban signal intelligence (SIGINT), imagery intelligence (IMINT), measurement and signatures intelligence (MASINT), communications intelligence (COMINT), electronic intelligence (ELINT) and other intelligence efforts
Conduct urban operations exercises
Integrate urban operations with other relevant environments
Coordinate maneuver in the urban environment
Coordinate multinational and interagency resources
Govern in the urban environment
Identify critical infrastructure nodes and system relations
Navigate in the urban environment
Plan urban operations
Provide common situational awareness
Provide fire support
Provide security during urban transition operations
Rehearse/war-game urban operations
Conduct urban noncombatant evacuation operations (NEOs)
Conduct U.S. domestic urban operations
Conduct urban combat search and rescue (CSAR)
Conduct urban operations during and after a WMD event
Consolidate success in the urban environment[a]
Disembark, base, protect, and move in urban environments
Engage in the urban environment
Orchestrate resources during urban operations
Shape the urban environment
Sustain urban operations
Transition to civilian control
Understand the urban environment
Achieve simultaneity in meeting requirements
Conduct training across multiple levels of war

[a] This requirement is derivative of the five phases of urban operations as outlined in Joint Publication 3-06, *Joint Urban Operations.* The phases, which overlap in time, are understand, shape, engage, consolidate, and transition. *Consolidate* can be succinctly described as protecting and building upon what has been gained via previous action. Relevant functions might include restoration of a stable and secure social and political environment and the repair of damaged infrastructure.

With this consolidated list, we now turn to the enumeration of existing JUT capabilities and their potential to satisfy the 34 requirements.

CHAPTER THREE

Determining Current and Planned Joint Urban Training Facilities, Simulations, and Other Training Capabilities

> We train a company team plus in Marnehuizen. It is not large enough to properly train a unit of greater size.
>
> Lieutenant Colonel Henk Oerlemans, Dutch Army, speaking of his army's 120-building urban training facility (paraphrase)

> The DoD Components, to the maximum extent possible, shall share training resources, ranges, maneuver areas, and other facilities and devices that have training or test potential.
>
> Department of Defense Directive Number 1322.18
> Subject: Military Training,
> September 3, 2004

Introduction

The shortfalls discussed in Chapter Four were identified by comparing the 34 requirements identified in Chapter Two against the capabilities that exist to meet those requirements. By "capabilities," we mean any assets that contribute to joint urban training. Such assets can be in support of horizontal, vertical, or functional training or integration exercises.[1] They might involve classroom instruction; live,

---

[1] "1. Horizontal training exercise: Build on existing Service interoperability training. 2. Vertical training exercise: Link component and joint command and staff planning and execution. 3. Integration exercises: Enhance existing joint exercises to address joint interoperability training in joint context. 4. Functional training: Provide dedicated joint training environment for functional warfighting and complex joint tasks" (U.S. Joint Forces Command,

virtual, or constructive training in field or simulated environments; or any combination of these.[2] Supporting capabilities include instructors, observer/controllers (O/Cs), opposing forces (OPFOR), and culturally savvy noncombatant role players. The character of capabilities ranges from human beings to urban structures and complexes, as well as targetry, props, and other support for live or simulated fires. Instrumentation that permits the recording or monitoring of exercises or the integration of training at different sites is yet another form of capability, as are simulators and simulations for individual trainees or groups linked to provide broader context. Environmental factors influence capabilities; at times, limits to the extent of damage allowed during training or demands that training sites be neat so as to present a "military appearance" impinge on the effectiveness of instruction, as do airspace restrictions and concerns about local political and social sensitivities or natural habitats.

This discussion concentrates on three capability groups that will play primary roles in the development of a JUT strategy: (1) purpose-built urban training sites (i.e., current and planned U.S. urban training sites and the capabilities found at them);[3] (2) current and projected simulations, simulators, and training involving synthetic environments (hereafter collectively referred to as *simulations*); and (3) innovative or novel urban training sites, which may offer benefits

---

"Western Range Complex Joint National Training Capability Horizontal Training Exercise," briefing, January 2004).

[2] "E2.1.11. Live, virtual, and constructive training: A dynamic training and operational environment, using live, virtual, and constructive simulations, that provides an interoperable, networked, training capability that includes mission-rehearsal. Live simulation involves real people operating real systems; virtual simulation involves real people operating simulated systems; and constructive simulation involves simulated people operating simulated systems," U.S. Department of Defense, Public, M&S Resources: Online M&S Glossary (DoD 5000.59-M). Online at <https://www.dmso.mil/public/resources/glossary/results?do=get& def=297≥ (as of September 27, 2005).

[3] Here, *purpose-built urban training sites* refers to what are commonly called MOUT (military operations on urbanized terrain) complexes. These are dedicated military training facilities for the conduct of live, manned, urban training, and are meant to be distinguished from similar civilian or private facilities and military simulation facilities. Note that human/ OPFOR assets do not have to be associated with a particular facility. In Chapter Seven, we discuss the notion of a "traveling OPFOR," as well as traveling noncombatant role players.

either in the generic sense or in cases of specific instructional needs (e.g., the use of ships, factory complexes, abandoned urban areas, closed military installations, commercially available sites or those leased by public institutions, amusement parks). We discuss our approach for each of these capability groups.

Purpose-Built Urban Training Sites

Approach

Compiling a list of all U.S. armed forces urban training facilities proved somewhat difficult, for two reasons. First, various Service commands had collations that differed in purpose and content. For example, some presented only those training sites on which funds were to be expended in the near term, while others failed to include smaller complexes used only occasionally. Second, the Services lack an agreed-upon definition for *urban training site* or *urban training facility*. Some sources included training resources with little more than a building façade through which soldiers threw training grenades, while others demanded greater numbers and more sophistication in structures comprising such facilities. Given these difficulties, we sought to identify all the sites that could legitimately be considered viable means to train sections, teams, aircraft crews, and larger units on urban-related skills. (Using this standard, building façades on grenade ranges and the like do not qualify for inclusion.)

The approach we took to identify the capability provided by purpose-built training sites was similar to that taken for identifying requirements: start comprehensively and then screen down. Starting out comprehensively gave us the greatest universe of facilities from which to select those with maximum JUT potential. In addition, a comprehensive list should prove useful to both joint and Service endusers who are planning training and seek to identify sites most pertinent to their needs and in reasonable proximity to home stations.

We relied on a wide range of sources in compiling the comprehensive list of facilities. Our ten-plus years of work in urban operations helped in expanding initial lists provided by the Office of the

Secretary of Defense–Readiness. Searches of the U.S. armed forces Non-Classified Internet Protocol Network (NIPRNET), as well as the Internet, expanded the roster and enhanced the information available on individual sites. Many of these sources contained references to other pertinent materials, allowing an inductive expansion of the initial source base. Additional lists provided by representatives from headquarters within the several Services and studies conducted prior to this effort further increased the number of facilities identified. These included materials from the 2001 J8 Dominant Maneuver "Joint Urban Training Facility Study" and the FY03 "Ongoing or Proposed Urban Operations Training Facility Investments." (Complete citations appear in the Bibliography.) Additional documents included doctrinal and other sources that referred to training facilities and at times provided further material on those facilities' characteristics and capabilities. Individuals interviewed during site visits (described below) provided additional data and theretofore unidentified references. Subject-matter experts, including Lieutenant General G. R. Christmas (USMC, ret.), Captain (Promotable) Robert Harward (USN), and Lieutenant Colonel Gregory McMillan (USA, ret.), reviewed the facility list and provided supplemental information.

Because we decided to err on the side of inclusiveness in compiling our ultimate list (with the exception of facilities of truly marginal value such as the building façades mentioned above), we ended up with an unavoidable inconsistency in the quality of information available on identified sites. At times, we included a facility even though we were unable to acquire the desired level of basic information about its size, scope, or other important characteristics. In other instances, we were unable to verify or augment the descriptive material available from incomplete, unclear, or fairly old sources.

Appendix E provides the quality of information available on each site that is included on the comprehensive list. We are confident that no major urban site was overlooked; this confidence has been reinforced by training experts who expressed belief that the products used were the best available at the time of writing.

From this comprehensive list, we selected those sites deemed to have the greatest potential to support joint preparation for urban operations. More specifically, the sites were selected for their uniqueness or because they possessed characteristics thought to be of value in determining what JUT resources a site should possess. For these sites, we decided to gather more information through site visits or (when not feasible) off-location interviews. To facilitate this approach, we designed a site survey instrument (reproduced in Appendix D), which we used as a tool to guide the data collection effort, either sending it in advance to sites we visited or using it as part of the interview process for off-location interviews.

The selected urban training sites (both purpose-built and novel) are listed below, along with the way the information was collected (by site visit or off-location interview). The tables in Appendix E include the more-comprehensive information obtained from the further investigation of these locations. (We wanted to make site visits to Camp Lejeune, NC, Ft. Bragg, NC, and several other facilities; unfortunately, we had to forgo those visits because of time constraints. We compiled information on their capabilities from previous visits and other sources.)

- Camp Pendleton, CA (site visit)
- Twentynine Palms, CA (interview)
- Yuma – Yodaville Training Range, AZ (site visit)
- Yuma – Yuma Proving Grounds "Little Baghdad" test range, AZ (interview)
- Nellis AFB, NV (site visit)
- Ft. Irwin, CA (site visit)
- Muscatatuck, IN (site visit)
- Joint Readiness Training Center, Ft. Polk, LA (site visit)
- Blackwater Inc., Moyock, NC, training facility (site visit)
- Ft. Knox, KY (site visit)
- 2nd Special Naval Warfare Group Training Facility, Norfolk, VA (site visit)
- Marine Corps Security Force Training Facility, Chesapeake, VA (site visit)

- Hurlburt Field, FL (site visit)
- Playas, NM (site visit)
- Dutch Army Oostdorp and Marnehuizen urban training facilities (site visits)
- British Army Copehill Down Village training facility (site visit) and Operational Training and Advisory Group (OPTAG) installation (interview)
- Singaporean Army Saremban Fiba urban training facility (visit)
- Bagram AFB, Afghanistan, urban training site (site visit)
- Ft. Benning, GA, McKenna MOUT site (site visit)

Basic Facility Types

All the purpose-built sites identified, in both the comprehensive list and the screened list, contain some combination of five site types:

- MOUT complexes
- Urban target ranges
- Shoot houses
- Aerial ranges
- Temporary or façade ranges

We discuss each of these facility types below before turning to an assessment of which types exist on each of the sites on our screened list.

MOUT Complexes

MOUT complexes generally have the greatest joint urban training and exercise potential. They usually consist of a collection of structures that together give the appearance of a small to medium-size village. Streets are clearly defined and often have curbs. Buildings, rarely of more than two to three stories, include doors, windows, and internal rooms. Many have additional supporting elements of infrastructure such as rail lines, streetlights, and signage.

Official MOUT complexes are designated as either collective training facilities (CTFs) or more-ambitious combined arms collective training facilities (CACTFs). There is no standard definition for a CTF; such a facility can be large or small. The average number of buildings is 27.[4] A CACTF is more rigorously defined; i.e., such a facility "consists of 2.25 square kilometers of urban sprawl with 20 to 26 buildings, roads, alleys, parking areas, underground sewers, parks, athletic fields, and command and control buildings. The actual size and configuration of the CACTF depends on the local installation site requirements. The CACTF is designed to support heavy and light infantry, armor, artillery, and aviation positioning and maneuver."[5]

Reserve and National Guard installations countrywide include urban training sites of varied size and capability. Many of these are underutilized during winter months and offer commanders of active-duty units alternative locations for training, the benefits of which include a lack of familiarization with the specific facility.

Urban Target Ranges

Urban target ranges are valuable resources for Service-level building-block training but are of limited value for most joint instruction. Such sites consist primarily of live-fire façades in which only the front surface of structures is replicated. They serve as areas for practicing various urban marksmanship skills and tasks, and they can include more-extensive capabilities such as shoot houses (see below). The lack of complete structures and the primary focus on targetry make such ranges very difficult to conceive of as useful for other than the most specific and lowest tactical-level joint tasks. Their primary use is for training individuals or small elements on skills that all participants should have mastered before coming to a joint training event.

Urban assault courses (UACs) are used to train individual soldiers, squads, and platoons on basic urban tactical tasks. A standard UAC contains five stations:

- Individual and team trainer
- Squad and platoon trainer

---

[4] Dominant Maneuver (DM) Division, J-8, Joint Chiefs of Staff, *Joint Urban Operations (JUO) Training Facility Study Phase III Final Report*, Washington, DC, 2001.

[5] Department of the Army, *Military Operations on Urbanized Terrain (MOUT) FM 90-10*, Washington, DC, August 15, 1979.

- Grenadier gunnery trainer
- Urban offense/defense trainer (two-story)
- Underground trainer[6]

"MOUT assault course (MAC)" is an older designation for a similar facility, one lacking the dynamic targetry and more-complex stations of a UAC such as shoot houses or underground trainers.[7]

While urban target ranges are by themselves not suitable for complex or joint training events above joint training level 1 (defined in Table 2.1), they may contribute to joint capabilities at an urban training site. For example, at Camp Pendleton, CA, the MOUT complex is collocated with a MAC. The MAC sits at one end of the MOUT complex, with its surface danger zones (SDZs) oriented away from the rest of the complex. This layout allows participants conducting sniper training to be positioned in buildings inside the MOUT complex while directing live fire toward targets in the MAC (because exemptions are granted from the standard range regulations).[8] While this particular site is used for training for Service and not joint tasks, it does provide an interesting example of synergistic benefits from collocation of different types of facilities. Similar innovation and multiuse design could well have joint training applications.

---

[6] John Le Moyne, Commandant of the United States Army Infantry School, "U.S. Army Urban Operations Training Strategy," briefing, February 2001, slides 49–55.

[7] Combined Arms MOUT Task Force (CAMTF) Study Group, *Urban Operations (UO) Resource Requirements and Combined Arms Training Strategy*, Volume V, *Final Report*, September 30, 2001.

[8] Interview/site visit with Terry Finch, Bill Ash, Andy Chatelin (Range Management), and Maj Bill Russel and Staff Sergeants Baker and McCarty, U.S. Marine Corps, by Christopher Paul and Barbara Raymond, Camp Pendleton, CA, May 27, 2004. The Dutch Army Oostdorp urban site has a very similar sniper training capability designed in the same manner (interviews and site visits with Lieutenant Colonel Henk Oerlemans and Lieutenant Colonel Johan Van Houten, Dutch Army, by Russell W. Glenn, the Netherlands, December 7–8, 2004).

Shoot Houses

A shoot house is a live-fire venue for Service training in room entry and clearing. A certified UAC should include a shoot house. The standard shoot house is a "complex single-story building with multiple points of entry. Units are trained and evaluated on their ability to move tactically (enter and clear a room; enter and clear a building), engage targets, conduct breaches, and practice target discrimination."[9] While the standard shoot house is a single-story facility, Special Operations Forces (SOF) require two-story shoot houses that allow users to rappel down exterior surfaces.[10]

Either kind of shoot house located as a standalone entity is suitable only for training of Service-level tasks or the lowest-level joint training exercises. It may be advisable to position shoot houses adjacent to or as part of larger MOUT complexes or other facilities adapted for urban training during future construction. Such action would support joint forces training of missions such as SOF takedown of the shoot house while other units isolate and secure the area.

Aerial Ranges

Aerial ranges allow aviators to train for urban combat. Such ranges are either instrumented for "no drop scoring" (a system that "scores" accuracy based on simulated release of ordnance and an aircraft's trajectory at the time of simulated release) or rely on live observation of inert munitions impacts for evaluation.

While there are numerous aerial ranges in the continental United States (CONUS), there are very few *urban* aerial ranges. The inert-drop, live-fire urban ranges in the U.S. armed forces inventory (e.g., the Yodaville range at Marine Corps Air Station Yuma, AZ, and the range at Mountain Home AFB, ID) lack the necessary size, complex-

---

[9] Combined Arms MOUT Task Force (CAMTF) Study Group, op. cit.

[10] U.S. Special Forces Command, *Tiger Team Report: Global Special Operations Forces Range Study*, McDill Air Force Base, FL, January 27, 2003.

ity, or scale or otherwise fall short of realistic urban training for aviators.[11]

Buildings at aerial ranges are generally not real, but rather are
façades, mock-ups, or other structures that provide the appearance of
a built-up area from the air but do not convey the same impression to
personnel on the ground. The two images of the Yodaville range
shown in Figure 3.1 demonstrate this point.

Yodaville was built from shipping containers and cluster-bomb
(guided bomb unit (GBU)) packing containers that were stacked,
welded, and painted. From the ground it looks like stacks of shot-up

Figure 3.1
**Marine Corps Air Station Yuma's Yodaville Range Viewed from the Air
and from the Ground**[12]

SOURCE: Photo courtesy Air Force Special Operations Command (left) and
Christopher Paul (right).

---

[11] A Nellis AFB initiative to build a large no-drop urban target-tracking facility with quite
sophisticated shipping-container structures included a small northern part with half-buried
containers meant to replicate tunnels or caves. While the concept has some individual points
of merit (e.g., the use of robotic vehicles as targets), the training payoff is questionable given
the facility's notably high cost and a prohibition on munitions drops in the "urban area"
proper (interview and site visit with Lt Col Lloyd "Ring" Ringgold, USAF, and Spencer
Anderson, by Russell W. Glenn, Nellis AFB, August 11, 2004).

[12] The aerial photo is from a briefing by Jim Stott, "AFSOC Range Initiatives," Air Force
Special Operations Command, briefing, May 17, 2004. Christopher Paul took the ground
image during a visit to the site.

shipping containers; from the air, it looks convincingly urban. However, similar future efforts will need improving. Issues of scale (e.g., the height of building floors and therefore the overall height of buildings is incorrect) and signature (replicating light, electromagnetic, heat, and other characteristics a pilot would confront when operating over a built-up and populated area) are among those that yet-to-be-built facilities should address.

Aerial ranges may also contain various embellishments such as lights (to increase visual realism for night operations) or threat emitters (such as "smoky SAMs" (surface-to-air missiles)) to simulate threats against exercise aircraft.

The air operations at aerial ranges have an inherently joint character; absent extraordinary communications issues, aircraft from multiple Services should be able to simultaneously operate over or in the proximity of such sites. Using an aerial range for joint operations involving ground forces is more problematic. Ground forces, including joint ground forward air controllers (FACs), will be extremely limited in how much they can meaningfully train using a range composed exclusively of facsimile buildings. Moreover, use of a live inert-drop aerial range complicates FAC and other ground element training because of the SDZs required for aerial ordnance, distances of such magnitude that they essentially preclude ground maneuver or positioning of fire control personnel in the target area because of the necessary safety standoff required when aircraft are dropping munitions, inert or otherwise.

These prohibitions on putting ground forces in the urban training areas have been partly overcome by linking an AC-130 Specter gunship simulator at Hurlburt Field, FL, with exercises at the U.S. Army's Ft. Benning, GA, McKenna MOUT site. Simulator pilots "engage" targets as requested by those on the ground, and their effects are replicated in real time. These results are reflected both in the simulator and at McKenna (the latter via O/Cs triggering smoke discharges to show the location of the strike). While imperfect, the interface does enhance training for both air and ground elements, and it points to ways for even more-realistic and effective interaction in the future.[13]

Temporary or Façade Ranges

Temporary or façade ranges offer little joint training opportunity on their own. This category includes ranges that might be elements of a UAC or a MAC: permanent or temporary two-dimensional building facsimiles that provide targets or urban context from either a single vantage point or any direction. Façade ranges are generally not realistic when viewed from all perspectives. They either are designed to be viewed from a single direction (a two-dimensional façade) or are exterior-only buildings, lacking interior walls or real doors and windows. The need to improve preparedness of combat services support (CSS) forces for combat as part of convoy operations has led to widespread construction of temporary façade convoy ranges—in some ways, a suitable match of capability to requirement given appropriate scenario definition. (However, because dismounting and moving into structures is called for under some conditions, great care must be taken to avoid improperly training personnel and units because of range limitations.) While façade ranges by themselves are unlikely to contribute to joint training requirements, the ease of constructing temporary façade ranges makes them a potential supplement to other resources when the intent is to broaden instructional scope through synergies obtained through collocation. In such instances, joint training may benefit from the impression of greater size that mock structures provide in some scenarios.

Simulation and Simulated Capabilities

To be as useful and realistic as possible, JUT exercises should incorporate elements that are rarely if ever sequentially—much less simul-

---

[13] Interviews and site visits with Scott E. Moore (Engineering Planner, Lockheed Martin Information Systems), Ron Miller (Distributed Missions Operations Lead, Lockheed Martin Information Systems), and Bill Taylor (Communications Engineer, Lockheed Martin Information Systems), Hurlburt Field, FL, October 29, 2004; and with Maj Raymond (Ray) C. Casher, U.S. Army, Ft. Benning, GA, November 1, 2004, by Russell W. Glenn.

taneously—incorporated in current training events. Such elements include airspace deconfliction, non–line-of-sight (NLOS) fire coordination, the use or replication of obscurants and nonlethal systems, and massed direct and indirect fires synchronized with air defense capabilities and noncombatant safety considerations. Further, urban training should involve variations in environmental conditions (heat, dust, wind, etc.), along with extremes in terrain and threat that a physical site can rarely provide, at least not at reasonable expense in dollars and time.

However, integrating these elements into live training poses significant challenges. For example, some elements can have an adverse impact on the environment. Others can impinge on participant safety. Also, some technologies that are worthy of exploration are still being tested, and thus prototypes are not available in the field. Further, there are often compatibility issues corresponding to the use of a particular site or suite of sites—not enough range size for indirect fires or close air support training, bridge weight-class limitations and road use restrictions, or the inability to hire and train sufficient noncombatant role players.

Simulations run alone or in combination with live exercises can mitigate many of the challenges confronted by decisionmakers in the field. They already do so in many operational and strategic-level training events, often serving as the primary supporting mechanisms for exercises addressing these levels of war. Yet developing virtual and constructive capabilities to support such training adequately has proven difficult when the events in question include significant urban components. In the immediate future and throughout the 2005–2011 period, it will likely be necessary to have man-in-the-loop complements to simulations capabilities for anything other than the most simple urban operations training events. Virtual environments and simulations must, no less than live events, accurately represent the urban environment and its demands for difficult decisions involving the full spectrum of urban challenges. Joint tactical events, including moving, sensing, communicating, and engaging, must all seem realistic and have foundations built on historical precedence, quality analysis, and an understanding of future challenges. A simulation must also capture a wide variety of behaviors—those of the friendly, enemy, noncombatant, robot, network, group, media, international, and regional communities, and more—before it can be considered a candidate for standalone support of urban operations exercises.

This section begins by describing many of the individual simulation and modeling systems available to the joint urban training community (JANUS, Joint Conflict and Tactical Simulation (JCATS), Integrated Unit Soldier Simulation (IUSS), OneSAF, Full Spectrum Warrior (FSW), Full Spectrum Command, Diamond, and MANA). Each is examined in terms of its near- and long-term application to urban operations training. We also explore enhanced versions of these systems, along with large-scale training systems that incorporate multiple simulations and can link to live exercises.[14]

Simulation-Based Training Systems

JANUS is a system-level, force-on-force (FoF) constructive simulation that is able to model some aspects of urban operations. Recent versions have added capabilities that represent precision weapons, acoustic sensing, robotic planning, command and control, air defenses, and urban clutter. The National Guard and Reserves use some specialized variations of JANUS for training responses to WMD attacks.[15] Unfortunately, it would be prohibitively difficult to rewrite the simulation to provide interior fighting, noncombatants, complex reactive behaviors, explanations, 3-D perspective views, or dynamic

---

[14] This study does not purport to list and describe all the myriad efforts involving training simulations, creation of synthetic terrain, or other tasks that have primary or other application to urban preparatory initiatives. Work on other simulations and development of new capabilities are forever ongoing. A small sampling of those not addressed here include joint operations on urban synthetic terrain (JOUST), Joint Calls for Fire Trainer, and the Warning Assessment Logic Terminal (WALT) structures and munitions penetration capability.

[15] See, for example, the 1998 DoD Tiger Team report available online at http://www. defenselink.mil/pubs/wmdresponse/; updated versions of WMD response simulators have been produced by Unitech-Mimic for homeland defense training (*Armed Forces Journal*, Training and Simulation issue, 2002). These specialized systems will not be considered as part of the training packages here. Similarly, such engagement-skills trainers as the small arms engagement trainer, the interactive call for fire trainer, and the Abrams trainer are not considered because of their open-terrain emphasis.

terrain. Accordingly, while JANUS continues to have analytic value, it does not show adequate training potential to be considered as a substantive urban training tool.

JCATS is a follow-on to JANUS that solves many of the problems noted above. A screen shot of a JCATS urban hostage rescue scenario is shown in Figure 3.2. The simulation is able to model building floors, interior fighting, and day or night operations with artificial lighting. It can model up to 10 parties (e.g., own force, allies,

Figure 3.2
JCATS Screen Shot of a Hostage Rescue Operation Showing Detections of Enemy (Red) and Noncombatant Entities (Green) by Forward-Deployed Unmanned Ground Vehicles (UGVs) (U.S. Dismounts in Blue Are Readying for an Attack from Behind Buildings)

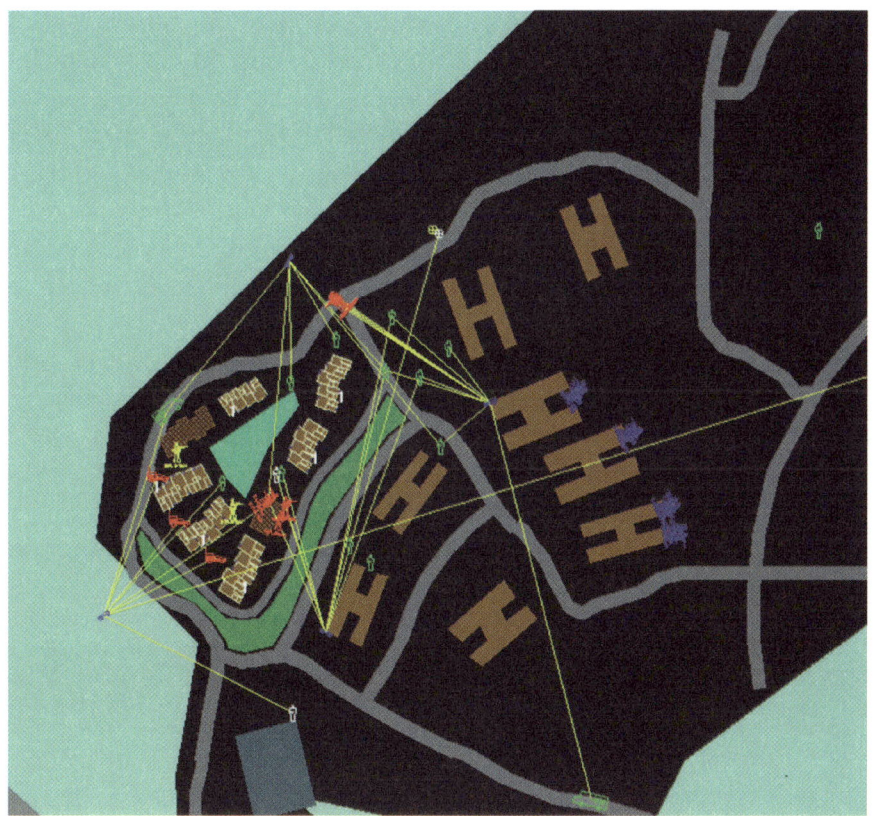

noncombatants, and adversaries), with rules for reactive behaviors. It is appropriate for joint operations, since it can handle small or large scenarios, links between Services, aggregation and disaggregation of units, and interoperation among sites using high-level architecture (HLA)-compliant protocols.[16] Its major shortcomings are the lack of a built-in 3-D visualization capability and the lack of complex-behavior input.[17]

Though considered primarily a tactical-level training aide, JCATS is capable of supporting training in both the tactical and operational levels of war. According to LTC Ken Bartlett of JFCOM, JWFC, in his November 2004 presentation "JCATS: Joint Urban Capabilities," the model is able to represent force interactions from the individual soldier to the JTF level. It can support visualization from the campaign view down to the individual building and street. Among the JCATS federations in use are the Live-Virtual-Constructive HLA Confederation and the Joint Multi-Resolution Modeling HLA Confederation at the Joint National Training Center (JNTC), and the Army Constructive Training Federation and Digital Battlestaff Sustainment Trainer. For most of these federations, JCATS emphasizes tactical-level operations, while other systems such as Air Warfare Simulation, Joint Semi-Automated Forces–Naval Forces, and Tactical Simulation (TACSIM) Intelligence Collection and Reporting, provide higher-level representations.

While considered an analytic rather than a training system, the Integrated Unit Soldier Simulation (IUSS) provides a small-unit analysis tool for examining the value of different items of equipment during urban missions. Figure 3.3 shows the 3-D image produced by

---

[16] See LTC Bill Robinson, "United States Joint Forces Command Joint Warfighting Center Modeling and Simulation Support to Homeland Security and Defense," Sim/C4 Program Strategy, October 2003, available electronically at http://www.mel.nist.gov/div826/msid/sima/simconf/proc/ftp/robinson.pdf, and Andy Bowers and David Prochnow, "JTLS-JCATS: Design of a Multi-Resolution Federation for Multi-Level Training," McLean, VA: MITRE Corporation, 2003.

[17] 3-D visualization with JCATS has been successfully demonstrated both at Livermore and at the Soldier Battle Lab. It is not part of the standard package.

Figure 3.3
Screen Shot of an IUSS Simulation Showing a Fire Team About to
Clear a Building

SOURCE: Image courtesy of U.S. Army Soldier Systems Center (Natick).

this system. The value IUSS offers during training is that it is able to model individual soldier physiological effects and complex behaviors at platoon level and below. As noted later in this chapter, some of these physiological and stress aspects are scheduled to be included in the OneSAF Objective System (OOS) set of models.

OneSAF is the most ambitious of the system-level constructive simulations intended for use in urban operations. This system has a long history of development, starting with the SIMNET (Simulator Networking) and SAFOR (Semi-Automated Forces) systems,

MODSAF (Modular Semi-Automated Forces in many variants—Dismounted Infantry, DISAF; Joint, JSAF; Medical, MED-SAF, etc.), and, finally, OneSAF Test Bed and OOS. This system has complex goal-directed behaviors, 3-D perspective display, and HLA compliance. As shown in Figure 3.4, it can model urban and open terrain with a reasonable level of detail. Like all the other systems, it does not have effective capabilities for explanation, although a special version of JSAF used in the Urban Resolve Exercise does have some after-action review (AAR) capabilities. These functions are described in more detail in the section on large-scale composite simulators.

Full Spectrum Warrior and Full Spectrum Command are two game-based training systems that focus on the squad leader and company commander levels, respectively. These systems, shown in Figure 3.5, have sound and special effects and demand a high degree of user participation. A version of Full Spectrum Warrior, developed by Pandemic Studios, is widely used as an action-oriented commercial game. Full Spectrum Command, developed by Quicksilver, Inc., is more leader- and tactics, techniques, and procedures (TTP)-oriented. Both of these appear effective for helping develop selected basic leader skills during urban training.

The constructive and virtual simulations described above are good for modeling combat. However, they do not capture many of

Figure 3.4
OneSAF Depictions of Urban and Open-Terrain Engagements

Figure 3.5
Full Spectrum Warrior (left) and Full Spectrum Command (right) Depictions

SOURCE: Images courtesy Pandemic Studios (left) and Quicksilver, Inc. (right).

the complex dynamics of crowd behavior and post-conflict operations. Two agent-based models that are more suited to this are Diamond and MANA (Map-Aware Non-uniform Automata). As Cares (2002) notes, in agent-based modeling, complex real-world systems are modeled as collections of autonomous decisionmaking entities, called *agents*.[18] Each agent individually assesses its situation and makes decisions based on its own set of rules. Agents may execute various behaviors appropriate for the systems they represent—for example, sensing, maneuvering, or engaging—and the collective actions then represent complex coordinated activities.

Neither Diamond nor MANA has detailed digital terrain, but both show complex emergent behaviors. Diamond is a high-level stochastic operations-other-than-war (OOTW) simulation developed at Britain's Defence Science and Technology Laboratory (DSTL). MANA was developed primarily with peacekeeping in mind and was produced by the Defense Operational Technology Support Establishment, New Zealand. Each of these simulations supports large numbers of agents with simple behaviors, and each has parameters for

_____

[18] Jeffrey R. Cares, "The Use of Agent-Based Models in Military Concept Development," *Proceedings of the 2002 Winter Simulation Conference*, available online at http://www.informs-cs.org/wsc02papers/123.pdf. It should be noted that there are also many other agent-based and gaming systems in use in the United States and other countries. The simulations described here were chosen for their widespread use and exemplary nature.

weapons, sensor, and communications systems. The primary value of these systems is in their ability to capture nonlinear phenomena and stochastic "swing" events (crowd dynamics, shifts in allegiance) that might be useful in training. Effects of this kind are extremely important for joint and coalition operations.

Enhanced Individual Simulations

To be effective in joint urban operations, these and other simulations need improvements in command and control (C2) representation, urban clutter, dynamic terrain, behavioral modeling, and many other aspects.

For example, specialized functions, e.g., dynamic communication links, can be modeled using statistical representations such as QUALNET and OPNET. QUALNET can currently model the behaviors of these functions on a single, high-powered personal computer (PC) if the number of moving nodes is limited to no more than a few hundred.[19] While the model is not able to simulate the passing of actual messages (it instead uses statistical representations of message size, priority, and routing), it does provide a means for showing the effects of terrain, interference, weather, bandwidth, jamming, and many other factors that add realism to a training experience. These simulations use terrestrial propagation models such as TIREM (Terrain Integrated Rough Earth Model), but they do not completely account for urban obstacles and multipath (a situation where signals bounce off various surfaces and arrive at the recipient at different times and amplitudes). As the forces move to network-centric warfare with Future Combat System (FCS) and other Service transformation programs (Sea Power 21, USMC Expeditionary Maneuver Warfare, Air Force Transformational Flight Plan), these communication models will become more important for effective training. In the longer

---

[19] Personal communication with Rajiv Bagordia, Scalable Network Technologies, Culver City, CA; for a tutorial on the model, see http://www.scalable-networks.com/pdf/QualNetTutorial.pdf.

term, the models will have to represent the Joint Tactical Radio System, AF Link 16, satellite radios, and many other systems.

With the advent of increased insurgent activity in Iraq and Afghanistan, much greater emphasis has been placed on field and experimental live-fire convoy training. Both Lockheed-Martin and Raydon Corporation have developed transportable simulators that provide 360-degree immersion in a virtual battlefield.[20] Both the Army and the Air Force are planning to use these systems.

Other specialized facets of urban operations training rely on dedicated models and representations. Nonlethal weapons, for example, require a simulation that can replicate their variable effects, effects that differ depending on target type, range, and weather conditions. Specialized models are also available for precision munitions, urban clutter, camouflage, active protection systems, acoustic sensing, and robotic system planning and mobility modeling. Many of these can be modeled using standalone simulations supported on a single PC.[21] They can be called on as needed to provide added realism and context during a training event.

Improvements to training system human interfaces are just as important as technology representations. Feedback systems based on explanations must be built-in from software inception rather than subsequently appended.[22] One of the higher-level training simulations that meets this requirement is the FBCB2/Tactical Decision-making intelligent tutoring system. This system, developed by Stottler-Henke, teaches planning and mission execution at brigade

---

[20] Sandra I. Erwin, "Dangerous Convoy Duties Prompt Expanded Training for Truck Crews," *National Defense*, December 2004.

[21] Examples of specialized simulations used in a local area net on a PC are described in Matsumura et al., *Exploring Advanced Technologies for the Future Combat Systems Program*, Santa Monica, CA: RAND Corporation, MR-1332-A, 2002.

[22] Some of the newer systems, such as Full Spectrum Command, have limited explanatory capabilities. They respond to queries about entity state, plans, and recent actions but are unable to present chains of logic or provide the reasoning behind behaviors needed for in-depth training.

level and below, using the tactical situations simulated by OneSAF Test Bed.[23]

Future training is expected to shift to increasingly sophisticated behavioral and organizational models for enemy, noncombatants, friendly-force members, and other groups. Improvements to agent-based models and to the Special Operations aviation regiment (SOAR) system in OneSAF should provide many levels of behavior—individual reactions to distinct events, goal-directed behaviors by individuals and groups, and crowd dynamics.[24] These models will be essential as training expands from combat to stability to peace operations. There should also be a general improvement in presenting situational realism and stress. One possibility in this regard is the addition of IUSS code to JCATS or OOS, thereby providing physiological-effects considerations to training supported by these simulations.

Of critical importance to urban operations is the development of high-resolution terrain appropriate to the modeling of dismounted movement, small robotic systems, and interior fighting. Robust simulation will require something along the lines of what are called automated building generation system (ABGS) and ultra-high-resolution building (UHRB) databases.[25] The OneSAF system has a UHRB environmental data model that is purported to be the most complete representation in any of the Army's stock simulations. The

---

[23] Patrick Chisolm, "Tutoring for Future Combat," *Military Training Technology: Online Edition*, September 8, 2003, available online at http://www.mt2-kmi.com/archive.cfm?DocID=219.

[24] For a description of SOAR and behavior-based modeling, see R. Michael Young et al., "An Architecture for Integrating Plan-Based Behavior Generation with Interactive Game Environments," *Journal of Game Development*, March 2004, and Scott Wood et al., "An Intelligent Interface-Agent Framework for Supervisory Command and Control," 2004 Command and Control Research and Technology Symposium, available online at http://www.dodccrp. org/events/2004/CCRTS_San_Diego/CD/papers/140.pdf. These capabilities have also been added to the Urban Resolve simulation using JSAF, described below.

[25] For a discussion of the difficulties of producing these inputs, see Dale Miller et al., *An Environmental Data Model for the OneSAF Objective System*, Lockheed-Martin Information Systems, available online at http://www.wood.army.mil/TPIO-TD/02F-SIW-082-FINAL.doc.

types of highly detailed images being produced by General Dynamics using the MetaVR visualization system are shown in Figure 3.6.

We noted earlier that the plan to embed training systems into all the FCS (and possibly some of the Stryker) vehicles should minimize the burden at the training sites. Embedding these systems should allow large-scale training without the need for added stations or semi-automated forces. However, the cost effect may be mixed, since extensive networking will be required to allow the high bandwidth needed to send and process HLA packets. Once completed, this embedded training interface should facilitate the linkage of live, virtual, and constructive simulations in large, joint, and coalition unit train-

Figure 3.6
Illustration of the FCS Training Database Being Developed by MetaVR

SOURCE: Image courtesy MetaVR, Inc.

ing. Also, many of the systems discussed (in particular, OOS, JCATS, and FSW) are being used by U.S. coalition partners, and some of these countries are also developing simulations similar in form and function to the specialized systems mentioned.

Large-Scale Composite Simulations and Simulators

Several current systems combine multiple simulations to provide many different levels of training. All these combination systems focus on joint and coalition operations, and many handle echelons up to theater and above. The Joint Theater-Level Simulation (JTLS) system, for example, has been linked to JCATS to cover the range from theater to small urban areas, and from peacetime operations to open warfare. Some limitations are present because the larger (up to 2,000 n mi by 2,000 n mi) terrain-box representations are hex-based rather than reliant on digital polygonal terrain, which is used by JCATS and other constructive simulations.[26] As a result, coordination of actions in the different simulations can be difficult and can result in errors. JTLS is currently implemented on a combination of Sun workstations and PCs.

Another example of a composite system is WARSIM, a suite of models currently in engineering development phases. This system has links to OneSAF and a logistics model (LOGFED) and focuses on next-generation C2 training up to the theater level. WARSIM relies on digital terrain and currently resides on a four-processor system. The simulation is a good candidate for high-power computing using clusters of processors.

The Joint Training Experimentation Program (JTEP) has just completed a linkup of the Close Combat Tactical Trainer (CCTT), JCATS, and a live exercise at an instrumented range as part of its second demonstration. This demonstration was intended to show a capability for future training for the National Guard but was said to be

---

[26] In hex-based representation, large six-sided chunks of terrain are used to abstractly characterize mobility, engagements, and other phenomena. This higher-level representation is quite different from detailed digital terrain models, in which elevation posts represent terrain slope, trafficability, and features at intervals as small as a few meters.

a useful battalion-level training experience in itself. The linkage of the live, virtual, and constructive simulations was accomplished using high-bandwidth links and DIS/HLA (distributed interactive simulation/high-level architecture) protocols. Some aspects of this demonstration involved urban operations; a future demonstration will be fully focused on urban terrain.[27]

JSIMS (Joint Simulation System), a system that could have provided a strong training environment, was recently canceled. This system, a combination of OOS and higher-level simulations, was intended to produce a joint synthetic battlespace. It was said to enable users to create or access a variety of simulation environments, including links to other simulation centers. It is expected that the majority of JUT sites will not have dedicated large-scale simulations or simulators, but that some networked, distributed capability along the lines of JTLS, WARSIM, JTEP, or JSIMS will be required to provide joint training at the higher echelons.

The largest demonstration of urban simulation is the Urban Resolve Experiment (see Figure 3.7), now in the first of three phases. This joint experiment, conducted by J9 in conjunction with the Institute for Defense Analyses (IDA) and the Joint Analysis Warfare Program, has demonstrated urban modeling of more than 1 million buildings and 100,000 noncombatants and vehicles.[28] The first phase concentrated on situation understanding using some 400 reconnaissance, surveillance, targeting, and acquisition (RSTA) platforms. The second phase will add communication and coordination of the forces, and the third stage (scheduled for 2007) will add urban engagement by linking human-in-the-loop simulators located at Ft. Benning. The system is built on JSAF and uses supercomputers located at Maui and Dayton, primarily for line-of-sight and logging functions.

---

[27] John Shockley et al., "The Joint Training Experimentation Program: Hot Wash from the Second Demonstration," 2004, available online at UO-FACT site.

[28] See Andy Ceranowicq and Mark Torpey, "Adapting to Urban Warfare," Interservice/Industry Training, Simulation, and Education Conference (I/ITSEC), 2004.

Figure 3.7
Screen Shot from Urban Resolve

SOURCE: Image courtesy Joint Forces Command.

Special Simulation Cases Considered Only for Longer-Term and Unique Urban Training Applications

Many systems may be too specialized, too expensive, or too high-maintenance to consider for widespread inclusion in JUT. For example, one application of OOS (actually a precursor thereof) is the CCTT. This system is able to replicate battlefield conditions using 3-D imagery, has operator controls similar to those on actual vehicles, and can include semi-automated forces (a form of constructive simulation with added behaviors). A platoon-size M1A2 tank capability in a mobile trailer costs about $5 million.[29] This capability is not yet specialized for urban operations, has combined arms only within U.S. Army (rather than joint) scenarios, and does not replicate links to

---

[29] Sandra Erwin, "On the Move, Combined Arms Training Available to Soldiers," *National Defense*, November 2000. Subsequent versions of this system, with computer-generated imagery produced by lower-cost hardware, are expected to be somewhat less expensive.

coalition forces. There is also an Army Aviation version called AV-CATT (aviation combined arms tactical trainer). A mobile version of this type of virtual simulation for training in convoy operations is now being produced by Lockheed-Martin.[30] It is claimed that this system helps crews communicate, maintain situational awareness, and acquire targets, all while moving along roads and in urban areas. AC-130 trainers have been successfully used to improve proficiency in ground-target designation, flight management, and target engagement.[31] These expensive and highly specialized simulators are expected to be located only at selected sites.

Some simulations require physical enclosures and linkages. One of the SAF versions located at Ft. Benning, the soldier station, has a roughly 10 ft by 10 ft walled space with projection on each of the four sides. This setup includes position location equipment and is said to cost $100,000 per station. Many variants of this "cave-type" immersion projection display system have been developed for a wide variety of commercial and military training applications.[32] These systems are considered to be practical parts of urban simulation suites only during the later portions of the 2005–2011 period because of their current high costs.

A MAPEX (map exercise) is a training event involving an interactive war game with participation by commanders and support element representatives. The focus in this type of exercise is on interactions between decisionmakers. A recent example is Joint Urban Warrior, conducted by the USMC and Joint Forces Command (JFCOM),[33] which consisted of a major war game and associated workshops, seminars, and planning events. It was used primarily for

---

[30] Scott Gourley, "Training for the Ambush," *Military Training Technology*, October 27, 2004.

[31] Roxana Tiron, "SOCOM a Trailblazer for Joint Training," *National Defense*, February 2004.

[32] Tim Shaw, "Full Scale Virtual Mockups," Penn State Applied Research Lab, briefing, undated.

[33] J9, Joint Experimentation Analysis Division, *Joint Urban Warrior 2004/Joint Urban Operations*, Final Report—Joint Experimentation Section, June 2004.

concept development and exploration. Combatant commands often conduct exercises with similar agendas, for example, Northern Command with its Ardent Sentry and Vigilant Shield events. The senior-leader seminar portions of such events are in particular known for having this focus. However, the conduct and outcomes of these events can also have training value, especially for such elements as fire-effects coordination cells. The support requirements for this type of simulation are significant. Various higher-level simulations described above (e.g., those supporting air, deployment, logistics, and coalition operations) need to link to exercise locations during similar undertakings and require staffing by specialists. The costs associated with this support during higher-echelon exercises could be considerable.

Some such systems (especially those of the 100,000-entity size currently demanded of joint operations at the theater level) cannot run without the support of supercomputers. Semi-automated forces (SAF) simulations, developed from the precursors of the Urban Resolve exercise, are run on a variety of supercomputer systems (Oak Ridge, Maui High Performance Computer Center, SPAWAR Systems Center, and others).[34] In the event that a very large-scale training exercise is to be run (simulating a counterattack on Seoul, for example), the local systems can be linked to a set of supercomputer simulations across the DREN (Defense Research and Engineering Network). This requires extensive interoperability protocols for time management along parallel compilers to ensure efficient operation on the multiple processors.

The future may offer even higher-realism simulation-user interaction, though advances in this area pose numerous difficulties. Such systems add aspects to increase the immersion process—e.g., realistic sounds, smells, and touch—along with natural language processing. One approach to this is the Institute for Creative Technology (ICT)

---

[34] Ted McClanahan et al., *Human and Organizational Behavior Modeling (HOBM) Technology Assessment*, MSIAC Project MS-00-0019/0028, July 2001, available online at https://www.moutfact.army.mil/whitepapers/whitepaper_behaviors.htm.

"holodeck," with 10 channels of sound and computer-generated avatars that respond to user commands and requests. Also interesting is the ICT Sensory Environments Evaluation Project, with recent work on a custom scent delivery collar[35] and on sophisticated, Web-based explanation tools. These efforts are very experimental in nature and highly manpower-intensive. They hold potential for introduction only in the latter years of the 2005–2011 time frame or beyond.

Important Research Directions in JUT Simulation

A variety of sources indicate six areas of substantial simulations effort over the next few years.[36] These areas highlight potential for specific progress but at the same time show the near-term shortcomings of simulation for high-fidelity/high-detail urban training experiences.

Very High-Resolution Terrain and Features

Many efforts are under way to improve the representation of buildings, obstacles, rubble, and clutter in urban areas. High-resolution models are important for accurate target detection and discrimination from clutter and noncombatants, the effects of weapons in interior and exterior fighting, and the movement of vehicles and dismounts in congested areas (especially small unmanned ground vehicles (UGVs)). Attempts to model structural and critical node effects, such as building collapse and interruption of electric power, are also under way. The models may use tools such as computer-aided design (e.g., the AUTOCAD program) for buildings and the Joint Integrated Database Preparation system for geospatial imagery, integrating these with the OneSAF ultra-high-resolution database. Additional models will also be needed for representing chemical, biological, and nuclear effects.

---

[35] ICT newsletter, *Selective Focus*, Summer 2004.

[36] Relevant sources include the FY04 DoD Master Plan for Joint Urban Operations, August 29, 2003 (draft), the Joint Concept Development and Experimentation Report for FY 2004, US JFCOM Report, and the Urban Operations Focus Area Collaborative Team Terms of Reference.

Behavioral Modeling

Enemy forces, noncombatants, and other semi-automated entities are expected to act realistically, both individually and in groups. Some actions, such as returning fire, moving out of danger, and performing scripted behaviors, can be produced using simple rule sets. When interrogated after the battle, these rule sets can provide tracking of conditions and actions and can therefore be used for explanation.[37] More-complex, goal-directed and cooperative/collaborative behaviors will likely require more of an agent and blackboard methodology such as that used in OneSAF.[38] Here, each agent can log its actions and beliefs on a virtual blackboard that can be seen and manipulated by other agents, resulting in complex reasoning and actions. Cooperative actions can also be developed as emergent behaviors resulting from simple rules in agent-based models. Many efforts such as the U.S. Marine Corps' Project Albert are under way to link these models to constructive simulations and virtual simulators.[39]

Engagement Calculations

It would be expected that decades of constructive model development would have made engagement calculations straightforward and well validated. This is true for open/mixed terrain and currently fielded sensors and weapons systems. It is less so for complex terrain and future, more-sophisticated technologies. For example, line-of-sight calculations in close terrain are said to consume 70 percent of the processing time for WARSIM. This is because there are many different features to check (e.g., buildings, vehicles, bridges, foliage) and various forms of obscuration to account for. Introducing clutter models, large numbers of moving entities, and three-dimensional fields of

---

[37] Explanation (also termed *explanatory artificial intelligence*) is thought to require coding in at the initial design of a system rather than as an add-on. As a result, explanation would require extensive rework or reprogramming for most of the simulations described.

[38] Scott Wood et al., "An Intelligent-Agent Framework for Supervisory Command and Control," 2004 Command and Control Research and Technology Symposium, San Diego, CA.

[39] Tom Lucas and Susan Sanchez, "High-Dimensional Explorations of Agent-Based Simulations," Operations Research Department, Naval Postgraduate School, briefing, date unknown.

view further complicates computer analysis. Special processors may be needed to handle the load in large urban settings as resolution improves. Similarly, target acquisition and weapon effects differ in urban environments, and simulations have to accurately depict these differences. Researchers have made specific recommendations for field data collection and model enhancements in this regard.

Communications Modeling

Network-centric warfare is the key element in the future force and the transformation methodology leading to its design. Simulation of network dynamics is difficult even in open and mixed terrain. It is extremely challenging in urban environments because of blockage (not in line of sight, NLOS), jamming, multipath, and other factors. Models such as QUALNET and OPNET have been integrated with constructive simulations, but more work is needed to characterize actual messages in a dynamically changing urban environment. It may turn out that specialized processors are needed to perform calculations about both communications and weather.[40]

Visualization

Several forms of visualization are needed for effective modeling. High-resolution, three-dimensional rendering of the visual scene—with all its perspective, scale, texture, clutter, lighting, and tactical cues—is needed to immerse a trainee in the scenario. Many efforts are being made to migrate this function from high-cost graphics units to low-cost graphical processing units (GPUs) used for gaming consoles. A second form of visualization is the use of diagrammatic tools to show network topology and status, planning processes (paths and danger areas, for example), and intelligence displays (with confidence, age, and accuracy of inputs). Specialized versions of the intelligence displays will also be needed to show critical infrastructure elements or cultural information (such as the activities and often changing sympathies of noncombatant groups). Because these areas are currently

---

[40] Personal communication from Don DePree, Boeing FCS Lead System Integration Team, date unknown.

undergoing intense development, it would be prudent to avoid the use of early-stage research systems in the near future and instead emphasize use of physical training environments and actual troops for tasks requiring high-fidelity/high-realism simulations. Greater proportions of these aspects can be subsumed into the simulation component as the simulations mature.

Tight Link Between Live and Virtual Simulation Training

The emerging capability of virtual simulations to model such complex and difficult training aspects as joint effects cell operations, tactical air controller links, and network management processes (on the high-abstraction side)—as well as dangerous tactical situations such as rotary-wing wire strikes, danger-close indirect fires, and armed UGV control (on the high-detail side)—make training linkages of paramount importance. OOS, JSAF, and JCATS have all demonstrated strong potential for real-time linkage between live training exercises and virtual or constructive components. Geographically distributed interactions and operations that depend on highly detailed "immersion" are especially suited to use of OOS and JSAF, while urban operations with simple rule-based actions are appropriate for JCATS.

Whether OOS, JSAF, or JCATS is being employed, the interaction of live, virtual, and constructive simulations has much to offer Service and joint force transformation. Development of and experience with virtual simulators will help the transition to embedded training, planning, rehearsal, and execution. Progress should also be especially helpful in coordinating training involving geographically distributed U.S. and coalition forces and interagency partners.

Near- and Far-Term Milestones in JUT Simulation

It appears that the next several years will see moderate use of simulations and simulators to augment live training. The majority of the effort should focus on part-task simulators for such functions as small-team operations, leader training, and contingency planning, along with some specialized complete tasks such as convoy operations and tactical air controller training. The farther term should see more integrated use of live and virtual training, greater amounts of tactical realism for urban engagements (interior fighting, dynamic terrain, reactive opponents, unmanned systems), and more characterization of networks and higher-echelon decisionmaking. As a result, the more-distant future should see much greater employment of simulations and simulators in training and operations. At no point in the foreseeable decades will simulations be expected to completely supplant live training, especially in dynamic situations in which the fog of war will play its role, but they should take over more and more of the routine and potentially dangerous aspects of training as time progresses.

Innovative/Novel Urban Training Sites/Capabilities

In addition to the dedicated MILCON facilities that form the backbone of the DoD urban training infrastructure, the possible uses and benefits of novel or alternative training sites must be considered. While not built to military specifications in most cases (the major exception being military and Coast Guard bases subject to Base Realignment and Closure (BRAC)), these sites are attracting increasing attention from defense analysts both inside and outside of government as DoD searches for ways to inject realism into the urban training provided to ever-larger numbers of troops (both active and Reserve). As is the case for dedicated MILCON installations, the alternative sites can range in size from individual buildings or exterior façades to large sections of urban sprawl. There are also a number of ways in which DoD can gain access to these sites. Direct purchase and ownership, rental from a private-sector owner, public-private partnerships, and short- or long-term leases are all available options.

In this section, we identify the different kinds of alternative training facilities available today and then assess their advantages and disadvantages. These alternative facilities fall into five principal categories: (1) abandoned or low-population towns; (2) closed (BRAC'd) military, Coast Guard, and other agency installations; (3) abandoned industrial infrastructure; (4) "other agency," private, or foreign urban training complexes; and (5) use of currently populated urban areas.

Table 3.1 details the types of facilities included within each major category.

Approach

Our method for identifying and categorizing alternative training facilities was much less formal than the approach we used to identify U.S. armed forces urban training sites. No comprehensive studies or databases have been produced to date that catalog all the physical locations of potential alternative training sites. Therefore, we had to rely on an ad hoc collection of sources to gain a rough picture of the

Table 3.1
Alternative Facilities Within Each Category

Category	Facilities
Abandoned or low-population towns	• Ghost towns • Low-population towns • Networks of abandoned towns
BRAC'd installations	• Army • Air Force • Navy • Marine Corps • Coast Guard • Other agency
Abandoned public and private infrastructure	• Ships as permanent urban training facilities • Mothballed ships temporarily used for urban training • Abandoned factories • Abandoned public-sector complexes
Foreign or "other agency" urban training sites	• Domestic law enforcement training sites • Dedicated MILCON foreign sites • Abandoned foreign towns • Privately owned "boutique shoot houses" • Private security-firm training complexes
Use of currently populated urban areas	• Terrain walks • Urban navigation • Urban simulated engagement • Urban live fire in populated areas • Use of vacant or condemned buildings • Use of empty public or private facilities

available alternative facilities. Interviews, sponsor and subject-matter expert input, site visits, and literature reviews all played significant roles in this effort, as did previous work by team members in the urban operations field.

Current Alternative Training Options

In the following, we identify current alternative training capabilities and consider the advantages and disadvantages associated with each. The discussion is not intended to be exhaustive but, instead, aims to provide a flavor of the current level of development in each of the five categories.

Abandoned and Low-Population Towns

The use of abandoned towns has moved beyond the concept phase into what might be considered the early test and development stage. Two possible urban training areas are being explored. The first is the largely abandoned town of Playas, in the southwest corner of New Mexico.[41] The second is an as-yet vaguely defined set of low-population towns that are spread throughout North Dakota.[42]

The Playas site is the more advanced of the two in terms of development and provides the prototype for this kind of alternative option. It encompasses 640 acres and includes 259 single-family homes and an apartment complex with 25 furnished units.[43] The town has no multistory dwellings. As of spring 2005, it was already being used to host suicide bombing deterrence and response command post exercises (CPXs) for mid-level local police and fire department personnel from across the United States. A limited number of U.S. Army

---

[41] For information on the size, layout, and composition of the Playas site, see Daniel H. Lopez, "Playas, New Mexico . . . Imagine the Possibilities," briefing presented to the Finance Committee of the New Mexico Legislature, June 8, 2004.

[42] Information on the network of abandoned towns in North Dakota was obtained during a telephone interview with Bill Goetz, chief of staff to the governor of North Dakota, November 24, 2004.

[43] Lopez, op. cit.

RSTA units are also conducting exercises at Playas. There is a rudimentary airstrip near the site.

An abandoned-town training area at Playas or a similar venue could conceivably accommodate company-size formations conducting FoF training with multiple integrated laser engagement systems (MILES) or Simunition ammunition. Live fire is probably out of the question, since the owners of the town would consider the structural repair costs prohibitive. Specialized hostage rescue vignettes could be accommodated in the town's apartment complex, where the furnished interiors inject an element of realism not present at many dedicated MILCON sites.

The instrumentation suite at Playas is limited and rudimentary in comparison to those at national training centers. There is currently no dedicated OPFOR or noncombatant actor population. This might change; the New Mexico National Guard has expressed interest in serving as a dedicated OPFOR to acquire realistic training otherwise deemed unaffordable by its leaders. It is not clear whether funding will allow this to come to pass. There is currently no organic pyrotechnics capability at Playas, but some could be added quite easily given that the site owners frequently conduct explosives demonstrations for visiting first responders at their test range in Socorro, NM. Pyrotechnics could conceivably be imported from Socorro if training customers requested it.

Joint air-ground exercises are feasible at Playas or a comparable site. The town's location in a remote part of southern New Mexico permits minimal airspace restrictions. U.S. Army small-scale unmanned aerial vehicle (UAV) testing has already been scheduled. Simulated bomb drops during ground force exercises should be feasible with minimal coordination (though means of measuring the effectiveness of aviation operations are not in place). A site like Playas could also support terrain walks in addition to formal FoF exercises. Such events would be especially useful if the architecture of the abandoned town site were modified to include walled compounds of the type that U.S. troops in Iraq and Afghanistan must at times isolate and clear.

The North Dakota abandoned-town model is somewhat different from the single-town site Playas offers. Early thinking about the North Dakota option indicated to us that it could involve a network of several low-population towns located in different regions of the state. This raises some interesting training possibilities that do not exist with single-town site options. A network of towns provides opportunities for mixing urban and rural training, exposing units to longer, more drawn-out operations often confronted in overseas theaters, in contrast to the limited-duration tactical urban vignettes currently practiced at home bases and combat training centers (CTCs). The rural-urban variability would also give units experience in selecting avenues of approach to built-up areas and in shaping the tactical environment during approach periods. The viability of various training options will depend on the number and location of the towns North Dakota (or other states) is willing to provide as training sites, as well as the layout of the towns and the terrain between them.

BRAC'd Installations

During the 1990s, the BRAC process went through four rounds of military base closures in the United States. A significant number of the affected installations have since been or are being converted into commercial or non-DoD governmental properties, but a fair number of vacated bases around the nation have still not been redeveloped. Some qualify as potential urban training candidates, and the 2005 BRAC list includes others that hold promise in this regard.

DoD has already recognized the potential training value of former military installations and has moved forward with implementation of this concept on a limited scale. The former George AFB (now known as the Southern California Logistics Airport) in Victorville, CA, has seen repeated use as a training and experimentation site by the USMC since the late 1990s. It has recently been used by both army and marine units preparing for duty in Iraq. A 200-acre area of the base, which includes large tracts of abandoned military housing, is available for urban operations training. (However, of late, local community members have been exerting pressure to cease use of the facility for such purposes.)

The U.S. Army's 1st Squadron, 14th Cavalry Regiment conducted a productive 10-day training rotation at George AFB late in 2002.[44] The 1-14 Cav is the RSTA unit for the Army's first new Stryker brigade. It was able to air-deploy into George AFB and employ a dedicated noncombatant player set of about 50 contract civilians. The training included urban reconnaissance and close-quarters combat.

Two major advantages of a George AFB–type BRAC'd facility are its size and its dilapidated condition. LTC James Cashwell, the 1-14 Cavalry squadron commander, put it best when he stated, "The advantage of George AFB is it is ugly, torn up, all the windows are broken and trees have fallen down in the street. It's perfect for the replication of a war-torn city. You can . . . then enter this complex old city—a wide variation of structures and multiple blocks, where at most MOUT facilities there are only a couple of blocks with maybe 20 buildings."[45]

Perhaps most important, George AFB and similar BRAC'd facilities offer opportunities for air-ground joint training events. The 1-14 Cav's rotation incorporated a Shadow tactical UAV that provided live feeds of the training area to company-level headquarters. A facility the size of George would potentially permit the use of fixed-wing aircraft (with simulated bomb drops) over at least a portion of the site. The presence of an airfield offers further opportunities for urban operations such as seizing and securing an airhead. Parts of such facilities might in the future be turned into low-fidelity areas for tactical air operations, while other areas are maintained at higher fidelity for ground forces training. Other factors, notably airspace coordination with commercial and other military facilities, will impact the feasibility of such use.

The overriding issue from a policy standpoint is the availability of BRAC'd bases for use as MOUT training facilities. Some are not

---

[44] J. R. Wilson, "Army Expands Home-Based MOUT Training," *Military Training Technology*, online edition, Vol. 8, Issue 5, December 1, 2003, http://www.mt2-kmi.com/archive_ articlle.cfm?DocID=361 (site accessed January 2005), pp. 1–3.

[45] Ibid., p. 2.

suitable for conversion to urban training status because of their location; size; urban character; legal, regulatory, and policy constraints; or other factors. Determining which are suitable candidates is beyond the scope of the current study, but such an investigation should be undertaken in a timely manner. Similarly, once authorities have identified future BRAC candidates, those candidates should be inspected for their potential training value as a matter of routine. It is notable that the U.S. Army is already investigating the suitability of Cannon AFB, which appeared on the spring 2005 BRAC list, as a possible large-scale urban training facility. Cannon AFB holds promise from the perspective of its proximity to units at Ft. Hood, TX, Ft. Carson, CO, and Ft. Riley, KS, and the potential for Ft. Bliss, TX, to assume the role of home station for units redeploying to the CONUS from Germany. That list and the Army decision to move its armor school from Ft. Knox, KY, to Ft. Benning, GA, could well result in considerable Service and joint interest vis-à-vis possible locations for urban training capabilities. (The move of the U.S. Army Armor School could enhance Ft. Knox's status as a possible home station for redeploying units, thereby making its MOUT facility a better candidate for expansion and use as a regional battalion-size training site.) The availability of sites for conversion to urban training use will, of course, depend on the ultimate outcomes of BRAC and redeployment review processes.

Abandoned Public and Private Infrastructure

The use of abandoned public and private infrastructure is largely conceptual at this time, although here too, recent innovation offers interesting opportunities and case studies. There are four subcategories within this category: (1) ships as permanent urban training facilities, (2) mothballed ships for temporary use, (3) abandoned factories or other large complexes, and (4) abandoned public-sector infrastructure.

Ships are potentially useful supplements to other urban training facilities for several reasons. First, their extensive networks of narrow passageways and corridors provide a reasonable facsimile of the subterranean environment American troops might encounter in subways, sewers, or alleys. Other types of urban training sites are hard pressed to replicate underground systems; where they do, the systems are generally very limited in scope and complexity. Second, because of the large amounts of machinery present on ships (pumps, generators, hydraulic lines, and electrical circuits, for example), troops training in this kind of environment can prepare themselves for various dangers and resource opportunities embedded in actual urban complexes.

Navy Sea, Air, Land (SEALs) special warfare units have, not surprisingly, conducted training on ships. While some of the training may have been of a joint character, it generally is joint within the SOF family of organizations. If docked ships (either mothballed or permanently decommissioned) are deemed worthy of urban training interest, the ships of the National Defense Reserve Fleet (NDRF) provide DoD with one readily available source of vessels. The NDRF is a reserve fleet of aging merchant ships that would be activated to supplement the Military Sealift Command's existing roster of fast sealift ships in the event of an emergency. NDRF ships are defense property under the control of the Maritime Administration; access for training purposes should therefore not be difficult to obtain. There are at the time of this writing 98 NDRF ships docked at Ft. Eustis, VA, 49 in Beaumont, TX, and 84 at Suisun Bay, CA.[46] Since these vessels are collocated in large clusters, several of them could be used simultaneously as different nodes in an "urban network" during an urban training exercise. The primarily metal-surfaced craft (high probability of ricochet) and their status as vessels with a potential for reactivation makes them dubious candidates for live ammunition training, but exercises involving MILES or some Simunitions-type rounds could well be within acceptable damage and risk limits. Most NDRF ships are not large vessels; urban training would, therefore, be limited to lower tactical echelons.

Abandoned factories, strip malls, or other commercial facilities offer yet another prospect for alternative training that could be simi-

---

[46] FAS Military Analysis Network, "National Defense Reserve Fleet," http://www.fas.org/man/dod-101/sys/ship/ndrf.htm, web page updated April 19, 1998 (site accessed January 2005).

lar to the kind of training carried out at George AFB or other BRAC'd facilities. As is the case with George AFB, commercial facilities could present more-realistic environments than many home-station training sites because of their size and worn-out state, conditions that generally mirror those found in many third-world cities.

Abandoned public infrastructure complexes constitute a broad subcategory that includes federal and state government schools, hospitals, and administrative buildings that have been closed. Some of these are in isolated rural areas and thus are lucrative candidates for socially insulated training sites. Perhaps the most notable current example of the potential offered by this concept is the Muscatatuck, IN, former school for the mentally challenged. The Indiana Army National Guard is assuming responsibility for this site. Its variety of building types, heterogeneous terrain, and near-perfect state of preservation (rooms still have virtually all their original furnishings) offer considerable value for use in supporting urban homeland security and international deployment training vignettes.

Muscatatuck is large and features a wide variety of structures, including multistory buildings. Although no dedicated OPFOR currently exists, the facility, when fully operational, will feature a number of permanent employees responsible for maintaining the facility, conducting research, or providing support to the staff. It is envisioned that they will all be incorporated into exercises as noncombatant role players. This civilian population is expected to include Purdue University college students (some language students will participate in training events) and local prison inmates. Muscatatuck also has a well-developed underground tunnel system of sufficient size to allow movement by, and concealment of, trainees, noncombatant role players, and OPFOR. There are plans to set aside part of the facility as a "laser friendly zone" into which aircraft can practice precision guided munition (PGM) targeting. A number of agencies have expressed interest in training at Muscatatuck, including Special Operations Command (SOCOM) and the Federal Emergency Management Agency (FEMA). In terms of capacity, Muscatatuck's built-up area could perhaps host up to company-team training in its current state (i.e., without augmentation by additional structures), but battalion task-force-level training seems readily within reach given upgrades that should not significantly (if at all) affect usage costs. There are tentative plans to introduce container express (CONEX) or similar additional structures to the complex that would significantly help in realistically disrupting lines of sight and would thereby provide additional training challenges.

Foreign or "Other Agency" Urban Training Sites

The use of foreign or "other agency" urban training sites is probably the most well-developed category within the universe of alternative urban training concepts. It is also the category over which the U.S. military has the least control, since all the sites in this category are owned and operated by either foreign military, private, or governmental non-DoD entities. In most cases, these owners do not see DoD as a core customer.

Probably the most intriguing training site option in this category is the privately owned shoot house. This is a relatively new phenomenon: Private investors build large, state-of-the-art complexes for close-quarters firearms training. These shoot houses go well beyond the standard firing range in that they incorporate advanced sound and light effects that closely replicate the sensation of being involved in a close-quarters engagement in various surroundings. The customer base for these "boutique" shoot houses is primarily wealthy civilians spending leisure time.

We identified two of these complexes in the United States—one operational, the other in the planning stages. The active shoot house is a 16,000-ft complex called Valhalla located near Telluride, CO.[47] The second shoot house is to be located near Fort Myers, FL, and is being built by private investor Steven Alexander.[48] Both projects are collocated with major resort hotels.

---

[47] See David Crane, "Valhalla Training Center LLC: The Future of Tactical Training Schools?" available online at david@defensereview.com (accessed July 8, 2004).

[48] Telephone interview with Dale Pruna, facilities manager of Hogan's Alley MOUT Training Facility, Quantico, VA, November 24, 2004.

These shoot houses could benefit fire-team or squad-level training by providing more sensory realism than is offered at most military
urban training facilities. While cost and capacity limit military applicability, there may be occasional value in using such facilities for specific purposes. However, the potential for other than very low-level
joint tactical training is likely to be limited.

Other facilities in this category—law enforcement training facilities, foreign MILCON sites, and private security-firm training sites—
do not seem to provide any capabilities that the U.S. military does
not already have access to at U.S. military sites. However, these sites
could have utility because they might provide more capacity for military training if the existing infrastructure becomes saturated. Proximity or the opportunity to conduct interagency training might also
promote usage during Service or joint events.

The most advanced foreign urban training sites appear to be
Copehill Down in the United Kingdom, Marnehuizen in the Netherlands, and Sarimban Fiba in Singapore. These sites support up to
company-level instruction and may not provide the level of amenities
found at some CONUS facilities. They may, for example, lack a
dedicated OPFOR or noncombatant role players.

One of the two largest (in terms of number of structures)
MILCON sites in the world as of early 2005 is the Dutch Army's
120 multistory facility at Marnehuizen.[49] The site allows use of
MILES only and has no permanent OPFOR or civilian role players;
the unit conducting training provides the OPFOR. A notable
strength of the Marnehuizen facility is its ability to host maneuvering
armored vehicles, including main battle tanks. Unfortunately, the
lack of role players currently limits the site's use as a venue for stability or support mission training in urban areas.

Abandoned overseas urban sites, parts of active indigenous urban areas, and purpose-built training facilities may offer training opportunities for deployed Service or joint forces virtually anywhere in
the world. Marines participating in Eager Mace 01 with Kuwaiti mili-

---

[49] RAND internal memo, "Visit to Netherlands Urban Facilities," December 7–8, 2004.

tary personnel benefited from urban environments available on Falayka and Bubiyan Islands in the Persian Gulf.[50] The commercially constructed urban training facility adjacent to the airfield in Bagram, Afghanistan (see Figure 3.8), was credited with saving "a lot of infantry lives" by the local combined and joint special operations task force (CJSOTF) command sergeant major, despite its being only three buildings in size.[51] Organizations from several nations regularly polished their room- and building-clearing skills at the facility, deliberately designed to replicate an Afghan compound. The construction of a high-rise training facility in South Korea offers similar opportunities in East Asia.

As the number of specialized SWAT and hostage rescue teams has grown within the American law enforcement community, so too has the number of training sites designed to prepare these units for urban contingencies. This set of training sites provides another potential urban training option for DoD organizations. The flagship domestic law enforcement MOUT site is the "Hogan's Alley" facility in Quantico, VA, where the FBI trains and exercises its elite hostage rescue team (HRT).[52] Hogan's Alley is already used as a supplementary training venue by Marine Corps units based on the East Coast, but the USMC's access to the site is fairly limited because of high demand by the FBI and other domestic law enforcement bodies.

Hogan's Alley has an actual training area of about five square blocks and is composed of actual structures, not just building façades. In addition to residential structures, Hogan's Alley has replicas of commercial buildings, including a motel, a bank, a theater, and a warehouse. Checkpoints can be set up on the streets leading into the

---

[50] "Eager Mace," online at http://www.globalsecurity.org/military/ops/eager-mace.htm (accessed January 11, 2005).

[51] Interview with William A. Howsden, MOUT Complex Manager, by Russell W. Glenn, Bagram, Afghanistan, February 16, 2004; and interview with CJSOTF, Afghanistan commander, deputy commander, and command sergeant major, by Russell W. Glenn and Todd C. Helmus, Bagram, Afghanistan, February 15, 2004.

[52] All information on the Hogan's Alley complex was provided during a telephone interview with Dale Pruna, November 24, 2004.

Figure 3.8
Bagram, Afghanistan, Urban Training Site

SOURCE: Photo courtesy Russell W. Glenn.

town site if necessary, and emergency vehicles can be brought into the area during training events to increase realism. However, Hogan's Alley is not equipped for the types of FoF or live-fire urban events that are conducted at MILCON facilities. The emphasis instead appears to be on the isolation of specific buildings or rooms and the subsequent assault of those specific areas to neutralize a criminal and/or rescue captives.

Collectively, this category of urban training capabilities seems to offer only limited potential for joint urban training, and even that instruction would be of value to only the lowest tactical echelons. The greatest benefits for U.S. military participation at such training events are likely to be those gained in establishing multinational and interagency relationships and lessons learned by individuals who can later become trainers in other venues.

Use of Currently Populated Urban Areas

A wide variety of different training events can be staged in cities, ranging from simple and unobtrusive terrain walks to urban live-fire exercises.

Basically, two kinds of training events are possible in actual cities. The first is a "survey and analysis" exercise, in which soldiers move through a piece of urban terrain to better understand the complexities of urban operations or in support of an exercise using the city in question for instructional purposes. This type of event would be low profile and would have at most a limited impact on the areas visited during the training. U.S. Army infantry officer courses have for many years conducted training in Columbus, GA, and pilots from various Services use urban sprawl to practice navigation over or adjacent to challenging urban environments.

The second type of training that might be conducted in a populated city would involve force maneuver and conceivably live fire within a restricted area, such as a vacant complex. For example, a deserted university campus, an abandoned amusement park, or an empty sports stadium could be used for a set period of time after provisions have been made to minimize the risk to civilians and infrastructure not involved in the exercise. USMC use of actual cities has long been a staple of the final phase of Marine Expeditionary Unit (Special Operations Capable) (MEU(SOC)) training. More recently, that service used the Oakland, CA, naval hospital complex and the cities of Little Rock, AR, and Boise, ID, to support training and experimentation.

Innovation has long been a key to effective training, and this is nowhere more true than with respect to urban training involving actual cities. The joint potential is considerable. Commanders can conduct inter-Service, interagency, and even multinational terrain walks, overflights, or other training as standalone instruction or in conjunction with military classroom, map, or headquarters exercises. The opportunities for coordination between SOF and regular forces and improved understanding of each other's capabilities provide the basis for better orchestration of assets under more stressful conditions. (The benefits derived from such coordinated urban operations in active theaters make joint training involving these capabilities highly attractive. Although considerably less desirable, use of contractors to replicate Special Operations capabilities is worthy of consideration if SOF commitments worldwide preclude participation by active, Reserve, or Guard units.) Integration of pilots and aircrews could facilitate improved intelligence, targeting, and logistical support; and other activities that are dramatically influenced by populated areas in ways often poorly understood under difficult situations would be actually confronted.

Potential Advantages of Alternative Training Options

Alternative JUO training sites offer potential advantages to DoD training managers and commanders. Not least are the cost and tactical advantages that are at times available at such locations. Many alternative sites can be used with limited or no expenditure of procurement funds. Rates for contract maintenance and event oversight may be lower than those for federal workers who serve as permanent staff at dedicated MILCON facilities. Clean-up and maintenance costs can be minimal. State prisoners convicted of nonviolent crimes provide labor in support of both these activities at Muscatatuck, so post-training repairs entail only the cost of materials.[53] Terrain walks and urban navigation exercises conducted in large cities present little risk of damage and negligible costs in most instances, providing an understanding of urban areas that is impossible to obtain otherwise and doing so at minimal cost.

Most "renewable" types of alternative training facilities would have clear operations and maintenance (O&M) costs associated with them. However, these costs may be more modest than those for dedicated MILCON facilities. The damage caused by exercises in abandoned towns, factories, strip malls, or other complexes destined for eventual destruction or complete abandonment may be considered acceptable wear and tear. Repairs, if necessary at all, may consist of

---

[53] Usage costs for the Muscatatuck facility had not been determined at the time of this report's publication. They are in any case likely to vary depending on the specifics of a given training event.

nothing more than returning specified items (e.g., damaged doors) to working condition to allow subsequent use during follow-on joint exercises. Rotating or sequencing training (e.g., switching areas used for ground urban operations and those involving inert bomb drops) could extend the usable life spans of such resources.

Some alternative capabilities offer greater environmental realism than do existing military training sites. The sound and lighting effects provided in the commercial shoot houses have been mentioned in this regard. Sites such as Muscatatuck or Playas offer structures with fully furnished interiors and complexes with the thermal, light, and electromagnetic signatures inherent in a populated urban area, benefits rarely replicated at purpose-built military training facilities.

Perhaps the greatest increase in realism that alternative training sites provide lies in the size and scope of the training area used. Whereas even the best urban operations training sites built specifically for military use (e.g., the Shughart-Gordon complex at the Joint Readiness Training Center (JRTC) or Marnehuizen in the Netherlands) have at most a little over 100 major buildings, abandoned towns and terrain walks through major cities offer military personnel the opportunity to conceptually grasp the complexity of clearing an urban area with hundreds of structures interlaced with alleys, commercial districts, underground passageways, highways, rivers, lakes, airports, and the many other elements commonplace in urban environments worldwide. And while MILCON sites train units for short, high-intensity sweeps through a small urban district, alternative sites can prepare military personnel for the long, drawn-out efforts that deal with the full spectrum of challenges a modern city offers—efforts that can last for days, weeks, months, or even years instead of hours.

Drawbacks of Alternative Training Sites

There are shortcomings to using alternative training sites that can attenuate the training advantages laid out above if steps are not taken to mitigate them.

Perhaps the most obvious downside of using some types of alternative training sites is that their availability to military units, especially on short notice, will be limited. Private boutique shoot houses, law enforcement training sites, corporate security training sites, and abandoned towns all have a variety of paying non-DoD clients who are eager to book training time and not always amenable to last-minute changes in schedule to accommodate military needs. As the terrorist threat to the U.S. homeland has grown in recent years, local police and fire departments have been sending greater numbers of their personnel to urban training sites for terrorist deterrence and consequence management training. This leaves less time available for military units to rotate through these facilities. Long-term planning is probably the only tool available for mitigating the availability problem.

Use of abandoned facilities such as factories, ships, or strip malls brings with it the risk of hazardous materials that may be present. Steps will have to be taken to ensure that threats are nonexistent or of negligible effect before training events are conducted. (Exceptions could include instances in which sites known to contain hazardous materials, such as dry cleaners and jewelry shops, are deliberately chosen for training involving such substances. These cases would naturally present special demands to ensure safe training and to avoid spreading contamination to adjacent areas.) Such sites should be purchased for training use only after careful investigation of ownership consequences. Many older structures were built before strict environmental laws were in place and thus contain substances that would not be used in construction today. As a result, significant abatement costs could be accrued by military Services or commands that buy such sites outright.

Physical access could be a deterrent as well. Some alternative sites are located in areas where the road and rail network is limited and airfields are either nonexistent or of limited capacity, making it difficult to move personnel and vehicles (especially armored vehicles) to the training site. In some areas, access may be seasonal. Some of the shoot houses, like Valhalla in Colorado, are also in fairly remote areas. All this means that a significant subset of the alternative sites will be readily accessible only to lighter forces, a condition that could mitigate or eliminate their joint training potential.

On the other side of the coin, those alternative sites that are located in populated areas with very dense road and rail infrastructure (e.g., populated cities, abandoned industrial sites, some foreign MOUT sites) come with their own set of risks for the military user, risks associated with the disruption of local civilian life and commerce and the possibility that this will engender public protests. The density of traffic increases the likelihood of vehicle accidents, both during training and during periods of arrival and departure. Public antipathy could force local curtailment of military training activity. Properly gauging public moods in the urban areas being considered as training venues and coordinating closely with local officials before any training agreements are made will assist in heading off such problems. USMC efforts at Yuma and Indiana National Guard initiatives regarding Muscatatuck are notable in this regard.[54]

---

[54] All of the alternatives presented in this chapter have advantages and disadvantages, of course. Even purpose-built training sites on military installations have safety, maintenance, periodic upgrade and redesign, civilian-use encroachment (both ground and air), and other costs, some of which are immediately evident, others of which will be apparent only with the passage of time. Comparisons of specific locations and types of alternatives should include rigorous investigation of such likely and potential "hidden costs."

CHAPTER FOUR

What Are the Shortfalls Between Requirements and Capabilities?

Introduction

Having enumerated JUT requirements and existing and planned JUT capabilities, we can now examine the shortfalls between what is needed to prepare the U.S. armed forces for urban operations and what capabilities exist. Eliminating, or at a minimum mitigating, the effects of these shortfalls is essential if America's joint force is to properly prepare for both near- and longer-term challenges.

It is important to stress that the focus of this study is not on urban training shortfalls within each of the Services and within individual joint units. Instead, we look at the U.S. armed forces as a whole and address where training capabilities require enhancing, such that JUT requirements are met. Individual commanders will have their own gaps, influenced by their missions, personnel turnover, time constraints, and the many other factors that leaders confront daily. Nevertheless, many of the capabilities sought as a result of this analysis should also benefit their efforts to prepare for future urban trials.

This chapter begins with a discussion of the challenges in trying to determine the training shortfalls. The chapter concludes by identifying the most significant shortfalls in both the near term (2005–2007) and the longer term (2008–2011).

Challenges to Determining Requirements/Capabilities Shortfalls[1]

At first glance, the process of identifying shortfalls seems very straightforward. In principle, we are simply comparing requirements against capabilities. However, several challenges make it less straightforward in practice. First, the requirements derived in Chapter Two range from very general to quite specific. The very general requirements (e.g., "govern in the urban environment") consist of innumerable sub-elements that are impossible to articulate comprehensively even for a generic case, much less for the complete set of all possible specific scenarios. Therefore, identifying all specific requirements is not a reasonable goal. They (and therefore the resultant shortfalls) must be defined in overarching rather than specific terms, with a focus on those most critical to success in reducing the total joint training shortfall.

Moreover, the requirements derived in Chapter Two are dynamic. Those drawn from even the most recent historical events will change character, some daily or weekly, others over periods of years. The example of improvised explosive devices (IEDs) illustrates the former case. Such weapons previously fell under the broader term "booby traps." Booby traps, explosive or otherwise, have long been the bane of ground force soldiers (and, to a lesser extent, sailors and aviators). Soldiers and marines in Vietnam suffered physical injuries from weapons that maimed and killed them and frustration because they had little opportunity to eliminate the people making and emplacing them. Many of the same challenges and frustrations exist for soldiers and marines in Afghanistan and Iraq. The dynamic character of the struggle both adds to the frustration and complicates finding a resolution.

---

[1] Early drafts of this study discussed shortfalls in terms of a requirements-less-capabilities gap, or delta, where requirements and capabilities are those presented in Chapters Two and Three. This means of presentation was later dropped when some reviewers found it confusing. The shortfalls presented here retain that context, however; i.e., they represent instances in which capabilities are insufficient to meet training needs.

IEDs evolve rapidly, changing in design, placement technique, triggering mechanism, and other aspects. They, like so much of urban operations, are quite rapidly adapted by the involved parties, in a process such as that presented in Figure 4.1. The result is that related gaps and resultant shortfalls constantly evolve as well. An effective training strategy must therefore be flexible in design and dynamic in application.

The tactical example of IEDs has operational- and strategic-level counterparts. Those responsible for developing campaign plans and those training forces to fulfill the requirements inherent in those plans face preparations that evolve not only over time but also over space. The threat and the social and geographic conditions facing Southern Command (SOUTHCOM) leaders during Operation Just Cause in Panama in 1989 differed dramatically from those confronted by Central Command (CENTCOM) in 1993 in Mogadishu. No less striking differences are present within a single combatant command's area of responsibility. Flexibility and dynamism in training are crucial to preparation at the operational and strategic levels as well as the tactical level.

The state of U.S. armed forces' urban training capabilities is also dynamic. Those resources that exist today (facilities, classroom courses, and simulations are but a sample) have to be maintained if they are to be an effective component of a joint training strategy. Virtually all will need regular upgrades to stay abreast of changing field needs.

How We Assess Shortfalls Between Requirements and Capacity

An organization might fail to meet a training requirement for any of the following seven primary reasons:

1. **Lack of capability.** No capability satisfies the requirement.

Figure 4.1
The Complex Process of Adaptation of IEDs

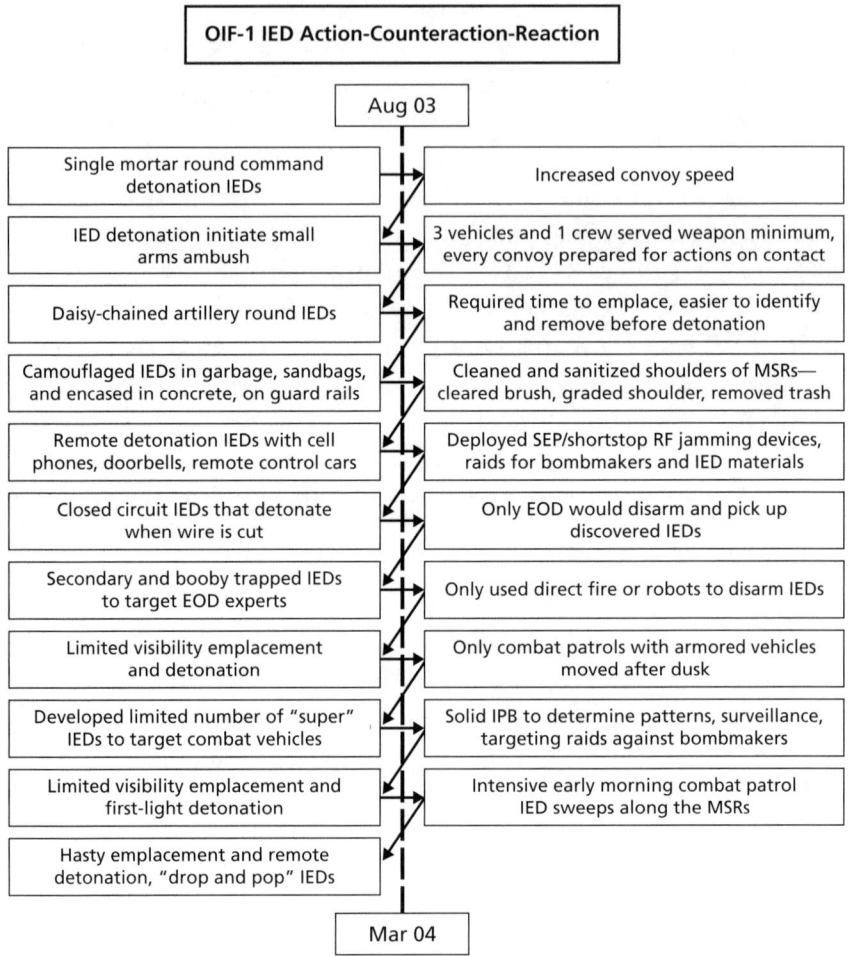

SOURCE: MAJ John Strycula, 4ID, U.S. Army.
NOTE: In 4ID sector of Iraq during OIF-1 rotation.
RAND *MG439-4.1*

2. **Inadequate throughput capacity.** While there are capabilities adequate to train for a requirement, an insufficient quantity of those capabilities is available to accommodate joint training demand.

3. **Lack of accessibility.** While there is sufficient capability and capacity, the capability is not available within the bounds of reasonable financial cost and travel time.

4. **Inadequate linkage or synchronization of capabilities.** Capabilities exist in sufficient capacity and accessibility, but they are geographically, functionally, or technologically separated to the extent that collective training requirements cannot be met. For example, live and virtual training capabilities exist so that a pilot can engage targets in urban areas through a simulator while ground-based fire support coordinators (FSCs) occupy the urban area replicated on the pilot's screen. However, there is no effective link that allows the FSC and the pilot to communicate in real time and credibly appraise the effects of their respective actions or measure the utility of the interaction itself. However, voice over Internet protocol (VoIP) technologies show considerable promise in this regard.

5. **Legal, regulatory, and policy constraints.** Environmental issues preclude using the full potential of otherwise effective JUT capabilities.

6. **Recognition of need.** If a requirement has only recently been identified, as may well be the case during periods of intense force commitment, the need to train for it may not have been recognized, regardless of whether a capability for such training exists.

7. **Training prioritization.** Unit commanders may choose to spend available training time or other resources on things other than JUT requirements.

The bulk of this study's JUT strategy development focuses on the first four of these reasons—lack of capabilities, inadequate throughput capacity, lack of accessibility, and inadequate linkage or synchronization of capabilities. These collectively address what capabilities are needed and how much of them will be sufficient to meet requirements.

Below, we discuss in detail some of the factors used in assessing whether particular capabilities exist and how we determine whether there is sufficient throughput capacity. Much of the discussion of accessibility and interlinkage takes place in Chapter Six, where we consider the costs of training modules identified in Chapter Five, and in Chapter Seven, where we examine the JUO training strategy itself based on the analysis done in Chapters Five and Six.

The fifth and sixth causes receive coverage only to the extent necessary to advise how they can impact training strategy development and resultant programs' execution. We discuss some of the environmental issues that play into our assessment of potential shortfalls below.

As for the seventh and final cause of shortfalls, much of the work in that regard has been done. Ongoing operations reinforce long-standing calls for both effective joint and Service urban training. Fortunately, savvy commanders and other leaders increasingly recognize the need to prepare for urban contingencies. There are, unfortunately, still too many leaders who do not understand the implications of recent and ongoing operations in urban environments for doctrine, training, acquisition, and force structure. We assume that readers of this monograph appreciate the importance of JUT, and we trust that they will encourage others to allocate the resources to conduct it. Previous RAND studies and reports from active theaters are among the sources that provide ample support for such preparation.

The following sections consider how well the available capabilities (as described in Chapter Three) meet the needs inherent in the requirements identified in Chapter Two. Where capabilities fail to meet requirements, the U.S. joint community must address the shortfalls if it is to adequately prepare the armed forces for future urban contingencies.

Whether a Capability Exists

In Chapter Three, we described the types of facilities (both purpose-built and novel) that the military can draw on to meet JUO training requirements. However, whether any one facility has the capability to actually address a requirement depends on a number of characteristics that fundamentally impact its potential as a joint training venue. We examine some of these briefly below.

Size/Scope

Facility size is an important measure of how useful a facility is in meeting requirements. While villages in Vietnam, Afghanistan, and elsewhere provide historical evidence that limiting complexes to only those with tens or hundreds of buildings would fail to provide the full range of environments in which joint forces need to train, small urban training sites still have value. There is little danger of a shortage in this regard because of the lesser cost of such sites, the ease with which they are constructed, and the possibility of using portions of larger sites to replicate those with limited capabilities. The same is not true of larger live training capabilities. While size can be measured in several ways—acreage of a MOUT site, total number of buildings, total square footage enclosed, and number of rooms, among others—it is a fundamental truth that there is a worldwide shortage of urban training complexes sufficiently large to properly support joint training. There is arguably no training site currently capable of supporting anything beyond the lowest-echelon JUT.

Urban Complexity

Real urban environments are inherently complex. Even small villages can have social networks and relationships with nearby rural and urban areas that are difficult to fathom. Social, political, and human infrastructure elements all pose their own challenges. Physical complexity differs in innumerable ways, including the following:

- Building function
- Number of floors in structures
- Subterranean features
- Furniture and other interior feature characteristics
- Construction materials
- Building spacing (urban density)
- Urban-canyon implications
- Line-of-sight disruptions
- Reflectivity and transmission properties of materials such as wires

Cultural variation is also important. Most traditional urban training sites have a northwest European layout, a legacy of the Cold War. Streets tend to have wide curbs, and buildings have Euro-familiar interior and exterior layouts. Ideally, urban complexity should differ from site to site, breaking away from this traditional mold of construction. In addition, the sites themselves should be altered to better replicate likely regions of deployment and to avoid having users become familiar with each location's layout. More-recently constructed facilities are attempting to steer clear of the Cold War model, but too often they do so only minimally. Very few present the density and mix of construction types and materials actually found in developing nations. Even fewer attempt to reproduce the filth and physical chaos that is characteristic, especially in times of crisis.

Unfortunately, according to several individuals we interviewed, some senior officers demand that sites be kept clean and orderly so that they present "a military appearance." Actual operations are not the time for soldiers to discover that not all homes look like something in a home and garden magazine. Trainees should have to deal with clutter, debris, and filth. They need to become familiar with regional building layouts and decor types likely to be found in potential operational areas. Infrastructures also vary. If a training site has electricity, it could also have lights; distractions such as TVs and radios, elevators, computers, working fuel pumps, electrocution hazards; and other features common in actual urban areas but virtually never found in training environments.[2] Buildings should have ventilation systems and other "mobility corridors" of potential value to defenders, attackers, or noncombatants. Running water, open sewers, wells, and rivers or streams add another layer of realism and complexity to an urban site. Similarly, pilots overhead need to confront the same light, smoke, reflection, and electromagnetic and other signatures that

---

[2] We are obviously not arguing for the presence of potentially fatal hazards, but replicating electrical shocks sufficient to provide a memorable jolt or odors that make one gag would provide long-retained lessons similar to that gained when a soldier or marine is struck by painful but nonlethal paint or other training rounds.

will favorably or unfavorably influence their missions over urban areas. Means of enhancing the replication of urban complexity in support of both Service and joint training is a potential area of significant growth.

Features that increase realism also increase the range of training requirements that can be satisfied at a site, thereby enhancing the value of repeated visits to a facility. Value is further added if part or all of the site is reconfigurable, with buildings themselves or walls within buildings being movable. Changing other features can help to offset an inability to alter buildings or other permanent features. Examples include adding or moving potentially temporary features such as shanty towns or targets; introducing vehicle check points (VCPs); altering areas occupied by representatives of a particular economic, social, or religious group; imposing political constraints such as those inherent in entering an embassy complex; and replicating the dangers associated with drug labs, jewelry stores, or dry cleaners.

Types of Forces Accommodated

The types of forces supported by a facility influence the quality of training that can be achieved there. The following types of units and functions (a non-comprehensive sample list) are potential candidates for JUT:

- Joint task forces of various sizes and makeup
- Air expeditionary forces
- Marine expeditionary forces and army corps
- U.S. Army divisions, corps, and units of employment (UEx and UEy)
- Theater support groups
- Combined arms
- Armored and mechanized maneuver
- Rotary-wing and fixed-wing aviation overflights
- Aircraft landings
- Air drops
- Seizing and securing air and sea ports of debarkation
- Operating from urban airheads and port facilities

- Fast rope operations
- Counterfire
- Countersniper
- Close-quarters combat (CQC)
- Amphibious operations
- Disaster relief
- Negotiations with factions, governmental representatives, or other urban-vital entities
- Demolitions and breaching
- Close air support (CAS)
- Live fire

Too many sites currently prohibit or severely restrict participation by tracked vehicles or aircraft. The age-old lesson that a force should train as it will fight seems forgotten, or it is at least subordinated to concerns about repair costs. Not only is such restricted training sub-optimal, it can instill the wrong lessons in participants. The design, funding, and manning of urban training facilities should account for the added costs of realism. An unwillingness to absorb these kinds of additional expenditures calls into question the very value of the training received at any such location.

Instrumentation

Instrumentation in support of urban training can be of several types:

- Monitoring instrumentation for observing and recording training for use during AARs
- Connectivity instrumentation that allows forces to participate in exercises with real, virtual, and/or constructive forces elsewhere
- Replication instrumentation, such as moving targets and munitions simulators

There are a variety of opinions about the value of expenditures on instrumentation. Some believe that the use of instrumentation for

AAR purposes results in too great a focus on individual performance and potential neglect of team and leader dynamics.[3] Others find the ability to accurately identify who was involved in engagements or incidents of fratricide vital to providing successful feedback. The high cost of installing, maintaining, and upgrading AAR instrumentation can significantly limit the number of buildings being instrumented, thus giving uneven feedback in the sense that "mistakes" are less likely to be discovered in areas lacking cameras.

There is no correct funding level applicable to all contingencies, but it is crucial that the tradeoffs involved be addressed when considering instrumentation expenditures. First, instrumentation that supports the extension of training realism will become increasingly essential. The ability to conduct artillery and aviation targeting in urban areas is fundamental to operational success, yet, as has been noted, training that effectively supports such activities is generally lacking. Similarly, little has been done in the way of reproducing the use of larger-caliber munitions (e.g., tank 120-mm rounds, antitank rockets or missiles, thermobaric weapons), leaving any but those who have actually experienced their use ignorant of the impact such capabilities have on structures, noncombatants, enemy targets, or friendly forces. Sensors, robots, and other emerging technologies are at times replicated for analytic purposes; only rarely are their effects and benefits incorporated to support training. Instrumentation that promotes progress in these areas and thus at once reduces training costs while providing enhanced operational preparation merits careful consideration on a case-by-case basis. Those capabilities that provide the linkage of different training sites and capabilities (such as the current tie between Hurlburt Field's AC-130 trainer and ground forces at Ft. Benning) fall into this category.

Some such technologies can be designed to support training AARs in addition to serving their primary functions of directly augmenting exercise realism. Potential economies of scale in this regard

---

[3] Interviews and site visits with Lieutenant Colonel Henk Oerlemans and Lieutenant Colonel Johan Van Houten, Dutch Army, by Russell W. Glenn, the Netherlands, December 7–8, 2004.

are worthy of consideration. Other AAR-related instrumentation should be reviewed in light of tradeoffs. The need for such equipment might be questioned in instances in which O/Cs accompany training units at lower echelons (e.g., squads).[4] The quality of feedback given by those providing oversight may reduce the call for instrumentation, and there may be little if any call when the focus is on larger units and higher-echelon decisionmaking. Since much JUT will focus on coordination, joint community funding for instrumentation of value primarily to lower-tactical-level training is an expenditure of questionable value. Exceptions in this regard include those activities that are inherently joint and for which feedback is essential for providing effective instruction, e.g., air delivery of munitions during CAS engagements. (The issue of instrumentation is discussed further in Chapter Eight.)

OPFOR and Noncombatant Role Players

Current urban training includes a wide range of OPFOR and non-combatant role-player options. Frequently, no permanent OPFOR or civilian actor staff exists, so a training unit must itself provide either or both if they are deemed necessary.[5] At the lowest levels of training, the opposition may not need to be represented by live individuals but can instead be replicated through the use of targets. Targetry can be more or less complex, ranging from stationary silhouettes to pop-up mannequins that can be dressed in different types of clothing to moving, reconfigurable targets or those that include thermal signatures. Battlefield-effects simulators that reproduce booby trap or other

---

[4] The Dutch Army uses this technique at both its Oostdorp and Marnehuizen urban training facilities (interviews and site visits with Lieutenant Colonel Henk Oerlemans and Lieutenant Colonel Johan Van Houten, Dutch Army, by Russell W. Glenn, the Netherlands, December 7–8, 2004).

[5] U.S. Joint Forces Command's Joint National Training Capability has made considerable strides in its short life span as it takes on the issues of training standards, OPFOR, and other elements critical to training support. Many of these initiatives are in the early stages of development or are otherwise incomplete. The discussion in this study describes outstanding requirements, existing capabilities, and initiatives with a firmly specified deliverable. It does not include efforts in the conceptual or other early phases.

enemy engagements can further enhance training realism. Threat signature emitters, such as "smoky SAMs" that visually simulate the launch of a SAM, have complements in systems that "squawk" a foe's air-defense radar signal. Well trained O/Cs can help training units get the most out of their training experience through instruction provided to OPFOR and noncombatant role players and oversight of their actions during training events.

Whether adversaries are represented by targetry, site-based OPFOR, or individuals brought in by the training unit, scenarios that can be supported vary from site to site. They include the following:

- Force on force (FoF)
- Force on targetry (FoT)
- Humanitarian assistance (HA)
- Noncombatant evacuation operations (NEOs)
- Peacekeeping/riot control
- SOF special scenarios
- Nuclear, biological, or chemical (NBC) attacks
- Information operations
- Scenarios requiring medical capabilities

Joint training, which is frequently focused at the upper tactical and higher echelons, should combine several of these scenarios during most exercises, the better to replicate the simultaneous offensive, defensive, stability, and support mission demands that urban areas will simultaneously present to joint headquarters and the units assigned to them. OPFOR and noncombatant role players need to confront those being trained with the same ambiguities, inconsistencies, and unfamiliar ways of reasoning that characterize actual operations. Operational- and strategic-level training should include representatives of multiple interest groups, individuals who are adept at playing friendly-force leaders off each other, who never allow situations to reach final resolution, and who represent the mix of selfish and selfless motivations characteristic of the exercise scenario region. Indigenous alliances and coordination between urban areas and regions ought to reproduce the challenges of simultaneous uprisings in 2004 Iraq. Noncombatant and OPFOR player coordination should reproduce reality, challenging exercise participants with complex coalitions of enemies united by antipathy to the friendly force and other rarely imitated behaviors.

Information operations are an area of particular concern. Current training at all levels (tactical, operational, and strategic) and of all types (live, virtual, and constructive) does not adequately represent information operations requirements, capabilities, or effects. There is a need for much greater sophistication in the quality and quantity of the intelligence a unit can access as it prepares to conduct urban actions. As articulated by LTG (USA, Ret.) L. D. Holder:

> The simulation support to relatively small live training facilities will have to be much more sophisticated than previously. . . . The smallest joint operations—SOF direct action, battalion-sized MEU non-combatant evacuations (NEO)—will be supported and affected by intelligence and other situational awareness factors that are far broader than in the past. Because operations and intelligence officers and their commanders will see their surroundings in greater detail, be subject to monitoring and direction from more higher headquarters and be able to "reach back" for almost any information support, the training support apparatus for all exercises will have to be extensive. [Further,] simulations and databases that represent all the actors, all the technical intelligence sources and all the elements of the joint force will have to be available. For example, a MEU commander or staff officer in a NEO operation would be able to call on regional experts anywhere in the world for advice, track the movements of TRANSCOM or SOCOM forces supporting the operation, and communicate with foreign military commanders and governmental agencies concerned with the operation.[6]

Other facets of information operations (e.g., psychological operations (PSYOP), public affairs, and deception) need similarly sophisticated representation in training at all echelons.

---

[6] LTG L. D. Holder (USA, Ret.) personal communication with Russell W. Glenn, June 12, 2005.

Live Fire

The range of live-fire activities allowed and the limitations imposed on live-fire activities vary considerably from site to site. There is variation in the types of live fire allowed; no one facility replicates direct fire, indirect fire, aircraft fire support, and naval gunfire in a manner suitable to the full range of JUT demands. Ground-based direct-fire training can include that employing live ball and tracer ammunition (green tip), short-range training ammunition (SRTA, also called blue tip), or Simunitions or other forms of man marker rounds (MMRs), blanks, or MILES gear.[7] Each has advantages and disadvantages. SRTA are lethal munitions that, therefore, cannot be used in FoF exercises, but they do less damage to walls and other training infrastructure than green tip ammunition does. However, the extent and character of this damage can be counterintuitive. Larger calibers may or may not cause more damage than other blue tip rounds. For example, 7.62-mm SRTA rounds (with a copper insert that improves ballistic properties) can do more damage to walls than the larger .50-caliber SRTA round that lacks the copper insert.[8]

Simunitions, MMR, and MILES laser all permit FoF engagements; each has its own advantages and drawbacks, but none realistically accounts for cover. MILES equates concealment and cover, a dangerous lesson for ground forces who sometimes demonstrate a frightening ignorance of the difference between the two. Trainers note that trainees often take risks in simulated combat that they would not take in real combat (creating "MILES heroes").[9] Simuni-

---

[7] See http://www.simunition.com/ for trademark and details on Simunitions; see http://www.xtek.net/catalogue/hrrequipment/utm02.shtml for details on UTM (ultimate training munition) MMRs.

[8] Lt Col Mark Axelberg, U.S. Army; Maj Everett Baber, U.S. Army; Sgt Maj Henry Legge, U.S. Army; Marty Martinson; and Capt Sven Myrberg, U.S. Army; interview with Brian Nichiporuk, Christopher Paul, and Barbara Raymond, Shughart-Gordon Urban Training Complex, Ft. Polk, LA, October 26, 2004.

[9] Interview with Major Scott Tatnell, Australian Army Liaison; LTC Christopher Forbes, Major Vern Randall, Captain Jason Tussey, and SFC Martino Barcinas, U.S. Army, O/Cs with Operations Group, JRTC, Ft. Polk, LA, by Christopher Paul, Brian Nichiporuk, and Barbara Raymond, October 26, 2004.

tions and other training rounds that perceptibly sting the recipient of a successful engagement move trainees closer to real combat. Range managers and trainers at several sites we visited advocate the use of such munitions during urban training because trainees know when they are hit, and they remember the event as a distinctively negative one. The principal drawback to these rounds is related to this advantage: Because the round has such heavy impact, it is more likely to cause an injury than is the case with other nonlethal training systems. This poses difficulties when training incorporates nonmilitary personnel (e.g., individuals representing noncombatants).

Further, all these training rounds lack the range and ballistics of real ammunition. While the difference is negligible at shorter ranges, at longer ranges it can introduce important consequences that can have notable negative impact on training lessons learned. Though many urban engagements are at 25 meters or less, a sufficient number are at greater distances. Training should realistically represent the characteristics and consequences of these longer-range engagements. In addition, unit members can learn the wrong lessons about providing supporting fire as fellow personnel move between buildings. Facility and exercise design can help to mitigate these effects (the Dutch Army, for example, ensures that adjacent buildings are close enough to allow for effective small arms supporting fires with weapons employing training ammunition). But in any case, steps should be taken to ensure that trainees leave an event understanding its shortfalls.[10]

Finally, live rounds penetrate drywall, corrugated metal, or other building materials, but training ammunition does not, potentially allowing ground personnel to inaccurately believe that such materials provide protection from small arms fires or air-delivered munitions fragmentation (reinforcing the "concealment is equal to cover" misconception). Future live, virtual, and constructive systems should seek

---

[10] Interviews and site visits with Lieutenant Colonel Henk Oerlemans and Lieutenant Colonel Johan Van Houten, Dutch Army, by Russell W. Glenn, the Netherlands, December 7–8, 2004.

to better reproduce munitions effects, both for ground forces and for those that might deliver munitions in their support.

Technological advances might address some of these issues in the 2005–2011 period. Global Positioning System (GPS)-augmented use of MILES that allows tracking of training and microwave capabilities are two areas currently being investigated.

Whether Throughput Capacity Exists

As noted above, it is not enough to merely have a particular capability on hand for use by U.S. joint force elements. It is also essential to have a sufficient number of these capabilities available, i.e., enough so that all personnel and organizations requiring training can obtain that training with the frequency necessary. Therefore, the problem is not only numbers, but also resource throughput capacity: How many such organizations can cycle through the capability in a given unit of time? Factors affecting throughput for a given facility include

- Number of days needed for a unit to complete training at a facility;
- Standard of training required;
- Quality of instruction provided (related to number of days needed, as training quality will influence the time required to achieve task proficiency at a given standard);
- Potential for simultaneous use (personnel or units training are complementary, or the training resource is designed to allow for independent but simultaneous use, e.g., separate STX training); this includes multi-echelon training, i.e., more than one organizational level participating in the same exercise;[11]

---

[11] For example, during visits to the USMC's Camp Pendleton, CA, urban training site, we witnessed two formations of troops training in the facility at the same time. One formation was a company practicing convoy ambush response within the built-up area proper; the other was a platoon from a different unit practicing snap-shooting (close-quarters combat target engagement) on one of the facility MAC ranges. Such simultaneous use is situation-dependent. In this example, range managers allowed sniper trainees to position themselves within the facilities buildings to fire into the adjacent MAC. However, a platoon-size formation of snipers training in this fashion closes the entire MOUT and MAC to use by other formations.

- Initial level of student expertise;
- Perishability of skill(s) being taught;[12]
- Availability of essential training augmentation (e.g., OPFOR, joint headquarters elements);
- Time necessary to maintain, adapt, or "reset" training capability between rotations;
- Amount of downtime required for cadre (e.g., leave, attendance at courses, deployments to active theaters).

Many of these are discussed further below.

Throughput capacity is sometimes difficult to measure. For example, many purpose-built urban training sites and their immediate environs provide simultaneous training for up to a battalion task-force-size organization. The British Army's Copehill Down Village (CDV, commonly referred to as Copehill Down), the Dutch Army's Marnehuizen, the USMC Camp Pendleton MOUT site, and the U.S. Army Shughart-Gordon facility all fall within this category (despite their significant differences in specific capabilities and quality of amenities). These sites are the largest and most sophisticated facilities in the world, yet none can effectively offer training to much more than a platoon and its full complement of supporting elements at any one time within the confines of the built-up area proper. The remainder of the task force is performing such tasks as isolation and providing supporting fires or other tasks that do not (or only marginally) require extended presence within space populated by buildings and noncombatants. Combat support (CS) and combat service support (CSS) elements are rarely committed to this area to the extent that they would be during an actual operation. Measures of throughput are very much a function of what a commander considers an adequate level of training expertise. The throughput capacity for a leader

---

[12] Individual and collective skill retention dramatically influences operational readiness and the frequency with which training should be repeated. Unfortunately, information regarding retention of individual military skills is scarce, and that regarding collective retention is virtually nonexistent. Both are areas in which further research could dramatically affect training design and force readiness. Though skill retention is generally beyond the scope of this study, a primer on it appears in Appendix F.

who is satisfied with only one platoon in a company actually training within the facility would be far greater than that for a leader who demanded completion of training in which each platoon in the company spends an extended period in the urban "box."

Throughput is also influenced by the quality of instruction available. Shughart-Gordon offers considerably greater sophistication in noncombatant behavior replication and other aspects of instruction than does the Camp Pendleton site, which is not surprising, given that the Ft. Polk facility is used Army-wide, while the Pendleton facility is rarely used by other than I MEF (First Marine Expeditionary Force) units. Increased sophistication means that more training requirements can be met per unit training time.

We sought data on the capacity and utilization rates of existing training sites as part of the information collection effort described in Chapter Three. (See the site survey in Appendix D.) Fewer such data are available than would be desired, making calculations an exercise in educated estimation. It is also worth noting that historical utilization rates do not directly convert to potential capacity values. One or more of the nine factors affecting throughput mentioned above may confound reported utilization rates.

Though throughput is difficult to measure with precision, it was evident that major U.S. urban training facilities were running at or above planned throughput rates in late 2004 because of their support of operations in Afghanistan and Iraq. This is a function of the time available between notification of deployment, actual movement to theater, and limited prior preparation of forces for urban operations.[13] Mission readiness exercises conducted at such facilities serve not as refresher courses to put the final edge on an already urban-

---

[13] There is also a lack of rigorous analysis regarding the periodicity of collective training (either civilian or military). While there are a limited number of training and education studies regarding *individuals'* knowledge or skill retention, there are virtually none that investigate *collective* (unit or organizational) skill retention. How frequently units should refresh specific types of training therefore relies largely on the commander's experience level and personal perceptions regarding his or her unit's readiness. Commanders, the Services, and the joint community would be well served by a study of collective skill retention, one that also accounts for such factors as personnel and leader turnover (unitwide and within crews or subunits).

ready force, but instead are often the means of training personnel and organizations in basic urban skills. This level of initial preparedness takes a toll, either through reduced throughput values or in a lower standard of urban training possessed by units as they pass too rapidly through the limited number of available training sites.

Environmental Restrictions and Encroachment

Environmental, safety, and other constraints limit the bounds of what can and will be accomplished through urban live training in the 2005–2011 period. Compounding the difficulty in a joint training context is the fact that regulations and usage constraints vary not only from site to site but also across Services. For example, U.S. Army range regulations preclude aircraft from firing live munitions directly over army ground forces; Marine Corps range regulations contain no such stipulation.[14]

Financial limitations can also act as a brake on the scope of live training. Urban sites in Bagram, Afghanistan; the Netherlands; Suffolk, VA; Ft. Polk, LA; and elsewhere have various means for economical replacement of door and wall components.[15] However, means permitting the use of live ammunition larger than 5.56 mm or shotgun pellets without extensive restrictions or causing unacceptable damage to training infrastructures do not exist. Use of artillery and air-delivered ordnance during realistic JUT involving both air and ground forces is infeasible, both because of the physical destructive

---

[14] Interview/site visit with Terry Finch, Bill Ash, Andy Chatelin (Range Management), and Major Bill Russell and Staff Sergeants Baker and McCarty, U.S. Marine Corps, by Christopher Paul and Barbara Raymond, Camp Pendleton, CA, May 27, 2004.

[15] Interviews and site visits with MOUT site manager William A. Howsden by Russell W. Glenn, Bagram, Afghanistan, February 16, 2004; Lieutenant Colonel Henk Oerlemans and Lieutenant Colonel Johan Van Houten, Dutch Army, by Russell W. Glenn, the Netherlands, December 7–8, 2004; and Naval Special Warfare Group Two representatives Steve Frisk and Stephen D. White by Russell W. Glenn, Norfolk, VA, September 10, 2004. Other sites lack even these basic means of supporting realistic training. The authors identified several urban training facilities that lacked even a rudimentary repair and maintenance budget. Training forces were therefore constrained both in the time available for training (because more time had to be dedicated to site preparation and/or cleanup) and in the amount of realism-related destruction they were permitted to inflict.

power of such ordnance and because of the dangers it poses for personnel on the ground. This constraint can be mitigated to an extent. Live-fire training can take place while avoiding unacceptable damage to training areas through the use of training munitions or by approximating the fires of heavier systems with smaller-caliber fires. However, such artificialities and current simulations fall far short of desirable realism standards in this regard.

The other major category of restrictions on what can and cannot be done at an urban training site falls under what DoD calls *encroachment*. DoD defines encroachment as the "cumulative result of any and all outside influences that inhibit normal training and testing." According to DoD, the eight encroachment factors are "endangered species habitat, unexploded ordnance and munitions constituents [post-use residual effects], competition for radio frequency spectrum, protected marine resources, competition for airspace, air pollution, noise pollution, and urban growth around military installations."[16]

Environmental concerns, complaints from local residents, and restrictions on the use of airspace above urban training sites can all place restrictions on JUT events. The exact nature of constraints and restraints differs dramatically from location to location and even at the same location over time. Some are unfortunate yet largely beyond the means of the U.S. armed forces to address. For example, flight profiles over Yodaville are significantly compressed because the Mexican government prohibits overflight of its territory.[17] Other restrictions also reduce the effectiveness of Service and joint urban training. For example, because of Yodaville's proximity to the Mexican border, the lights of the site attract illegal immigrants, who believe it is an actual town that would be a source of water. Regardless of their nature, environment-related constraints will continue to present a constantly changing palette of challenges, plague training realism, and

---

[16] General Accounting Office (GAO), *Military Training: DoD Report on Training Ranges Does Not Fully Address Congressional Reporting Requirements,* GAO-04-608, June 2004, p. 1.

[17] Interview with Marine Corps Air Station Yuma representatives Lt Col Bill Sellars, CWO3 Deana Sherrill, Major Rascon, and Major Pearce (first names not available), by Christopher Paul and Brian Nichiporuk, Yuma, AZ, July 7–8, 2004.

impose additional costs for related studies and mitigation. Live training and environmental issues are increasingly in tension both domestically and internationally.

Identifying the Shortfalls Between Requirements and Capabilities

We used the issues raised above, historical study, interviews with serving officers of all Services, and recent reports from active operations to identify the shortfalls most critical to adequately preparing the U.S. joint force for urban operations. Closing these gaps while maintaining and adapting existing capabilities will dramatically enhance the nation's readiness for the urban contingencies of today and those of the years to come. In the following, we list shortfalls for the near term (2005–2007) and the longer term (2008–2011) for both physical facilities (purpose-built and novel) and simulations. (The shortcomings noted for the immediate 2005–2007 term also apply to the longer 2008–2011 term unless otherwise noted.)

Many of the capabilities already available to the joint force serve to at least partially address current and emerging requirements. Other gaps fall outside the purview of joint preparation; responsibility for their closure lies with Service training authorities. The requirement "control maneuver in urban areas," for example, includes many tasks for ground, air, and SOF units that support joint maneuver but for which the Services properly feel they should retain responsibility. Therefore, only those tasks relevant to the joint aspects of the requirement receive attention here. Similarly, the specific shortfalls of interest when considering all 34 requirements identified in Chapter Two include only those pertinent to joint operations (and, by extension, joint training).

Avoid Fratricide

In the near term, physical facilities capabilities are insufficiently realistic to replicate through-wall, structure collapse, and other casualties that would in reality be caused by close fire support, grenade fragmentation, or ground fires. Also, facilities are not large enough to replicate possible detections of SOF elements by friendly forces at greater ranges.

Resolutions in current terrain databases tend to be good to only a few meters, insufficient for urban areas in which several meters can mean the difference between one entryway or another or even an entirely different building. Larger simulations have been built that provide better resolution over urban areas encompassing tens of square kilometers (e.g., Urban Resolve), but these require supercomputer support. Most simulations of urban areas have to be constrained to squad- or platoon-level operations with a few dozen buildings if they are to operate at reasonable speed. Yet it is larger-scale and more-detailed databases that are needed for joint operations.

Communicate in the Urban Environment

For the near term, current purpose-built sites lack the density and verticality to sufficiently replicate radio, UAV/UGV control, visual, and GPS line of sight (LOS) interference; in addition, smoke, noise, structure density, and debris are lacking in many sites. Simulations suffer from the same shortfalls.

Conduct Airspace Coordination

Training sites for conducting airspace coordination lack the size necessary to replicate the density of aircraft (rotary, fixed-wing, and UAV) in airspaces above and in the vicinity of urban areas of operation. In addition, facilities lack the capability for execution of suppression of enemy air defenses (SEAD) before or during air support. Moreover, ground-based aircraft controllers are required to be remote from drop sites for safety purposes. No training facility currently has the capacity for simultaneous munitions engagements by multiple aircraft flying realistic flight profiles.

Simulator links to MOUT sites are currently very limited (e.g., there is only one AC-130 simulator with connectivity to the Ft. Benning McKenna training facility). There is a need for the capability to link large numbers of aircraft simulators to represent simultaneous operations over the same urban terrain.

Synchronize Joint Rules of Engagement

Aside from SOF/regular ground force exercises, there is very little joint training currently executed at available urban training facilities, thus practice of joint rules of engagement (ROE) is limited. We also find an inability to conduct live air-ground engagements and a lack of inter-Service urban aviation training, which preclude refining, practicing, or gaining sufficient familiarity with current ROE. Simulations and simulators generally lack ROE feedback (i.e., they do not reflect whether engagements are acceptable or violate standing ROE).

Conduct Stability Operations in the Urban Environment

At current facilities, live urban training lacks sophistication in how well it represents noncombatants. This is in no small part attributable to the fact that the numbers of noncombatant actors are rarely sufficient to represent those found in actual environments. Better efforts are called for in replicating noncombatants vastly outnumbering combatant forces during live training. In addition to simply increasing raw numbers, trainers should consider the clever use of mechanisms such as bottlenecks that restrict friendly-force movement and timely concentrations of civilian role players. (There is no excuse for simulations not reproducing more realistic ratios of noncombatants to friendly and enemy forces when civilian play is called for.) Also, the various interest groups and organizations (both noncombatant and enemy) are underrepresented. The built-up area is the focus of training, with live, virtual, and constructive exercises generally failing to consider the urban area as but one element in a larger regional context. Scenarios also tend to emphasize urban combat as the primary mission, to the detriment of challenging participants with simultaneous demands involving FoF combat, stability operations other than those involving combat, and support tasks.

Simulations in the near term also generally lack scenarios other than those involving conventional FoF combat situations. In addition, current scenario generation is time-consuming, error-prone, and limited in scale. The few standardized scenarios are based on existing urban training sites. Few extend the region sufficiently to represent the urban area's role in a larger region or to encompass the full spectrum of demands likely to be confronted during joint operations.

Unfortunately, the longer term does not offer much promise for significant improvement vis-à-vis urban simulations capabilities. Larger simulated scenarios with bigger geographical areas and more entities will be difficult to generate, especially when the simulations are attempting to replicate the large numbers of interactions between systems that are required for effective training but that are too costly to reproduce in live training. Rapid generation will require special tools for adding systems, producing plans, operating at different levels of abstraction, and synchronizing forces. Some tools will be available with OOS, but these are still too limited to meet desired levels of realism and are likely to remain so through FY2011.

Conduct Support Operations in the Urban Environment

Some of the same issues confronted in addressing urban stability-operations training influence preparation for support operations. For example, facilities lack sophistication in how well they represent noncombatants. Once again, the numbers of noncombatant actors are rarely sufficient to represent actual environments, and the various interest groups and organizations (both noncombatant and enemy) are underrepresented. Again, the city is too often considered in isolation from the surrounding region, and combat operations dominate scenarios.

As for simulations in the near term, the picture is similar to what we find for conducting stability operations. Simulations lack scenarios other than conventional FoF combat simulations. The few standardized scenarios are based on existing urban training sites alone. Few extend the region sufficiently to represent the urban area's role in the larger area or to encompass the full spectrum of demands likely to be confronted during joint operations. Longer-term training challenges are likewise similar to those for stability operations.

Conduct Urban HUMINT Operations

Facilities lack size, and as a result, urban exercises fail to replicate the duration of training that would cause trainees to gain a sense of the tempo and character of urban operations. There is also insufficient granularity in representations of different enemy and noncombatant factions, as well as inadequate replication of intelligence capabilities at all relevant echelons. Urban simulations lack anything other than generic noncombatant behaviors, including anything beyond a super-ficial interaction with friendly force elements (if that); there are lim-ited or no mechanisms to provide realistic HUMINT input.

Conduct Urban SIGINT, IMINT, MASINT, COMINT, ELINT, and Other Intelligence Efforts

For this particular requirement, facilities are too small, and the ways in which the varied structural, population, and other relevant densi-ties are replicated limit most intelligence inputs to scripted inserts. There is virtually no participation by the variety of strategic assets on which intelligence personnel could actually draw. The limited size of facilities and the limited time allocated to urban training preclude the level of understanding of urban environments necessary to develop the expertise needed to appropriately use intelligence sources or the information provided. Finally, the range of threats, noncombatant factions, and physical diversity is insufficient.

Sensor models almost universally employ the Night Vision and Electronic Sensors Directorate (NVESD) Acquire algorithms for tar-get detection and engagement, which inadequately account for mod-eling of street clutter, obscurants, and camouflage. Once again, the capabilities necessary to realistically represent urban densities, spread, and levels of complexity are likely to be beyond simulations fielded before 2011.

Conduct Urban Operations Exercises

Few joint urban exercises are currently conducted at any echelon, and most of those that do take place are limited to superficial ground-air interaction or small-unit, regular force/SOF missions. The few higher tactical- or operational-level events that do exist fail to adequately replicate the complexity of the urban environment and the need for greater interagency (e.g., PVO and NGO) cooperation. This lack of realism is exacerbated by the lack of operational- or strategic-level instruction on urban operations or civil governing responsibilities in U.S. armed forces schools (except for an occasional elective course).

There are also numerous near-term simulation shortfalls. For example, no simulation allows appropriate replication of even FoF combat. Most simulations are of commercial game quality and therefore lack adequate sophistication or consideration of larger implications such as those involving stability or support requirements. Specifically, training systems need to have extensive feedback about student performance, along with explanations of how, where, when, and why performance was adequate or otherwise. Full Spectrum Command (FSC) has shown the ability to provide some of these forms of feedback, but it does not explain why or how. More-extensive feedback is planned for Urban Resolve using JSAF (with the Future After Action Review System, FAARS), but supercomputers will be needed to store the volumes of data produced.

The longer term offers both positives and negatives with respect to simulations. For the far term, AARs should also tell why something did not occur, recognize common errors in reasoning and show the consequences, and graphically trace the processes involved. None of the simulations now operational have these capabilities, but RAND and ICT are jointly developing new tools for achieving some of them. Unfortunately, progress is limited because needed algorithms not designed into the simulation from the start are extremely difficult, if not impossible, to add later.

Integrate Urban Operations with Other Relevant Environments

Facility and simulation capabilities are the same in the near term for this requirement. There is virtually no training—live, constructive, or virtual, including that conducted in classrooms—that incorporates environments other than those immediately adjacent to the focal area. Offensive, defensive, stability, and support implications and how they would influence regional or other areas are not encompassed in training.

Coordinate Maneuver in the Urban Environment

Facility size limits maneuver coordination to the lowest tactical levels and generally includes very limited CS or CSS components. Current simulations lack the terrain detail necessary to simultaneously represent interior and exterior fighting, mounted and dismounted operations, and a wide variety of equipment, including ground and air robotic systems. Some individual personnel, vehicle, and aircraft simulators approach acceptable levels of realism.

There are both facility and simulation capability shortfalls in the longer term. In terms of facilities, maneuver will involve a greater number and diversity of systems as Services incorporate transformation forces while maintaining legacy capabilities. The greater variety of kinetic and non-kinetic effects will correspondingly require replication. Longer-term requirements for simulations include the need to model currently emerging tactics and other technologies (e.g., loitering and nonlethal weapons, penetrating radars and hyperspectral sensors, and autonomous unmanned systems). These new technologies and associated new tactics will require more-detailed modeling of terrain, communications, and decision-aiding systems.

Coordinate Multinational and Interagency Resources

There are no established procedures in place for multi-echelon interagency training involving all relevant parties, including NGOs and PVOs. There is also no live, virtual, or constructive capability that replicates any but the most fundamental interagency challenges; noncombatant actors rarely represent the range of civil or social demands.

Govern in the Urban Environment

The shortfalls for governing in the urban environment are similar to those for coordinating multinational and interagency resources.

Identify Critical Infrastructure Nodes and System Relations

Facilities shortfalls in the short term include a lack of classroom instruction or higher-echelon joint urban exercises, which means that the understanding of urban physical and social infrastructure is limited. Also, determining second- and higher-order effects receives little if any attention, and the role of infrastructure nodes and linkages is even less well understood from the standpoint of stability and support operations or transition responsibilities than are offensive and defensive combat operations. No simulation sufficiently incorporates inanimate and social-node considerations or their interactions.

Navigate in the Urban Environment

Purpose-built facilities are too small to challenge navigation at other than the lowest tactical levels in the near term. They have insufficient size, density, and dispersion to cause confusion between air and ground elements during air support coordination, and they have insufficient density and verticality to sufficiently disrupt GPS signals. Inadequate numbers of noncombatant and OPFOR combine with lack of facility size to preclude replicating realistic conditions.

Current simulations offer no significant navigation challenges for a joint force. Manned and unmanned system mobility replication requires extremely detailed terrain models with dynamic changes to craters, mouse holes, and rubbling. Again, this level of detail is thus far lacking.

For the longer term, we find both facility and simulation capability shortfalls. The increased introduction of UGV and UAV will add to navigation requirements, and both size of area and complexity of internal spaces present at training sites will be critical factors, as will the variety of obstacles and human interference.

Simulations providing an economical way of testing new or proposed technologies need to replicate the full range of navigation, obstacle, and human tampering factors (e.g., terrain databases will need to be even more detailed in the far term to account for the advent of autonomous or semiautonomous ground robotic systems, Land Warrior technologies, and new high-resolution sensors). Less-than-meter resolution will be needed, with tagging of building features to indicate composition, weapon effects, and clutter level. Fast-response input will also be needed, with training and rehearsal taking place almost immediately after the outbreak of hostilities in an area.

Plan Urban Operations

Instruction and exercises involving the full spectrum of planning considerations exist nowhere in Service or joint training capabilities. Higher-level training events consistently fail to adequately represent other-than-military agency and noncombatant capabilities or interests. Simulations provide no planning challenges above the lowest, ground-specific echelons.

Provide Common Situational Awareness

Lack of size and low density in live training currently preclude adequately challenging situational awareness development above the lowest echelons. Training scenarios have to cover the span of intensity from stability and support operations to major combat operations. In the near term, it may be sufficient for these scenarios to be sequences of linked tasks. However, reproducing rapid changes in conditions such as clutter, quality visual acquisition and intelligence, threat capabilities, and own equipment performance is rarely possible. While FSW and FSC both allow for the quick change of conditions, they are limited to small-scale scenarios.

In the longer term, simulations will have to account for emerging sensor improvements while properly reflecting continued LOS difficulties; in addition, replication of 2008–2011 capabilities should have integrated deployment, fighting, resupply, and other components. Controllers will need to be able to quickly develop and edit potential scenarios.

Provide Fire Support

Current facilities fail to meet the requirement of providing fire support, because they have either inadequate representation or no representation of indirect fire accuracy or effects. Urban fire support training is further hindered by the inability to position FSCs realistically on ranges so as to allow actual incoming munitions (i.e., within the built-up area itself) and a very limited capability to link live training with simulated effects.

Simulations currently lack sufficient linkage between existing simulators and live training. Geographically distributed simulations with human operators should have a delay time of less than 200 milliseconds to minimize distraction and errors. In the near term, this has been accomplished on only a very limited basis by using OOS with T1 or equivalent lines and moderate numbers of entities. Delays will likely increase as the number of entities rises and the complexity of actions increases. Weapon models are designed to represent the vulnerability of a target in open or defilade positions, but few models characterize rubbling effects, firing through walls, or the dynamics of interior fighting. Many of these effects will be extremely important during coordination with tactical air and gunship operations, as well as during combined arms ground force operations.

While there are no additional facility shortfalls in this area in the longer term, there are some longer-term shortfalls for simulations. Delays will be further increased because of the use of unmanned systems, the passing of images, and additional requirements to replicate new technological capabilities. In addition, sensor and weapon models will have to be updated to reflect new penetrating sensors, updated in-flight weapons, and nonlethal devices. Finally, unmanned systems will need to be represented by explicit programming of planning and negotiation routines.

Provide Security During Urban Transition Operations

Current facilities cannot provide security during urban transition operations because they are not large enough to realistically portray OPFOR infiltration threats and other challenges related to the non-linearity of many urban operations. Similar shortfalls impede the inclusion of noncombatant theft and other lesser threats to mission accomplishment. Actual urban areas are rarely used to exercise threats to airfields and ports. The simulations that exist fail to encompass operations security and force protection implications at other than the lowest echelons.

In the 2008–2011 period, there will be a need to represent future threats to information technology (IT) systems in simulations (e.g., threats to computer software and efforts to deceive sensors).

Rehearse/War-Game Urban Operations

The shortfalls for rehearsing/war-gaming urban operations—both near-term and longer-term—are similar to those identified above for conducting urban operations exercises and coordinating multinational and interagency resources.

Conduct Urban Noncombatant Evacuation Operations (NEOs)

Up to now, facilities have failed to replicate the size and scope of urban complexity, including appropriate levels of nuance when attempting to demonstrate the effects of noncombatants on operations. There are no simulations to support NEO-specific scenarios.

Conduct U.S. Domestic Urban Operations

Current urban training involving domestic contingencies includes the military in only a peripheral role, if any. Few U.S. military sites support such training. Domestic urban simulations are primarily oriented toward downwind hazard or other WMD weapon-specific models. There is little interface between military and police crowd-behavior software. U.S. military representatives do participate in local or regional preparedness initiatives (e.g., the Los Angeles County Terrorism Early Warning Group periodic meetings), which include exercises. But much more can be done. For example, there is a call for

- Improved knowledge of U.S. domestic emergency capabilities by military personnel;
- Greater participation of DoD representatives in domestic planning and exercises to enhance interpersonal relationships and identification of areas in which military capabilities can best fill civil shortfalls;
- Practicing the integration of military forces with civil assets in the field and during emergency operation center decision-making.

Simulations in the longer term will need to model nonlethal systems that have area (rather than point) effects.

Conduct Urban Combat Search and Rescue (CSAR)

Facilities are not large enough, dense enough, or complex enough to adequately represent challenges inherent in urban CSAR above basic extraction procedures. Simulations fail to provide scenarios for this mission. Future training, including simulations, will have to replicate UGV CSAR.

Conduct Urban Operations During and After a WMD Event

In the short term, no facility supports relevant training at other than the lowest tactical levels. The available simulations are planning- rather than training-oriented. There is a need for training to prepare forces, planners, and leaders for domestic as well as international WMD contingencies.

Consolidate Success in the Urban Environment

Current field training in support of consolidating success in the urban environment is essentially limited to combat operations and crowd-control activities. Higher-echelon exercises oversimplify consolidation and transition challenges; there is insufficient detail or variety in non-combatant/third-party representation; inadequate participation by other federal agency, NGO, PVO, and other interest representatives; and a failure to fully represent second- and higher-order consequences of decisions and actions. No simulation represents consolidation or transition implications.

Disembark, Base, Protect, and Move in Urban Environments

Training rarely includes pre-combat (e.g., CSS operations through developing-nation urban aerial and sea ports of debarkation(APODs and SPODs)). No simulations exist that support early phases of ur- ban operations. There appears to be an inherent assumption that such operations will take place in passive environments. The bombing of the U.S.S. *Cole* in Yemen and extensive coordination involving New York City docks during World War II provide tactical and strategic examples with the types of considerations that should be incorporated in current and future training. U.S. Air Force and U.S. Navy training facilities should include replications of developing-nation APODs and SPODs that can provide realistic training for SOF and ground forces assisting in their seizure and subsequent security and those that will conduct unloading, transloading, and other logistical tasks at these nodes.

Engage in the Urban Environment

Urban-environment engagements other than those involving small arms and FoF are too rarely replicated in live, virtual, or constructive training. Combined arms training involving mechanized and armor vehicles is prohibited at many facilities, and there is very limited ability to realistically replicate indirect and air support. (See the comments above under "Provide Fire Support."). No simulation adequately replicates all forms of urban engagement realistically, and there is a lack of combined arms and joint fires representation. There is virtually no accounting for a broader concept of engage (e.g., PSYOP, use of civil affairs) or of the difficulties inherent in coordinating Service, joint, multinational, interagency, and indigenous intelligence efforts; regular-force/SOF operations; multi-Service fires; and similar highly complex command and control issues.

In the longer term, new kinetic and non-kinetic systems for live, constructive, and virtual training will need to be incorporated.

Orchestrate Resources During Urban Operations

The fact that facilities are not large enough to contain the full complement of combat, CS, and CSS elements of all combatants, as well as representative types and quantities of noncombatants, precludes the level of realistic training needed to meet orchestration challenges. In addition, higher-echelon CPXs do not have sufficient resolution. Simulations similarly fail to represent the full palette of joint combat, CS, and CSS capabilities or those of adversaries, noncombatants, or other interested local, regional, or broader international parties.

Shape the Urban Environment

Training and exercises rarely have the resolution needed to test and provide feedback on shaping efforts. Activities other than traditional information operations are seldom included in training, and there is no evidence of efforts to introduce a level of sophistication that would intimate a shaping campaign. There is a similar lack of replication in simulations.

Sustain Urban Operations

Urban CSS considerations rarely exceed a superficial level of involvement, regardless of the training venue.

While there are no additional facility shortfalls for the longer term, there are some longer-term shortfalls for simulations. Specifically, future simulations should include the introduction of emerging systems and the difficulty of tailoring packages for legacy and transformation forces. Logistical support for UGV, new sensor, and other technologies will also be required.

Transition to Civilian Control

The shortfalls for this requirement are similar to those for "Consolidate Success in the Urban Environment."

Understand the Urban Environment

Current training limits the focus of understanding the urban environment to low-level tactical combat in most instances. Higher-echelon training events seldom include significant considerations beyond those influencing combat missions; the complexity of urban areas and the difficulty of obtaining accurate intelligence are poorly replicated. Simulations rarely provide anything beyond the immediate tactical situation.

In the longer term, training will need to represent improved sensor technology in live, virtual, and constructive training.

Achieve Simultaneity in Meeting Requirements

The shortfalls in achieving simultaneity are similar to those listed under "Orchestrate Resources During Urban Operations."

Conduct Training Across Multiple Levels of War

Upper-echelon exercises involving urban areas include only limited consideration of strategic implications; operational-level considerations seldom encompass the regional or larger influence of the urban areas concerned. The focus is generally on combat operations (i.e., the JP 3-06, *Joint Urban Operations,* engage phase) to the detriment of the understand, shape, consolidate, and transition phases. Few training events include effective consideration of tactical operational-level relationships. No simulation more than superficially replicates interlevel war considerations.

Summary

"Train as you fight" and "make the practice harder than the game" are long-standing informal standards for military leaders preparing troops for operations. Soldiers returning from victories during Operation Desert Storm in 1991 found their training at Ft. Irwin's National Training Center (NTC) tougher than anything they had confronted during combat against Iraq's military. The same cannot be said today. The U.S. armed forces are thus far unable to replicate the challenges their soldiers, sailors, marines, and airmen meet in the towns and cities of Afghanistan and Iraq.

Several of the reasons for this shortcoming are immediately evident in the gaps between identified JUT requirements and existing live, virtual, and constructive training capabilities. The most evident is lack of size. Training in complexes of 25, 50, or even 150 buildings is inadequate preparation for tactical actions in which structures number in the hundreds, if not thousands or tens of thousands. That quantity of buildings implies correspondingly greater numbers of people, vehicles, infrastructures, and other elements that imbue actual cities with a complexity that is altogether lacking in current live exercises. Simulations supporting virtual and constructive training are, unfortunately, similarly overly simplistic. Regardless of how many buildings they might replicate, the notional behaviors of opposing forces and noncombatants fall far short of reproducing the range of actual interactions and the scope of potential higher-order effects of each action and decision.

Analogous oversimplification likewise inhibits the effectiveness of urban exercises attempting to replicate the operational and strategic levels of war. These exercises fail to adequately challenge participants by focusing almost exclusively on combat operations, marginalizing the influence of agencies other than DoD, effectively ignoring noncombatant support requirements or their attitudes toward the friendly force, and glossing over governing responsibilities. While much improvement is also necessary in both Service and joint tactical-level training, preparation at this stratum by and large employs the accepted building-block process of first schooling the components and then educating larger units of which they are a part. The same cannot be said for readying those who participate in higher-level training events. Service and joint schools rarely address governing responsibilities, interfacing with indigenous populations, or urban concerns in general.

"Train as you fight" is no longer adequate when preparing for urban operations. Winning battles is but one element of success, and often not the dominant one. Joint urban training must prepare the American armed forces for the entirety of conflict's spectrum, for the complete hierarchy of tactical to strategic, and it must integrate these many parts into a single whole, for that is what awaits its students overseas and, potentially, at home.

CHAPTER FIVE

Deriving Joint Urban Operations Training Modules

"We need to try to develop some type of MOUT training facility for aviators," said [Col Greg Gass, commander in the 101st Airborne Division (Air Assault)]. It does not mean that such facilities do not exist, but they do not replicate cities the size of Najaf or Karbala, in Iraq, with populations of at least 600,000 people, according to Gass. Most importantly, it is the conventional forces that need to get this training, he added. "It will pay off in the end."

Roxana Tiron, "Pilots Spurring Training, Tactics Revolution,"
National Defense, June 2004

Training shall resemble the conditions of actual operations to the maximum extent possible.

Department of Defense Directive Number 1322.18
Subject: Military Training, September 3, 2004

Introduction

Chapter Two provided JUT requirements, and Chapter Three identified the current capabilities available to the joint community for urban training and sought to identify those that will be forthcoming in the near term through 2011. Chapter Four identified near- and longer-term shortfalls by comparing the requirements against the capabilities. Uncovering outstanding shortfalls allowed us to determine what further capabilities would be required for the nation's armed forces to be ready for both ongoing and pending urban undertakings.

This chapter explains how we derived the modules that simultaneously include existing capabilities identified in Chapter Three (including those pending in future years out to 2011) and those needed to close the shortfalls identified in Chapter Four. We begin with a discussion of why we chose a modular approach. We then present the original list of candidate modules and assess them in terms of their capability to close shortfalls. Finally, based on that assessment, we eliminate the modules that do not adequately apply to the development of a JUO training strategy to produce the final set to be used in constructing that strategy. All the removed modules have pertinence to Service or very limited joint applications, but their loss does not reopen any shortfalls closed in the original development of the modules.

Why a Modular Approach?

The most obvious solution to developing a JUO training strategy—developing a single, overarching "best answer" for the designated period of application (2005–2011)—suffers from one overriding deficiency: The seemingly optimal solution in 2005 might prove far less effective in the later years of consideration. The past decade of urban operations studies has provided many lessons, a primary observation of which has been that the complexity and dynamism of urban environments guarantee ever-differing and always-evolving conditions and challenges. The extent of change is striking even when episodes are separated by only a few years. Panama City in 1989, Mogadishu in 1993, and Baghdad in 2003 seem to have at least as many dissimilarities as commonalities. Even urban areas within the same nation at a given moment in time—for example, Baghdad, Basra, and Mosul in Iraq—present commanders with dramatically differing trials. Any strategy developed to prepare the U.S. armed forces for urban operations has to account for this constantly changing collection of opportunities and challenges.

A modular approach overcomes this deficiency by providing the flexibility and adaptability that are essential. Instead of each individual training site or simulation being a module in and of itself, a training module consists of *categories* of facilities or simulations.[1] This limits the number of modules to a manageable size. For example, training sites such as those at Camp Lejeune, NC, and Ft. Knox, KY, and the Joint Readiness Training Center at Ft. Polk, LA, differ in detail but have many similarities. Therefore, they are all associated with a single module (which is discussed in greater detail below). Defining modules in terms of categories also permits adaptation over time. Periodically editing module definitions will incorporate evolutions in field conditions, which means that a strategy that relies on a set of modules will not become invalid. Users can also adapt modules to account for change as capabilities change—as the joint community develops new training technologies, software, methods, or doctrine. Any financial impact of module modification can likewise be incorporated into an updated version. Thus, a training strategy that incorporates a given module can be adjusted, and the new costs associated with the strategy can be readily determined.

Further, it is apparent that a modular approach offers benefits beyond those directly related to the construction of a DoD-wide urban training strategy. Several are readily apparent:

- Modules will facilitate joint organization and Service unit development of similar strategies. A modular approach supports combatant command, Service component, and subordinate unit short- and longer-term training programs aimed at enhancing urban preparedness and, similarly, urban training coordinated

---

[1] As used in this report, a *training module* is a capability or system of capabilities associated with a type of facility, simulation, or other joint training resource. Examples include battalion and larger purpose-built facilities, terrain walks in actual urban areas, and classroom instruction. Each module meets multiple training requirements to varying degrees. Thus, one can construct a joint training strategy, program, or event by identifying the objectives to be achieved and combining modules that collectively provide the means to meet those requirements.

by multiple combatant commands and other headquarters echelons.

- Modules have value during the design and conduct of joint training exercises at virtually all echelons.
- Modules similarly serve to support the design and execution of contingency and campaign plans, war gaming, and rehearsals in support of such plans.
- Modules have potential value in analytic as well as training applications, including those involving the testing of communications and sensor network configurations.

Unlike an inflexible strategy that assumes fixed operational conditions, a modular methodology is responsive to differences in available resources, missions, and conditions in a given theater of operations. Modules provide commanders with a "tailorable" tool adaptable to their needs and to changes that occur over time.

The modular approach also provides a means of effectively normalizing various means of accounting through the common measure of training requirements. Differences in types of funds and their sources are subsumed; all modules are defined in terms of capabilities. Their value is measured by how much they allow an organization to meet one or more joint training requirements. Thus, a commander can compare and combine very dissimilar alternatives to meet his or her particular requirements by considering the resources most readily available, funds available, and other pertinent variables. Joint urban training, like all training, is not an end in itself. It is a means to the end of prepared individuals and units. An overarching training objective is to achieve that end at the least cost in dollars and time.

Before we discuss how we derived the first cut at training modules, two points about the assignment of specific training sites, simulations, or other capabilities to a module type merit further discussion. The first deals with the point highlighted above: Not all purpose-built sites of near equivalent size are identical (nor would we want them to be—variation means that units are exposed to different challenges when using different facilities). The Zussman facility at Ft. Knox differs in capability from Shughart-Gordon at the JRTC and the MOUT site at Camp Lejeune. All, however, are considered as examples of platoon purpose-built facilities.[2]

Similarly, not every simulation purporting to provide tactical-level training for leaders is equivalent in terms of effectiveness or the aspects of leadership it emphasizes. We have already noted that the extreme alternative of having a separate module for each training site or other capability is infeasible; there are well over 100 purpose-built facilities alone. Conversely, grouping similar capabilities into modules provides a means of developing a strategy while preserving the desirable variation in those capabilities. An individual wishing to determine whether a specific facility fully meets all aspects of a module definition, fails to attain the standard in some regards, or exceeds those expectations can easily adapt his or her strategy, training program, or exercise to compensate for the specific characteristics of a given facility, simulation, or other capability.

The second point deals with the specific examples used to describe each module. We expect that there will be little disagreement with most of these cases—that is, with the capabilities assigned to a particular module. However, our interviews revealed that there is some controversy with respect to the size of unit that can be trained at selected sites. As noted above, units too large to train *within* the built-up portion of a training complex can and do train *in and around*

---

[2] A *platoon purpose-built facility* is defined as one similar to a Module 1 training site but sufficient for a joint force up to and including a platoon-size ground force element. The ground force need not include a full complement of supporting forces/arms. Module 1 (*battalion and larger purpose-built facility*) is in turn defined as a purpose-built facility capable of handling a joint force up to and including a battalion task-force-size ground maneuver element and all supporting combat, combat support, and combat service support components. It is fully instrumented with live-fire capability in selected structures or areas, including convoy live fire. Blue tip, Simunitions-type, or paint-ball fires are permitted throughout, as are all types of ground vehicles. The facility's size must be sufficient to house all Blue Force (BLUEFOR) ground components, OPFOR, and noncombatants with sufficient physical separation and breaks in LOS to prevent the training unit from seeing the outer boundaries of the facility. Construction and size must allow OPFOR to infiltrate BLUEFOR positions with reasonable probability of not being detected. The number and character of buildings is such that multiple structures could be confused for each other from the ground and air. The complex includes at least one area or structure that can be used as a helicopter landing and pickup zone. All structures permit interior training, and the site includes a subterranean system. For a complete set of joint training module definitions, see Appendix A.

the part of the facility that includes a notional village or town. Some would argue that this and similar facilities should be considered capable of handling up to a battalion for urban training (given that the built-up area and surrounding terrain support a unit of that size). We do not disagree that some missions involving a village of even a few tens of structures could require a battalion or even larger unit to isolate the built-up area, interdict reinforcements, contain unrest, provide support to noncombatants, and conduct whatever tasks within and among the buildings prove necessary. We have also repeatedly emphasized that urban operations often include many considerations and therefore considerable terrain beyond the boundaries of a built-up area itself. The standard used here in no way seeks to undermine this more-encompassing definition of urban operations.

However, to consider an organization ready to conduct operations in densely populated built-up areas when its personnel have been trained on only the periphery of such sites would be a disservice to both Service and joint communities. All aspects of urban training need to be incorporated in the development of a joint training strategy; therefore, our focus in defining site-related modules is on the densely urbanized portions of those sites (the built-up areas). It is in such environments that American military personnel are likely to meet their greatest challenges in the field. A joint training strategy should reflect that. Shughart-Gordon and similar sites are not capable of containing an entire battalion task force within the primary facility while replicating field conditions military personnel are likely to confront in the field.

The expansion of these core facilities' boundaries (via façade, empty interior, and other structures along approach routes and elsewhere) can help to some extent, providing a richness rarely achieved at other sites. A battalion could simultaneously occupy its organic and supporting manpower within this larger urban environment. However, one or more of the definitive elements would be missing for many of the trainees (e.g., the ability to enter all structures and receive realistic training value in so doing). For this reason, we take a conservative approach to module definition. When it comes to live training in urban areas, size has a quality all its own. (It is noteworthy in this regard that the Dutch Army considers its Marnehuizen facility—the largest purpose-built urban training facility in the world—capable of handling only slightly more than a company team within the boundaries of the structure-filled area.)[3]

The Training Modules

In this section, we describe our first cut at the training modules. We then describe the screening process used to get to the final list. Finally, we present the final list that was used to build the JUO strategy described in Chapter Seven.

First Cut at Training Modules

The initial list of training modules comprises a compilation of capability types that might reasonably be of potential value in constructing a strategy. This initial list allows readers to see that capabilities with seeming application were not overlooked. That is not to say that every *possible* module appears in this list, but every *feasible, nontrivial* capability should be encompassed somewhere in it. A number of the modules that for one reason or another failed to make the cut will still be of value to those conducting Service component urban training, and some modules that do not meet broader demands are still of potential value to low-level tactical or very specialized joint training. Identifying these modules may help commanders, other trainers, and research analysts testing various simulations or technologies, even if the modules fall short of meeting the needs of strategy development.

Table 5.1 presents the initial list of 39 training modules.[4] The text following describes the modules and provides examples of exist-

---

[3] Interview with Lieutenant Colonels Henk Oerlemans and Johan Van Houten, Dutch Army, by Russell W. Glenn, Oostdorp and Marnehuizen urban training sites, the Netherlands, December 7–8, 2004.

[4] We have sought to include all relevant components in our module design. The level of specificity given physical urban training facilities, both purpose-built and those adapted from other existing resources, is a function of two factors. First, the congressional interest

Table 5.1
Initial List of 39 Modules

No.	Modules
	Purpose-Built Facilities
1	Battalion and larger purpose-built facility
2	Company purpose-built facility
3	Platoon purpose-built facility
4	Modular purpose-built facility
5	Façade-based facility
6	Commercially manufactured portable training facility
7	Hybrid facility
8	Air-ground facility
9	Shoot house
	Use of Populated Urban Areas
10	Terrain walks
11	Urban navigation
12	Urban simulated engagement
13	Urban live fire in populated area
14	Use of vacant buildings in populated area
15	Use of buildings scheduled for demolition
16	Use of public facilities during hours of closure
	Alternative/Other Training Concepts
17	Use of abandoned domestic urban areas
18	BRAC'd installations[a]
19	Ships as permanent urban training facilities
20	Mothballed ships temporarily used for urban training
21	Abandoned factories and surrounding urban infrastructure
22	Abandoned/constructed overseas urban areas
23	Use of existing other-agency and commercially available urban training facilities
24	Classroom instruction
25	Conduct of combatant command or JTF headquarters, large-scale schools, or multi-echelon/interagency exercises
	Simulation Capabilities
26	Tactical behaviors in and around structures
27	Higher-echelon planning and coordination
28	Joint, multinational, and interagency operations

that initially spurred this study has its origins in queries regarding military construction issues. We have therefore ensured that issues related to these concerns are addressed. Second, most collective urban training conducted during the 2005–2011 period will employ such facilities, a situation substantiating the need to investigate alternatives thoroughly.

Table 5.1 (continued)

No.	Modules
29	Specialized-technology simulation
30	Scenario-variant generation
31	Physiological and other stress simulation
32	Geographically distributed joint simulation
33	Environmental degradation and urban biorhythm
	Training Support Elements
34	Infrastructure trappings
35	OPFOR
36	Noncombatant role players
37	Targets to support urban training
38	Instrumentation/connectivity
39	Joint-force headquarter(s)

[a]Installations subject to BRAC.

ing facilities, where available. Many of these modules have application to operational challenges beyond those of urban environments. This is a sometimes less-than-obvious benefit of analyses that involve urban matters: Much of the training and other preparation for urban contingencies applies to portions of the conflict spectrum well beyond operations in villages, towns, and cities. In view of the greater densities and increased complexities found in urban areas, a force prepared for action in built-up areas can more often than not readily adapt to other environments. The reverse is less often the case: Preparing for missions in deserts, jungles, or mountains leaves significant gaps in Service and joint readiness to conduct urban operations.

Purpose-Built Facilities

1. **Battalion and larger purpose-built facility.** This module is capable of handling a joint force of battalion task force or larger ground maneuver elements. It is selectively instrumented with live-fire capability in some structures or areas, including on-site or nearby convoy live fire. Engagement with small-caliber paint, chalk, or similar training rounds is permitted throughout. All types of ground vehicles are permitted in the facility and in the

vicinity of the facility. Size must be sufficient to simultaneously contain BLUEFOR ground components (including all combat, CS, and CSS components), OPFOR, and noncombatants. The facility allows sufficient physical separation and breaks in LOS to permit nonlinear operations and infiltration of BLUEFOR positions by OPFOR squad or larger units with reasonable probability of not being detected. The number and character of buildings is such that multiple structures could be confused with each other from the ground and the air. The complex includes at least one area or structure that can be used as a helicopter landing and pickup zone. Most structures permit interior training. The facility includes subterranean system and rooftop training. (Example: Proposed USMC Twentynine Palms facility)

2. **Company purpose-built facility.** This module is the same as Module 1 but is sufficient for a joint force with ground maneuver elements of between company team and battalion task-force size with all supporting components. (Examples: None exist as of March 2005; the closest approximations are Victorville/George AFB and the urban training facility at Ft. Lewis, WA)

3. **Platoon purpose-built facility.** This module is the same as Module 1 but is sufficient for a joint force with a ground maneuver element of between platoon and up to company team size. The ground force need not include a full complement of supporting forces/arms. (Examples: Shughart-Gordon, Ft. Polk, LA, considered in isolation from its extended surroundings (see Module 7 below); Zussman, Ft. Knox, KY; urban training sites at Camp Pendleton, CA, Camp Lejeune, NC, and Ft. Benning, GA)

4. **Modular purpose-built facility.** This module is constructed with CONEX containers or similar components capable of supporting interior and exterior training. It is unique to its location and can be either permanent or movable. (Example: Bagram, Afghanistan)

5. **Façade-based facility.** This module is metal-framed (scaffolding-type construction) with a plastic, tarpaper, or other façade. It is capable of supporting exterior training only. (Example: Israel

Defense Forces (IDF) facility in central Israel, constructed of plastic fabric on robust metal scaffolding frames)

6. **Commercially manufactured portable training facility.** This module consists of prefabricated modular components built on-site. It generally has CONEX-type construction. Each unit's interior is identical, or the variety between units is quite limited. (Examples: Facilities constructed by Blackwater, Anteon, and similar companies)

7. **Hybrid facility.** This module is the same as Module 1, but it is built in segments of varying quality (perhaps collocated with an extant facility comprising one or more of the segments). Most structures support interior training. This module could have multiple Module 2 or 3 components as a core and could also use adjacent combinations of Module 4, Module 5, existing actual urban areas, or other module types. It could include a Yodaville-style aviation range as an adjacent or contiguous element. Another example would be sites created by modifying underutilized or abandoned port and airfield facilities to support preparation for seizing and securing of developing-nation ports of debarkation. (Example: Shughart-Gordon, Ft. Polk, LA, when immediate environs are included)

8. **Air-ground facility.** This module permits simultaneous operations by fixed-wing and rotary-wing aircraft, UAVs, SEAD, and FACs. Some of those operations may be constrained (e.g., by a safety standoff between the drop zone and FACs) or provided in simulated or notional form (e.g., SEAD). (Example: Yodaville, Yuma, AZ, although it does not satisfy all definitional elements)

9. **Shoot house.** This module consists of live-fire facilities that allow for the use of green tip (ball or tracer) ammunition. It is capable of supporting at least fire-team-size elements. (Example: Naval Special Warfare Group Two, Chesapeake, VA)

Use of Populated Urban Areas

10. **Terrain walks.** This module allows for training conducted in populated urban areas with the purpose of familiarizing students with the challenges associated with such areas. The value of terrain walks is not limited to tactical-level instruction. The complexities inherent in physical and social infrastructure, air-to-ground targeting, and even seemingly mundane issues such as the difficulty of navigating in cities devoid of signage or recognizable landmarks should be made evident to those employing units and other resources in built-up areas. Hue and other examples from history remind us that higher headquarters too often impose grossly unrealistic tasks on tactical units because they do not understand the difficulties of urban operations. (Example: Training conducted by the U.S. Army Infantry School Career Course in Columbus, GA (adjacent to Ft. Benning, GA))

11. **Urban navigation.** This module allows for air and ground urban navigation conducted over and within populated urban areas. (Example: USMC overflights of Southern California urban areas)

12. **Urban simulated engagement.** This module allows for air or ground urban targeting but no live fire. (Example: Exercises conducted by USAF over DeRitter, LA)

13. **Urban live fire in populated areas.** This module allows for live-fire engagement of targets, generally limited to sniper engagements into bullet traps or short-range engagements in a structure or otherwise circumscribed area. (Example: USMC MEU(SOC) training during qualification exercises)

14. **Use of vacant buildings in populated areas.** This module consists of vacant buildings (or vacant portions of partially occupied buildings) in populated areas. Because buildings are occupied during working hours or will potentially be used in the future, there is a need to minimize the extent of damage inflicted on them. (Example: USMC MEU(SOC) training during qualification exercises)

15. **Use of buildings scheduled for demolition.** This module consists of buildings scheduled for demolition. Thus, the extent of destruction or damage that results from training is not a significant factor. (Example: None available)

16. **Use of public facilities.** This module consists of using subways and other public facilities during periods of non-use (e.g., the DC Metro in early morning hours). (Example: USMC MCWL using the streets of Little Rock, AK, and Boise, ID, to test urban infiltration techniques)

Alternative/Other Training Concepts

17. **Use of abandoned domestic urban areas.** This module uses previously occupied civilian built-up areas. (Examples: Playas, owned by the New Mexico Institute of Technology; abandoned towns in North Dakota; Muscatatuck, IN; abandoned towns in river flood plains)

18. **BRAC'd installations.** This module uses military posts and bases now in caretaker status. (Example: George AFB, CA)

19. **Ships as permanent urban training facilities.** This module uses former USN, USCG (U.S. Coast Guard), commercial, or other vessels to support urban operations training. Ships can be employed unmodified ("as is") or modified to better replicate urban areas on land. (Example: None available)

20. **Mothballed ships temporarily used for urban training.** This module uses inactive vessels for urban operations training. Such vessels include mothballed/abandoned/destined-for-scrap ships. (Example: None available)

21. **Abandoned factories and surrounding urban infrastructure.** (Example: None available)

22. **Abandoned/constructed overseas urban areas.** This module uses international sites similar to those described in Modules 1–3, 17, and others. (Example: Built-up areas on selected islands off the coast of Kuwait)

23. **Use of existing other-agency and commercially available urban training facilities.** This module entails the use/expansion of existing law enforcement, fire, or other agency training facilities. (Examples: Federal Bureau of Investigation (FBI) Academy Hogan's Alley, Blackwater training compound)

24. **Classroom instruction.** This module consists of training conducted in a classroom environment. (Examples: Urban operations training electives at the U.S. Army Command and General Staff College, Ft. Leavenworth, KS; electives previously offered at the Naval War College, Newport, RI; and classes and courses taught by joint and component headquarters as professional-education sessions or parts of other instructional programs)

25. **Conduct of combatant command or JTF headquarters, large-scale school, or multi-echelon/interagency exercises.** This module consists of major urban exercises focused on command, control, or decisionmaking issues. Like other training in the strategy presented here, and as evident in the definitions of the levels of JUO training introduced in Chapter Two, this module should demonstrate a building-block approach. Exercises at the pinnacle of the module will be highly complex, replicating the many and varied demands of urban operations. None currently reproduce the difficulties inherent in coordinating air, maritime, ground, and SOF component theater fires, intelligence activities, information operations, and logistics, including passage of personnel and supplies through urban APODs and SPODs. Service and joint headquarters at multiple echelons should practice the command and control linkages and simultaneous use of urban areas of operation, control made more difficult by the fact that such towns and cities also house the daily residences and workplaces of thousands or millions of members of the indigenous population. JTF and other headquarters similarly need to synchronize their activities with, support, or coordinate information campaigns with Special Operations foreign internal defense (FID), civil affairs (CA), PSYOP, and other missions in and

around urban areas. (Examples: Exercise Urban Challenge, Urban Resolve, and Joint Urban Warrior)

Simulation Capabilities

26. **Tactical behaviors in and around structures.** This module involves the comprehensive replication of urban operations from the lowest tactical levels (i.e., movement and behaviors of individual BLUEFOR, OPFOR, and noncombatants) up to and including battalion task-force operations. It includes realistic representation of operations across the spectrum of conflict, inside and outside all structures and above and below ground, and of transition between environments (e.g., from exterior to interior, floor-to-floor, and firing from interiors to engage a full range of exterior targets). (Examples: JRTC and McKenna sites, which use OOS variants (DISAF, JSAF, OTB (objective test bed)), JCATS, and Full Spectrum Warrior to model interior fighting at tactical levels)

27. **Higher-echelon planning and coordination.** This module involves the comprehensive replication of higher-level tactical urban challenges (e.g., maneuver brigade or JTF and all supporting elements, including air, space, and national assets/inputs). It includes realistic simulation of subordinate units, adjacent units, and other entities (e.g., civilian government representatives and multinational representation) as necessary to support decision-making and presentation of the full range of three-block war contingencies. (Examples: Defense Modeling and Simulation Office (DMSO) Joint Theater Level Simulation, JFCOM Urban Resolve, DMSO JOUST, and Lockheed-Martin WARSIM programs all support command and staff training)

28. **Joint, multinational, and interagency operations.** This module involves the comprehensive replication of operational and/or strategic-level urban situations capable of supporting joint, multinational, and interagency participation, including live participation by representatives of organizations representing these capabilities. (Examples: DMSO Integrated Live/Virtual Joint Urban

Operations Range has Army, Navy, Air Force simulations and MOUT sites linked; USAF Joint Close Air Support Training and Rehearsal Program links Services)

29. **Specialized-technology simulation.** This module consists of the simulated representation of conditions too dangerous, expensive, or technologically advanced to exercise with live training. Virtual or constructive sub-modules include the following:

- Cockpit or simulator representation of realistic urban signatures and hazards for fixed- and rotary-wing training (e.g., a pilot over an abandoned town views heat, electromagnetic, dust, light, and other signatures as though the area were occupied by a civilian population). Replication also includes antiaircraft fire, wire obstacles, UAV and other aircraft in the vicinity, targets in close proximity to proscribed targets, and other hazards encountered when flying over an urban environment. (Examples: Aviation Combined-Arms Tactical Trainer (AVCATT) at Ft. Rucker used for rotary-wing training, Army Research Laboratory (ARL)/Penn State CAVE systems used for 3-D visualization)

- Similar representation for armor and other vehicle training, including attacks by insurgents, mines, happenstance appearance of innocent civilian vehicles, domestic animals, and other micro-events. (Examples: UAMBL CCTT used for armor training, Lockheed-Martin and Raydon simulators specializing in convoy operations training, ARL/Penn State CAVE systems used for 3-D visualization, Project Albert MANA and Diamond agent-based models emerging as noncombatant representations)

- Air-ground coordination that reproduces targeting issues, accuracy and reliability data, SEAD, the presence of friendly and noncombatant personnel, and similar urban conditions. (Example: USAF AC-130 simulators at Hurlburt Field, FL)

- Representation of scenarios and robotic, sensor, FCS, or other technologies under development or existent and either too dangerous or too expensive to employ during live training. (Examples: UAMBL OTB, RAND JCATS, and Sandia UMBRA models all specialize in robotic applications, but none are used for training yet)

30. **Scenario-variant generation.** This module consists of simulations providing repeated training "trails" with similar but sufficiently varying scenarios to provide trainees the opportunity to learn from mistakes while precluding their playing the system rather than the scenario. (Examples: Full Spectrum Warrior and Full Spectrum Command are able to quickly change conditions; other models are more cumbersome)

31. **Physiological and other stress simulation.** This module involves the realistic replication of urban influences at all levels of simulation (e.g., personnel exhaustion, logistics consumption rates, stress, and interaction with civilians with heterogeneous behaviors that evolve over time). (Example: Natick IUSS models fatigue, casualties, and stress but is not a trainer; some of these capabilities are to be incorporated into future OOS versions)

32. **Geographically distributed joint simulation.** This module provides for compatibility with air, sea, and ground systems participating at dispersed locations in live, constructive, or virtual mode to replicate a "seamless" reality in which the consequences of all actions are accurately and stochastically replicated. (Examples: Geographically distributed OOS variants used widely for this, e.g., Urban Resolve, MC02, BFIT)

33. **Environmental degradation and urban biorhythm.** This module involves the representation or the imposition of realistic conditions on the exercise environment (e.g., the interruption or degradation of communications, sensor signals, UAV transmissions, ground-ground and air-ground visibility, and changing ambient conditions—vehicle traffic, density of on-street foot traffic, holidays, market days. (Examples: Urban Resolve phase 1 models some aspects of biorhythm; most constructive models allow control of atmospheric conditions; some jamming is modeled in the

Scalable Technologies Qualnet model, the Soldier Battle Lab C4ISR Sim Lab at McKenna, and the Defense Advanced Research Agency (DARPA) FCS Comms program; none of these are comprehensive)

Training Support Elements

34. **Infrastructural trappings.** This module provides lights, sewers, running water, and electronic signatures. (Example: Zussman training site, Ft. Knox, KY)

35. **OPFOR.** This module provides OPFOR personnel armed similarly to a friendly force during FoF urban training. (Example: Ft. Polk, LA)

36. **Noncombatant role players.** This module provides personnel playing the role of civilian urban residents. (Example: Ft. Polk, LA)

37. **Targets to support urban training.** This module involves two- or three-dimensional targets that respond when engaged with green tip or other ammunition. (Examples: Pop-up dummies in Bagram, Afghanistan; Naval Special Warfare Group Two shoot house, Chesapeake, VA)

38. **Instrumentation/connectivity.** This module uses cameras, communications, or other capabilities in support of training enhancement, linking physically separated training sites/ simulators/players during scenario execution or AARs. (Examples: Yuma, AZ; many other facilities)

39. **Joint force headquarter(s).** This module involves a joint planning, coordinating, and directing organization. (Example: JFCOM standing JTF)

The Initial Screening

These initial 39 modules were then screened with additional filters. Only those that passed through all gradations of this sieving process merit possible inclusion in the ultimate training strategy design. The categories of filters through which the initial set of modules was screened are:

- Does the module meet a sufficient number of JUT requirements? If so, does it provide the force with a sufficient level of proficiency?
- Are there environmental, ergonomic, or other considerations that make use of the module impractical?
- Is the module cost-effective in terms of dollars and time spent in its application to training?

In short, does a module provide sufficient joint training effectiveness to merit continued consideration as a component of a U.S. joint urban training strategy? We discuss the first two criteria below, and we deal with cost-effectiveness in Chapter Six.

The first filter follows directly from the work presented in the previous two chapters. Since each module is a system of training capabilities, the final list must contain every capability needed to meet the immediate and longer-term requirements that will prepare the nation's armed forces for joint urban operations. Comparing the module list against the training requirements developed earlier, then, should both confirm that the list is sufficient to meet this need and identify those modules that fail to adequately address JUT preparation.

We assessed the modules in terms of the building-block approach that the U.S. armed forces and those of many of its allies use to construct training both within a given echelon and over multiple levels of command. Individuals' basic training generally incorporates physical fitness, drill, weapons familiarization, and other blocks of instruction, each important in and of itself and as a component of providing soldiers, sailors, marines, and airmen with the minimum capabilities necessary for them to serve effectively when assigned to a unit in the field. Commanders similarly build readiness by training the various units in their command so that each part can meet the functional standards that, when combined, constitute operational readiness. A ship's captain, for example, ensures that maintenance personnel, radar technicians, gunners, and the many other subcomponents of his command are trained adequately to serve the greater whole, an effective vessel on the seas. Pilots could conceivably learn to fly simply by being put in a cockpit with a qualified aviator who attempts to teach the neophytes all aspects of their duties simultaneously. The result would be very costly in terms of time, lost aircraft, and casualties. Services instead train prospective pilots by teaching skills in components that establish foundation competence in navigation, instrument reading, radio communication, and operation of the many other systems that together constitute an aircraft or aid in its operation. Only after a prospective pilot demonstrates foundation-level proficiency is he or she given more-complex training and, eventually, responsibility for an airframe.

Each module was assigned one of four ratings, indicating how well it fills each of our 34 joint urban training requirements (using our own definitions):[5]

- C. Denotes achievement of a "crawl" standard of readiness, defined as attainment of foundation skills necessary for developing more-advanced skills or combinations of skills. Being able to establish basic air-ground communications in an urban environment, for example, is essential for coordinating CAS. A module supporting a "crawl" measure of ability has to support development of base-level skills translatable to application under actual operational conditions in the field.
- W. Denotes achievement of a "walk" standard of readiness, defined as achievement of greater sophistication in task accomplishment and the ability to coordinate several "crawl"-level or other "walk"-level skills in accomplishing a mission. Having the skills to communicate ground-to-air, transmit target grid coordinates or successfully provide laser designation, and conduct

---

[5] These definitions, with additional explanatory discussion, are given in Appendix A. We conducted four iterations of assigning C, W, R, or S to each module-requirement combination to obtain the results shown in Appendix G.

accurate post-strike battle damage assessment (BDA) would together constitute a "walk" measure of preparedness. A module supporting attainment of a "walk" measure requires managing several skills under realistic field conditions sequentially or simultaneously, as demanded by the situation.

- R. Denotes achievement of a "run" standard of readiness, defined as accomplishment of complete operational preparedness (combat readiness, for missions involving combat action). A "run" status implies proficiency in all supporting tasks and the orchestration of those tasks to accomplish assigned missions. Being able to successfully coordinate CAS under any feasible conditions, including situations in which one or more alternative means of doing so are impractical (e.g., talking a pilot onto an urban target given the failure of GPS and laser designation equipment), would constitute reaching a "run" measure of readiness. A module supporting attainment of "run" status must provide sufficient challenge to replicate the most adverse operational conditions.
- S. Denotes that the module "supports" meeting a training requirement. A support module cannot fulfill the needs of the requirement under consideration by itself, but its use adds realism, provides additional challenges, or otherwise enhances another module in the attainment of a C, W, or R rating. The addition of a support module cannot, in and of itself, raise a module's capability to meet a training requirement from a lower to a higher rating (e.g., C to W).

This "crawl," "walk," "run" evaluation method supports a building-block training approach; in fact, it is the way some trainers refer to such a methodology and should therefore be familiar to most readers. There is, of course, an inherent equivalency implied in this scheme: All joint training requirements are inherently considered of equal importance in our approach. This is acceptable, since the primary use of the matrix is as a tool to support development of a JUO training strategy. While some might argue that one or more requirements are two, three, or even several times more valuable than others, such arguments would almost inevitably presuppose that a specific scenario or type of contingency was being considered. Since the strategy designed in Chapter Seven covers the full spectrum of military operations, from urban support in an environment with no human enemy to major FoF combat in a WMD-contaminated environment, any weighting system would be specious. Commanders and staff officers seeking to develop contingency plans or training to meet specific scenarios are free to weight training requirements and, as appropriate, modules to fit their situations.

Whether a module merits a C, W, R, or S is based on what a trainer should reasonably expect to be able to achieve if the module has all the characteristics overtly presented in its definition and any others that would inherently be part of such a capability. Therefore, the rating is a measure of potential. It does not represent the actual degree to which all specific cases that fall under a given module will meet the requirements shown (e.g., every facility that is considered to be within Module 3). As noted previously, some Module 3 facilities (e.g., the Camp Pendleton urban training site and the Zussman training complex at Ft. Knox) will meet module-definition conditions to only a limited extent. Some may have capabilities that exceed those specified in the module definition and might therefore merit a higher ranking (e.g., an R rather than a W) for one or more requirements than would a less-inclusive site.

The results of this comparison are reproduced in their entirety in Appendix G, which presents a matrix that maps the 39 modules against the 34 joint training requirements, with a C, W, R, or S assigned at each intersection. This matrix indicates how well each module meets each training requirement.

Table 5.2 synthesizes the results detailed in Appendix G, allowing the reader to readily identify which modules appear to be star performers and which lag the greater pack. Leaving S entries unchanged, we assign numerical values of 1, 2, and 3 to C, W, and R modules, respectively. Given that there are 34 requirements, the maximum

Table 5.2
**Initial List of 39 Modules Ranked in Terms of How Well They Meet
JUO Requirements (Highest Score Possible = 102)**

No.	Module	Score
	Purpose-Built Facilities	
1	Battalion and larger purpose-built facility	84
2	Company purpose-built facility	55
3	Platoon purpose-built facility	44
4	Modular purpose-built facility	32
5	Façade-based facility	30
6	Commercially manufactured portable training facility	31
7	Hybrid facility	81
8	Air-ground facility	31
9	Shoot house	16
	Use of Populated Urban Areas	
10	Terrain walks	39
11	Urban navigation	26
12	Urban simulated engagement	29
13	Urban live fire in populated areas	18
14	Use of vacant buildings in populated areas	32
15	Use of buildings scheduled for demolition	41
16	Use of public facilities	52
	Alternative/Other Training Concepts	
17	Use of abandoned domestic urban areas	90
18	BRAC'd installations	91
19	Ships as permanent urban training facilities	34
20	Mothballed ships temporarily used for urban training	33
21	Abandoned factories and surrounding urban infrastructure	40
22	Abandoned/constructed overseas urban areas	84
23	Use of existing other-agency and commercially available urban training facilities	34
24	Classroom instruction	45
25	Conduct of combatant command or JTF headquarters, large-scale school, or multi-echelon/interagency exercises	73
	Simulation Capabilities	
26	Tactical behaviors in and around structures	38
27	Higher-echelon planning and coordination	43
28	Joint, multinational, and interagency operations	41
29	Specialized-technology simulation	18
30	Scenario-variant generation	1
31	Physiological and other stress simulation	1
32	Geographically distributed joint simulation	4
33	Environmental degradation and urban biorhythm	1

Table 5.2 (continued)

No.	Module	Score
	Training Support Elements	
34	Infrastructure trappings	2
35	OPFOR	10
36	Noncombatant role players	22
37	Targets to support urban training	1
38	Instrumentation/connectivity	3
39	Joint force headquarter(s)	27

score a module could achieve would be 102 (or 34 times 3).[6] As an example, the Module 1 score of 84 out of 102 is the result of 16 "run" evaluations (3s), 18 "walks" (2s), and 0 "crawls" (1s). If the module does not meet the "crawl," "walk," or "run" criterion for a particular requirement, it can either support other modules (receive an S rating) or not meet the requirement (be assigned a score of 0). The numerical effect is a score of 0 in either case.

The differences are quite dramatic in several cases. Five modules stand out as exceptional in meeting the demands of joint training requirements:

- Module 1: Battalion and larger purpose-built facility (84)
- Module 7: Hybrid facility (81)
- Module 17: Use of abandoned urban areas (90)

---

[6] The reader might question why we did not apply a numerical evaluation system originally in the modules rather than a requirements comparison process. The explanation given for using a "crawl," "walk," "run," "support" metric provides the basis. Numerical entries are appropriate for less-subjective evaluations, ones in which quantitative comparisons are straightforward, such as comparing costs (the cost of attaining requirement A is $X, while that for meeting requirement B is $3X would be an extreme example; not all applications need be this clear cut). This is not the case here. Assigning numerical values implies not only a relative ranking (e.g., a module attaining an R ranking meets a joint training requirement better than does one receiving a C or a W). Numerical values imply a proportional measure of relative value. For example, using 1, 2, and 3 to represent C, W, and R might lead a reader to infer that a module receiving a 3 is three times better at meeting a requirement than is one with a value of 1. Using numerical values for anything other than helping to demonstrate relative rankings more clearly (as we do here, with one set of quantities for C, W, and R to deliberately avoid implying proportional relationships) implies a specificity that is simply misleading given the nature of modules and training requirements.

- Module 18: BRAC'd installations (91)
- Module 22: Abandoned/constructed overseas urban areas (84)

Three others form a more widely dispersed but still well-above-average set:

- Module 2: Company purpose-built facility (55)
- Module 16: Use of public facilities during hours of non-use (52)
- Module 25: Conduct of combatant command or JTF headquarters, large-scale school, or multi-echelon/interagency exercises (73)

There are also some obvious underperformers in terms of meeting JUT requirements. (Recall, however, that these modules may still have value in meeting some Service or very specific joint training needs.) Several of the modules with very low numerical values have to be granted special dispensation because they contain many S ratings and therefore have value in conjunction with other modules that they support. These tend to be located toward the bottom of the matrix in the "simulation capabilities" or "training support elements" categories of modules. Allowing them to remain on the list (but viewing them again in later considerations) means that only modules 1–25 are susceptible to elimination at this stage. Two of those merit exclusion:

- Module 9: Shoot house (16)
- Module 13: Urban live fire in populated areas (18)

Nine other modules perform somewhat better but are still well below par:

- Module 4: Modular purpose-built facility (32)
- Module 5: Façade-based facility (30)
- Module 6: Commercially manufactured portable training facility (31)
- Module 11: Urban navigation (26)
- Module 12: Urban simulated engagement (29)

- Module 14: Use of vacant buildings in populated areas (32)
- Module 19: Ships as permanent urban training facilities (34)
- Module 20: Mothballed ships temporarily used for urban training (33)
- Module 23: Use of existing other-agency and commercially available urban training facilities (34)

Do any of these modules warrant elimination when analyzed more closely? The evidence suggests that from a micro-perspective, several are subsumed in other modules at the level of focus of this study. Modules 4 (modular purpose-built facility), 5 (façade-based facility), and 6 (commercially manufactured portable training facility) could have value in the design of unit training, but for an overarching strategy they are potentially a part of Module 7 (hybrid facility) and can be used to complement modules such as 1, 2, and 3 (battalion and larger purpose built facility, company purpose-built facility, and platoon purpose-built facility, respectively).

Only one of the group of nine modules in question meets a "run" standard for any module, and that is an anomaly (see Appendix G). Module 11 (urban navigation), not surprisingly, merits an R in meeting the "navigate in the urban environment" joint training requirement. That a module and requirement are almost identical explains the strong ranking. While specific training on urban navigation is crucial, this is generally true for Service component or lower-level joint rather than upper-echelon preparation. Map exercises, computer terrain replications, and urban terrain walks can address the requirement at these higher levels. Also, each module among those with the highest scores at the highest levels has at least two R ratings; all but two (Modules 2 and 16) have at least eight. Taking these considerations into account, the 11 lowest-scoring modules and "urban navigation" are eliminated from further consideration for inclusion in a JUO training strategy.

While it is obvious that all of the modules qualifying for exclusion have value in training individuals and organizations for urban contingencies, this preparation applies to what would take place at lower tactical levels or during Service training. Elimination from our final list does not mean that these modules should not be incorporated in appropriate Service training events. The complete set of modules appears here to enable us to identify any that might serve this purpose or that of supporting specialized or very specific joint urban training (e.g., small SOF unit use of shoot houses). The matrix in Appendix G can be adapted for other-than-JUT strategy development, according to individual needs. Users may choose not to eliminate one or more of the modules that did not pass muster here.

Also, some capabilities within modules that scored well in this first screening will fall far below the potential that the C, W, and R ratings suggest. The Playas, NM, training facility (Module 17, an abandoned domestic urban area, which scored a 90) may have value for some Service or specialized joint urban instruction. However, because of its lack of high-rise buildings, low structure density, and limited size, it will be of little value in meeting most larger joint unit or headquarters training requirements. As noted repeatedly, each specific capability, whether a facility, simulation, or other, must be evaluated on a case-by-case basis given the demands of the joint commanders' training needs. In the case of Playas, while it scores well, the facility itself falls well short of the potential value represented by the definition of Module 17.

Further analysis of the modules included consideration of the following:

- Which of those in the "simulation capabilities" and "training support elements" categories (both having modules with a large number of S ratings) failed to demonstrate adequate inherent value (i.e., the module itself meets training requirements) or support value (i.e., the module is significant in enhancing the quality of preparation provided when used in conjunction with other modules)?
- Which modules inherently presented insurmountable environmental challenges or related problems (e.g., noise pollution or extensive terrain damage) of sufficient severity to make their use infeasible?

- Which modules presented unacceptable internal social challenges (e.g., every possible example poses unavoidable danger to participants)?

Several modules in the simulation capabilities and training support elements categories supported training requirements to the extent that they inherently merited consideration for inclusion. Simulation capabilities that assist in developing tactical behaviors in built-up areas and those promoting the development of planning and coordination in urban environments are notable in this regard. Noncombatant role players similarly have considerable standalone merit in addition to their value in a support role. Only one of these modules, Module 31 (physiological and other stress simulation) failed to both independently meet requirements and provide adequate support for joint operations. Such capabilities have proven vital to Service and specialized analyses of given situations or during the use of particular technologies (e.g., pilot stress during urban engagements involving numerous noncombatants), but application to joint training in a more general sense falls short of what is needed for inclusion in a JUO training strategy.

With respect to the other two areas of consideration, environmental and internal social issues, further examination revealed none worthy of exclusion. However, there are specific examples of capabilities that pose such problems, some to the extent that they might limit a particular site or simulation from broad usage in support of joint urban training. Yodaville, mentioned in Chapter Three, is a good example. This air-ground urban training facility in Yuma, AZ, suffers from its proximity to the Mexican border. Denial of overflight rights during training severely restricts the directions from which some aircraft can approach targets. Further (as mentioned previously), illegal aliens sometimes enter the site, thinking the lights represent an actual town in which they might acquire water or meet other needs. Neither of these limitations precludes the conduct of joint or Service training, but they can in some cases limit the value of such events.

The Final List

On the basis of the above analysis, we created a final list of modules for inclusion in the JUO training strategy. Appendix G shows the final list in detail. Table 5.3 repeats Table 5.2 but deletes the modules that were not retained. To avoid confusion, we have left the numbers associated with retained modules as they were in the initial compilation.

Table 5.3
Final List of Modules Retained

No.	Module	Score
	Purpose-Built Facilities	
1	Battalion and larger purpose-built facility	84
2	Company purpose-built facility	55
3	Platoon purpose-built facility	44
4	Modular purpose-built facility	32
5	Facade-based facility	30
6	Commercially manufactured portable training facility	31
7	Hybrid facility	81
8	Air-ground facility	31
9	Shoot house	16
	Use of Populated Urban Areas	
10	Terrain walks	39
11	Urban navigation	26
12	Urban simulated engagement	29
13	Urban live fire in populated area	18
14	Use of vacant buildings in populated area	32
15	Use of buildings scheduled for demolition	41
16	Use of public facilities	52
	Alternative/Other Training Concepts	
17	Use of abandoned domestic urban areas	90
18	BRAC'd installations	91
19	Ships as permanent urban training facilities	34
20	Mothballed ships temporarily used for urban training	33
21	Abandoned factories and surrounding urban infrastructure	40
22	Abandoned/constructed overseas urban areas	84
23	Use of existing other-agency and commercially available urban training facilities	34
24	Classroom instruction	45
25	Conduct of combatant command or JTF headquarters, large-scale schools, or multi-echelon/interagency exercises	73

Table S.3 (continued)

No.	Module	Score
	Simulation Capabilities	
26	Tactical behaviors in and around structures	38
27	Higher-echelon planning and coordination	43
28	Joint, multinational, and interagency operations	41
29	Specialized-technology simulation	18
30	Scenario-variant generation	1
~~31~~	~~Physiological and other stress simulation~~	~~1~~
32	Geographically distributed joint simulation	4
33	Environmental degradation and urban biorhythm	1
	Training Support Elements	
34	Infrastructure trappings	2
35	OPFOR	10
36	Noncombatant role players	22
37	Targets to support urban training	1
38	Instrumentation/connectivity	3
39	Joint-force headquarter(s)	27

In closing, to paraphrase Donne, "No module is an island." All are de facto interrelated through the inability of any single one to meet all training requirements for any but the most limited of joint training scenarios. Technological constraints, distance to a facility possessing a given module's characteristics, variations in mission and training requirements, throughput capacity limitations, and cost constraints will limit the value or availability of some modules during joint commanders' efforts to ready their forces for urban operations. The development of any joint training strategy must include consideration of such obstacles.

CHAPTER SIX

Cost Analysis

As discussed in Chapter Five, we used three screening criteria to identify the final set of modules to use as building blocks in constructing our JUO training strategy. The first two—whether the module met a sufficient number of JUT requirements and, if so, whether it provided the force with a sufficient level of proficiency, and whether there were environmental, ergonomic, or other considerations that made the use of the module impractical—were discussed in Chapter Five. The third criterion—cost-effectiveness in terms of dollars and time spent in applying a module to training—is the subject this chapter. Regardless of how effective a module is in addressing requirements, it will be rejected if it does so at prohibitive cost.

This chapter assesses the cost of each of the modules that met the screening criteria. These cost estimates are independent of use (i.e., whether Service only or joint training) and therefore reflect the expenditures needed to build, maintain, and use a training facility, not accounting for any potentially unique demands associated with joint training.[1] We conclude by offering some observations (from a cost perspective) of the issues associated with each module as a prelude to crafting a JUO training investment strategy, which is discussed more fully in Chapter Seven.

---

[1] The exception to this is Module 38 (instrumentation/connectivity), which includes the cost of capabilities necessary to link facilities together in support of combined live, virtual, and constructive urban training events.

Methods and Assumptions

We followed standard DoD procedures in conducting the cost assessment.[2] The costs of the modules are derived from a combination of engineering data, parametric analysis, analogy, and interviews with subject-matter experts.[3] For analytical purposes, certain costs that are generally common to all modules (e.g., local transportation) are not included, nor are minor costs that would not be germane to the conclusions derived from the assessments, such as those associated with coordinating use of a facility. Operational training costs for such items as controllers and role players are provided separately, while those for items such as range safety and scheduling are not included, since they are generally encompassed in base operation budgets regardless of the range used or the type of training conducted. If more urban training were done, less of another type of training would be done; thus, the net is a zero-sum. Finally, the joint training tasks are not ammunition- or equipment-intensive, so these costs are also not included.

As a starting point, we constructed a comprehensive cost-breakdown structure, which we then modified as needed to accommodate the specific characteristics defining each of the modules. Ultimately, the assessment focuses on the life-cycle cost categories of investment (nonrecurring) and sustainment (recurring). Because de-

---

[2] See, for example, *Department of the Army Cost Analysis Manual*, U.S. Army Cost and Economic Analysis Center, May 2002; *Economic Analysis Manual*, February 2001; and similar departmental publications.

[3] Data sources will be provided within the discussion of each module. However, some sources were generally relied on throughout. These include Ann Miller, Robert Book, Pete Kusek, "Analysis of Alternatives for Providing Joint Urban Warfare Training Capabilities," Center for Naval Analyses Research Memorandum CRM D0009201.A2/FINAL, Alexandria, VA (hereafter referred to as the CNA study); Dominant Maneuver (DM) Division, J8, Joint Chiefs of Staff, *Joint Urban Operations (JUO) Training Facility Study Phase III Final Report*, 2001 (hereafter referred to as the J8 study); *Combined Arms MOUT Task Force (CAMTF) Study Group Urban Operations Resource Requirements and Combined Arms Training Strategy*, Volume V, *Final Report*, Appendix 1, Tab F, "Form DD-1391" (hereafter referred to as the CAMTF study); and DA Form 1391 as contained in justification data submitted to Congress in FY2003 and FY2005, Department of the Army Budget Estimates for Military Construction, Army.

tailed costs were not available for many of the modules, we used aggregate recurring and nonrecurring costs. More than one generally comparable cost source was available for some modules; in those cases, we used blended costs in developing our estimates. Each module is assessed on a life-cycle basis, using standard factors for discount rates and inflation derived from the Army's Force ;and Organization Cost Estimation System (FORCES) model website.[4] The *DoD Facilities Costs Factors Handbook* was used as a source of data and methodology.[5] To the extent possible, all costs are computed on a constant FY2004 dollar basis and then discounted to their net present value. The data presented in the tables in this section are limited to the future years chosen for this study (2005–2011). However, we used a 30-year life cycle for the comparison of the costs presented later.

Table 6.1 shows the factors we used to convert cost estimates to constant dollars and then discount them. The tables in this chapter for each of the modules show constant discounted dollars for each year of the analysis.

---

[4] For discounting, midyear constant-dollar discount factors were assumed, as costs are likely to occur in a steady stream throughout the year. Although inflation indices exist for various appropriations, those specifically associated with military construction are used as needed, as shown in Table 6.1, to convert current dollars to base-year constant dollars.

[5] For purpose-built facilities, we used the definitions and methods in the handbook. Construction provides a complete and usable facility capable of serving the purposes for which it was designed. Project costs (e.g., design, supervision, inspection, and overhead) are not included in the handbook construction cost factor, but we were able to estimate and include them in nonrecurring costs based on other sources. (The handbook calculates these costs at 20 percent.) Information on area cost factors (Headquarters, Department of the Army (HQDA), TM5-800-4, May 1994) is available and will either increase or decrease engineering estimates to put them on a normalized basis of 1 to make them independent of geographical area. (The Twentynine Palms and Muscatatuck modules were kept as geography-specific. Budget estimates for a specific base would need to reverse this process to be specific for a particular geographical area.) Facilities sustainment provides resources for maintenance and repair activities necessary to keep an inventory of facilities in good working order (Ibid., p. 6). We use facility-specific but not location-specific cost factors. In other words, estimates are general estimates that would need to be changed to reflect sustaining in a particular region of the United States. For the most part, sustainment is calculated as a percentage of construction cost (1.1 percent, derived from the handbook) and used as a constant dollar figure for the out years.

Table 6.1
Converting Current Dollars to Constant Discounted Dollars

Fiscal Year	Current Dollars	Inflation Index	Constant Dollars (2004)	Discount Factor	Discounted Dollars (Present Value)
2000	1,000	0.9525	1050		
2001	1,000	0.9696	1031		
2002	1,000	0.9774	1023		
2003	1,000	0.9872	1013		
2004	1,000	1.0000	1000	1.0000	1,000
2005	1,000	1.0130	987	0.9882	975
2006	1,000	1.0282	973	0.9651	939
2007	1,000	1.0457	956	0.9424	901
2008	1,000	1.0655	939	0.9203	864
2009	1,000	1.0869	920	0.8988	827
2010	1,000	1.1086	902	0.8777	791
2011	1,000	1.1308	884	0.8571	758

While we are confident about the validity of the underlying data for the rough order-of-magnitude costing and economic analysis that follows, it is important to stress that costs would need to be refined for program/budget (planning, programming, budget, and execution system, or PPBES) or funding appropriations purposes. As discussed later, the variable costs used for comparison would at a minimum need to be aggregated to total cost for program/budget purposes. However, the broad conclusions derived from the rough order-of-magnitude cost and economic analysis below would not change even with such refinements.

The next section summarizes the individual assumptions and underlying cost data used to calculate cost estimates for each of the retained modules described in Chapter Five.

Cost Analyses for Purpose-Built Facilities

Module 1: Battalion and Larger Purpose-Built Facility

We used the cost estimates associated with the potential upgrading of the Twentynine Palms facility to a large-scale MOUT facility to represent costs for Module 1.[6] We expect that this facility would support realistic training for up to brigade in a comprehensive environment that should accommodate urban maneuver by a battalion-plus combined arms team in the built-up portion of the site.[7] As designed, the greatly expanded facility will

- Support either FoF or live-fire training using ball-and-tracer ammunition;
- Include adequate air- and land-maneuver spaces to facilitate marine air-ground task force (MAGTF) shaping operations in areas surrounding the urban objective;
- Be of sufficient size and complexity to enable integrated urban air-ground training;
- Replicate C2 challenges associated with operating in an urban environment;
- Allow for the use of multiple urban operations scenarios;

---

[6] Density of structures, size, and other factors are critical to cost. The examples shown here could be scaled up or down, as appropriate (*Large Scale MOUT Feasibility Study*, Appendix A, 2004). The term *MOUT facility* is still in common usage in referring to urban training sites despite doctrinal movement toward use of the term *urban operations*. MOUT was a marine and army doctrinal term not formally recognized by joint or other Services doctrine. The term focused on the manmade construction aspects of built-up and densely populated areas to the exclusion of the people that occupy the terrain, a fundamentally critical urban component.

For base planning, we assume 25 buildings is the minimum number for a platoon-size facility. A company-size facility has a minimum of 70 structures, while 300 constitutes the minimum for a battalion or larger complex. These are planning factors developed primarily for cost computations in support of this analysis. The lay of the ground, density of structures, and other features would influence any specific facility's capacity in this regard. By this standard, the 900 buildings of the primary town would allow training inside and outside the structures for at least the better part of a brigade under most conditions. How often a full brigade will have the opportunity to train simultaneously on a given site will be a function of many factors, including the site's accessibility for the brigade.

[7] See the concluding pages of this chapter for a revised analysis of this option, one involving somewhat different assumptions regarding usage levels and throughput rates.

- Incorporate technology enhancements to support instrumentation, AAR, position-location information, and digital data links to the combined arms C2 training upgrade system.[8]

Specific components envisioned for this mega-MOUT complex include

- A 20 km x 20 km maneuver area;
- A 2 km x 2 km primary town (approximately 900 buildings);
- A ground live-fire area;
- An indirect-fire area;
- A convoy live-fire training area;
- An outlying airfield;
- Nearby port and industrial facilities;
- A nearby residential village.

Figure 6.1 illustrates the general layout of the proposed large purpose-built complex. Construction was initially planned to occur in six phases over multiple successive years. However, the USMC has already invested in portions of the convoy live-fire component. For this assessment, therefore, per-year nonrecurring costs are derived from total costs from all phases of development and construction. These are presented in six equally divided increments.

Table 6.2 summarizes recurring and nonrecurring costs for this facility in constant FY2004 dollars. We estimated the total expenditure to be $328 million by FY2011, a value that includes completion of construction and annually recurring costs.[9] Most of the total is nonrecurring (one-time) costs. Nonrecurring cost estimates cover the components listed above and also include standard costs for overhead

---

[8] MOUT Initiatives briefing given to RAND by Lt Col Richard D. Hall, Future Plans Officer, Operations & Training Directorate, MAGTF Training Command, Marine Air Ground Combat Center, Twentynine Palms, August 2, 2004.

[9] Subsequent estimates put the total cost for this facility and supporting infrastructure at $450,000 to $500,000.

Figure 6.1
Twentynine Palms Mega-MOUT Concept Plan

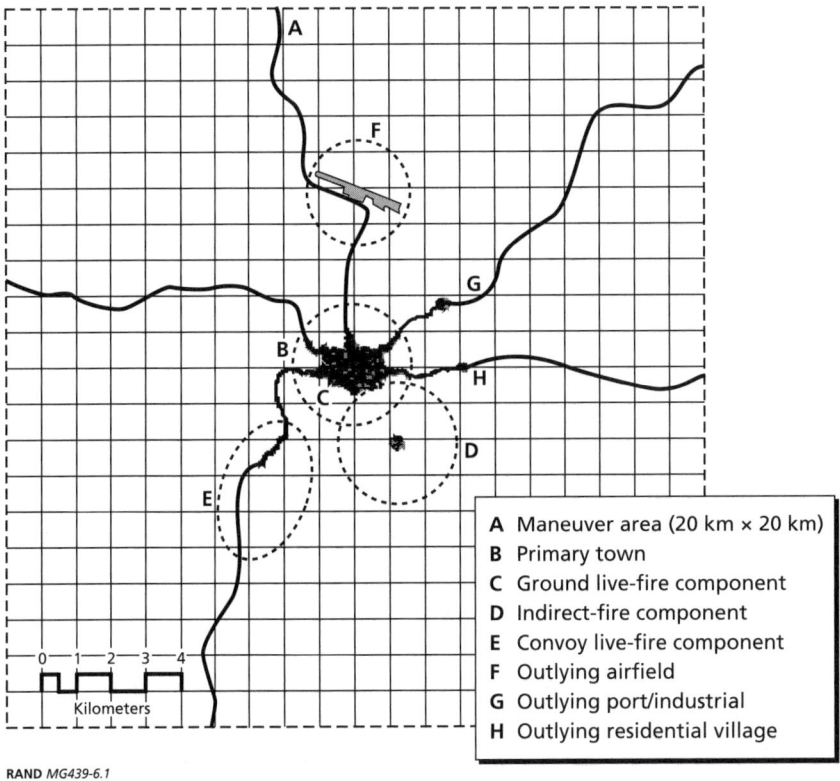

A Maneuver area (20 km × 20 km)
B Primary town
C Ground live-fire component
D Indirect-fire component
E Convoy live-fire component
F Outlying airfield
G Outlying port/industrial
H Outlying residential village

RAND *MG439-6.1*

Table 6.2
Cost Estimates for a Battalion or Larger Purpose-Built Facility
($ thousands)

Type of Cost	FY2005	FY2006	FY2007	FY2008	FY2009	FY2010	FY2011	Total
Nonrecurring	52,870	52,870	52,870	52,870	52,870	52,870		317,220
Recurring		561	1,096	1,606	2,091	2,552	2,991	10,897
Total	52,870	53,431	53,966	54,476	54,961	55,422	2,991	328,117

(e.g., design costs and contingency). Annual (recurring) facility sustainment costs are based on a standard rate of 1.1 percent for an urban combat training area (non-fire).[10]

Module 2: Company Purpose-Built Facility

We used cost estimates derived from several facilities to inform the cost analysis assessment for the company purpose-built facility module. One is the combined arms collective training facility (CACTF) MILCON estimate provided in the CAMTF study. The projected costs for the CACTFs planned at Ft. Lewis, Ft. Wainwright, and Schofield Barracks in Hawaii were used as a second range of estimates for CACTF-like facilities.[11] The third set of cost estimates is based on Dutch experiences at the Marnehuizen training site.

The Dutch Army Marnehuizen facility (120 structures) was designed to accommodate battalion-level training. Subsequent analysis determined that realistic training could be provided to no more than a company team in the built-up area of the complex. The Ft. Lewis CACTF is a 50-building complex, while those at Ft. Wainwright and Schofield Barracks are planned as 24-building complexes. The CAMTF CACTF is envisioned as a 20- to 26-building facility. Because few detailed cost data for the Marnehuizen facility were available, we based our assessment primarily on the CAMTF-study CACTF cost data and the budget estimates for the three Army posts.[12] Parametric estimates are made to "scale up" the CAMTF CACTF to reflect the additional costs associated with a larger, 70-building complex, which is more in line with the other purpose-built

---

[10] *DoD Facilities Cost Factor Handbook*, Version 2, April 2000.

[11] These estimates are from DA Form 1391 as contained in justification data submitted to Congress in FY2003 and FY2005 Department of the Army Budget Estimates for Military Construction, Army.

[12] Construction costs for this three-year-old facility are estimated at $17.5 million, but this may or may not include related infrastructure and targetry/instrumentation costs.

facilities.[13] The CAMTF study describes a CACTF designed to allow commanders a means of evaluating unit urban-operations proficiencies. Training can include branch-specific lane training, combined arms platoon situational training exercises, company team situational training exercises, and battalion task-force field training exercises.

Although many in the Army consider a CACTF facility suitable for training a battalion in preparation for urban contingencies, we do not agree, for the reasons repeatedly outlined in previous chapters. We further find a 20- to 26-building facility insufficient to train a company team if "train" means simultaneously conducting all team-related functions within the built-up area of such facilities. The costs associated with building, operating, and maintaining a CACTF are the basis for evaluation here, but only after they have been extrapolated to account for these functions from the perspective of the aforementioned 70-building site (i.e., a CACTF-type facility, but with approximately three times the number of buildings specified).

The following characteristics of the CAMTF CACTF, notionally illustrated in Figure 6.2, are used as the basis for cost estimation:[14]

- 20 to 26 buildings (1.5 km x 1.5 km);
- Tunnel and sewer system;
- Reconfigurable shantytown;
- One three-story building;
- Three one-story buildings;
- Industrial area;
- Electricity and potable water;
- Control facility;
- Breachable walls;

---

[13] CAMTF study; David Harris, "Support to the Warfighter: Fort Lewis Gets Major Urban Warfare Site," available online at http://www.hq.usace.army.mil/cepa/pubs/jan03/story16.htm; notes from a visit to Netherlands urban facilities, Russell Glenn, December 7–8, 2004.

[14] *Department of the Army Training Circular 90-1* (1 Apr 2002) contains a description of a home-station CACTF and a depiction of a 24-building CACTF similar to that in the CAMTF study. The specific details and layout of a CACTF may vary from one installation to another.

Figure 6.2
Combined Arms Collective Training Facility

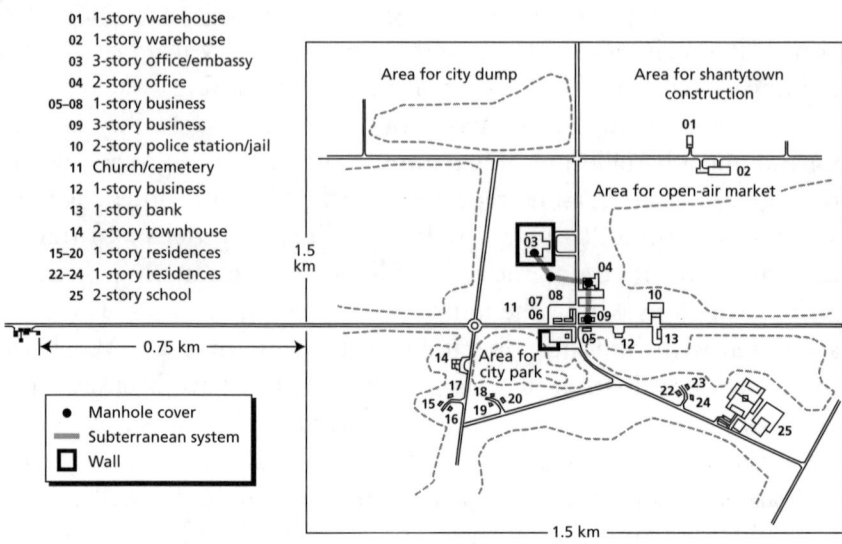

01   1-story warehouse
02   1-story warehouse
03   3-story office/embassy
04   2-story office
05–08 1-story business
09   3-story business
10   2-story police station/jail
11   Church/cemetery
12   1-story business
13   1-story bank
14   2-story townhouse
15–20 1-story residences
22–24 1-story residences
25   2-story school

● Manhole cover
▬ Subterranean system
□ Wall

RAND *MG439-6.2*

- Props and furniture;
- Video capture/edit/replay capability;
- AAR area.

The CAMTF CACTF also incorporates targetry that can be moved or modified, replicating precision and non-precision human and vehicle target systems.

Table 6.3 summarizes the costs for a 24-building CACTF, estimated to total almost $36 million through FY2011. Nonrecurring costs include those for constructing the CACTF ($8.1 million) and supporting facilities ($15 million) and miscellaneous contract and contingency costs ($1.4 million). The nonrecurring cost estimate also includes expenditures for procuring instrumentation ($6 million). Annual (recurring) facility sustainment costs are again based on a standard rate of 1.1 percent. Finally, the costs reflect an additional estimated $3 million in FY2010 to upgrade the instrumentation and

Table 6.3
Cost Estimates for a 24-Building Company Purpose-Built Facility
($ thousands)

Type of Cost	FY2005	FY2006	FY2007	FY2008	FY2009	FY2010	FY2011	Total
Nonrecurring	31,119							31,119
Recurring		330	323	315	308	2,934	293	4,503
Total	31,119	330	323	315	308	2,934	293	35,622

do restoration as needed. This is based on historical evidence pointing to a requirement for significant technology refreshment and refurbishment about every five years.[15]

The CACTF planned at the Ft. Lewis training facility is estimated to cost approximately $25 million to $35 million, an outlay generally consistent with the $31 million CAMTF CACTF nonrecurring cost projections. Like the CAMTF CACTF model, the project at Ft. Lewis will incorporate an extensive instrumentation package. However, there are differences, notably that the Ft. Lewis CACTF will potentially have approximately 50 buildings on 85 acres rather than the 20 to 26 buildings that define a CAMTF CACTF.

The budget estimates for facilities at Wainwright and Schofield and the generic CACTF form a natural three-point estimate for a 24-building company-size facility. At the high end is the CACTF estimate of about $31 million; at the low end is the Wainwright estimate at about $13.5 million; the midpoint is the Schofield estimate at about $23.9 million (all in constant 2004 dollars, with geographic cost factors removed). The midpoint estimate is used as a base to calculate the additional cost to scale up to the size and cost of the 50-building Ft. Lewis facility. The resultant average cost of an additional building is about $534,000. This gives an equation of the form

---

[15] Similar costs were assessed for all the real property modules at five-year cycles after completion of all initial construction. We estimated costs at $1 million for small modules, $3 million for medium modules, and $5 million for large modules. While these costs might not always appear in the data in these tables (because the tables end at FY2011), they were included in the 30-year life-cycle estimates.

$$\text{CACTF (of } x \text{ buildings)} = \$23{,}900{,}000 + (\$534{,}000)(x - 24),$$

where x is between 25 and 100. This is simply a rough estimate of what it might cost to scale up a CACTF-like facility to a size more conducive to effective training of larger company- or company-team-size units. Using this equation, a CACTF-like facility of 70 buildings is estimated to have nonrecurring costs of $48.45 million. Total costs for a 70-building company-size purpose-built facility are nearly $54 million, as shown in Table 6.4.[16]

Module 3: Platoon Purpose-Built Facility

Shughart-Gordon is a 29-building mock city complex and is one of three live-fire MOUT complexes at the JRTC, Ft. Polk.[17] (The others are Self Airfield and the Word Military Compound.) We used the nonrecurring costs associated with the Shughart-Gordon facility to represent the platoon purpose-built facility module. The Shughart-Gordon mock city replicates a main assault complex and includes a church, a hospital, several multistory buildings, and an underground

Table 6.4
Cost Estimates for a 70-Building Company Purpose-Built Facility
($ thousands)

Type of Cost	FY2005	FY2006	FY2007	FY2008	FY2009	FY2010	FY2011	Total
Nonrecurring	48,450							48,450
Recurring		514	502	490	479	3,101	457	5,543
Total	48,450	514	502	490	479	3,101	457	53,993

---

[16] The use of this equation for a CACTF of more than 50 buildings extends beyond the limits of the data on which it is based (the underlying values are based on the 50-building Ft. Lewis urban training site costs). However, the estimate of $534,000 seems reasonable, based on the cost of individual buildings in the DA Form 1391 budget estimates, including an appropriate slice of infrastructure and instrumentation/targetry costs.

[17] Today, Shughart-Gordon is larger than the 29-building core complex used in the cost assessment for this module. It now includes several miles of simulated urban "sprawl" along the access roads leading to the village complex itself, which consists of mostly one-story aluminum garden-shed-type structures.

tunnel and sewer system. Air assaults and fast rope operations can be conducted at Shughart-Gordon. The core facility covers a 7-km^2 area that includes four live-fire buildings for platoon-size live-fire training with short-range training ammunition. The city's water tower acts as a C2 facility/observation platform. Computerized targets and audio-visual systems are controlled from an administrative facility that includes an AAR theater.

Based on the Shughart-Gordon experience, the total cost to build and sustain a platoon purpose-built facility is estimated at nearly $13 million and is summarized in Table 6.5. The nonrecurring portion of the total cost, derived by scaling[18] to the nonrecurring cost quoted for the combined three MOUT complexes, is $10.4 million.[19] As in previous modules, the two key components of recurring costs are annual sustainment (1.1 percent) and instrumentation upgrade costs at about the five-year mark. However, instrumentation upgrades

Table 6.5
Cost Estimates for a Platoon Purpose-Built Facility
($ thousands)

Type of Cost	FY2005	FY2006	FY2007	FY2008	FY2009	FY2010	FY2011	Total
Nonrecurring	10,358							10,358
Recurring		110	107	105	102	1,855	98	2,377
Total	10,358	110	107	105	102	1,855	98	12,735

[18] Nonrecurring construction costs for the three MOUT complexes combined totaled $13.2 million. Shughart-Gordon costs are derived by scaling the number of buildings at Shughart-Gordon to the total number of buildings at the three complexes (29, 7, and 5 at Shughart-Gordon, Self, and Word, respectively). A separate project for an urban assault course (UAC) and two live-fire villages (a total of about 16 buildings) is budgeted for FY2005 at Ft. Polk at a total cost of about $3.7 million, with another $1.2 million in targetry/instrumentation, for a total of about $4.9 million. Army Training Circular 90-1 describes a UAC as providing squad- and platoon-size units with a facility at which to train and evaluate urban operations tasks. Other budget estimates for similar facilities describe the UAC as a squad-size facility. Our estimate for the platoon-size facility will be based on the Shughart-Gordon data as understood for both size and cost.

[19] "Military Operations on Urbanized Terrain Facility," available online at http://www.jrtc-polk.army.mil/JRTCExercise/MOUT.HTM, posted July 16, 2001.

for this module are estimated at $2 million, less than the $3 million assumed for the other, much larger training facilities.

Other facilities of this size have similar approximate costs. For example, a USAF MOUT project was bid in 2003 at $9.7 million (2004 dollars). This project included a mock city (13 buildings), a mock weapons storage area, and a mock aircraft alert area. The Navy Special Warfare Command was authorized in FY2005 to build a maritime MOUT training complex for use by SEAL teams. Adjusted for geographical construction costs, the cost of this project is about $7.2 million, including site costs and targetry.

Module 7: Hybrid Facility

As described in Chapter Five, this module is envisioned as a facility with capabilities comparable to those in Module 1 (battalion or larger purpose-built facility) but constructed in segments or comprising a mix of existing facilities, each providing some portion of the overall desired capabilities (e.g., an abandoned-building complex, purpose-built structures, and shipping-container mockups). We used the cost estimates associated with the 70-building CACTF developed in Module 2 as the basic core element with which to build a cost estimate for the hybrid module. Added to this are costs associated with reconfigurable buildings of the kind often used on movie sets and costs associated with reconfigurable shipping containers assembled into urban structures.

The reconfigurable-movie-set buildings can be moved from one location to another between exercises or even during phases of a single exercise. The CNA study projects construction costs of $1.54 million to provide sufficient movie-set-reconfigurable buildings to simulate a Ft. Lewis–size CACTF (50 buildings). The CNA study also estimates that costs to replicate a Yodaville-size facility (178 buildings) would approximate $5.54 million.[20]

The U.S. Army currently has two operational disposable mobile MOUT units to support training requirements for troops based in

---

[20] Yodaville in Yuma, AZ, is described more fully in Module 8.

Kuwait and Afghanistan. Each unit is designed for training on platoon-size objectives (encompassing the facility itself and its immediate vicinity) and consists of shipping containers configured into three buildings with instrumentation, exercise control, and AAR capabilities.[21] Discounting for labor and shipping costs unique to these combat environments, the U.S. Army Project Manager Training Devices (PM TRADE) estimates that a similarly constructed mobile MOUT for CONUS-based training would cost approximately $2.6 million (in 2004 constant dollars).[22]

Table 6.6 shows our cost estimates for the hybrid facility with a 70-building company purpose-built CACTF as its core. Two reconfigurable-movie-set complexes augment this, one comparable in size to the 50-building Ft. Lewis CACTF and the other comparable to the 178-building Yodaville air-ground training facility. The facility includes three disposable mobile MOUT units; calculations assume

Table 6.6
Cost Estimates for a Battalion-Size Hybrid Facility ($ thousands)

Type of Cost	FY2005	FY2006	FY2007	FY2008	FY2009	FY2010	FY2011	Total
Nonrecurring	63,000							63,000
Recurring		669	653	638	623	10,030	594	13,207
Total	63,000	669	653	638	623	10,030	594	76,207

---

[21] The units using these facilities are frequently less than platoon-size. In keeping with the previous discussion of urban training facility capacity, we would consider these three-building sites inadequate to train a platoon.

[22] Telephone interview with PM TRADE representative Barbara Raymond, December 1, 2004; and "Mobile MOUT Training System—System Description," PM TRADE briefing, n.d. In a subsequent communication, the PM TRADE representative indicated that it was appropriate to revise the mobile MOUT estimate to "a range of $2.6M to 2.8M" as "shipping containers [were] at a premium" in the aftermath of the December 2004 tsunami disaster.

replacement of these once during the time period under considera-tion.[23] These together produce a 300+ building battalion-size facility estimated to cost $76 million.

Module 8: Air-Ground Facility

Cost estimates for this module are based on the Yodaville air-ground training facility. The site is currently available to users at no cost and is regularly employed during joint training exercises. The CNA study reports that Yodaville was built for $539,000 in 1998. Imitation roads were subsequently added at a cost of $2.16 million. Other costs related to construction (e.g., personnel, sustainment) are unavailable from the USMC but were estimated by CNA to be approximately $1.4 million. According to facility personnel, no maintenance has been performed on this facility since its initial inception (i.e., Yoda-ville O&M costs are currently $0).

A cost estimation for a new purpose-built air-ground facility similar to Yodaville would have to include additional costs not ac-counted for above. Most significantly, it would be necessary to con-sider the cost of shipping containers, which were originally provided free of charge to the USMC. CNA estimated a range of container costs, which vary depending on current market supply and demand factors and associated transportation costs to move the components to the facility. CNA concluded that shipping containers like those used at the Yodaville site (or other similarly located facilities) would cost between $4.6 million and $36.8 million.

Based on the CNA estimation of "moderately" priced shipping containers ($18.4 million), a manned and operational air-ground purpose-built facility (not very distant from a seaport) with purchased containers would have a nonrecurring total cost of approximately $22.5 million. The cost would increase with distance from seaside ports. For example, costs for these same moderately priced shipping containers could approach $44 million in Salt Lake City, UT. For

---

[23] Mobile MOUT units in Afghanistan and Kuwait are entering their third year of use for training, with no serious damage to-date (Thomas Crosby, Ph.D., PM Trade, CDR, USN retired, phone interview with Barbara Raymond (RAND), December 1, 2004).

either location, CNA estimates an annual sustainment (recurring) cost of $41,400, based on cost factors reported in the *DoD Cost Facilities Handbook*. Total estimated costs (nonrecurring and recurring) for this module through FY2011 in the two locations are shown in Tables 6.7 and 6.8.[24]

Brigadier General Omer C. Tooley of the Indiana National Guard alludes to the challenges (and, by extension, costs) associated with aviation engagements of urban targets at Muscatatuck and elsewhere because of the potentially extensive damage that even small ordnance can inflict. General Tooley thinks that it "would be of huge R&D (research and development) interest to develop aerial-delivered

Table 6.7
Cost Estimates for an Air-Ground Purpose-Built Facility Fairly Near a Seaport ($ thousands)

Type of Cost	FY2005	FY2006	FY2007	FY2008	FY2009	FY2010	FY2011	Total
Nonrecurring	22,500							22,500
Recurring		40	39	38	37	36	35	225
Total	22,500	40	39	38	37	36	35	22,725

Table 6.8
Cost Estimates for an Air-Ground Purpose-Built Facility in Salt Lake City, UT ($ thousands)

Type of Cost	FY2005	FY2006	FY2007	FY2008	FY2009	FY2010	FY2011	Total
Nonrecurring	44,000							44,000
Recurring		40	39	38	37	36	35	225
Total	44,000	40	39	38	37	36	35	44,225

[24] CNA study; unpublished trip report by RAND JUT team, July 26, 2004; *DoD Cost Facilities Handbook*, Version 2.0, April 2000. Readers may note that the values in Table 6.7 differ from the CNA estimates for a Yodaville-size facility. The calculations for this module are for a facility with capabilities equivalent to Yodaville, of which the number of buildings is only one.

training ordnance that minimizes terminal damage."[25] While the obstacles to such development are considerable (attaining acceptable damage levels, minimizing risk to personnel on the ground, and accurately replicating munitions trajectories are but three of the challenges), development would potentially reduce O&M costs, expand potential construction options, and allow maintenance of lights and other features to realistically present urban area signatures during training.

Cost Analyses for Use of Populated Urban Areas

Module 10: Terrain Walks
There are no traditional nonrecurring or recurring costs associated with this module. The only costs incurred would be those to transport personnel to and from the terrain walk location.

Module 15: Use of Buildings Scheduled for Demolition
Use of a building scheduled for demolition is probably a one-time/limited-usage training option. There is no investment cost associated with this module, assuming that the owner agrees to such use of his property and that a suitable building(s) can be found near units to be trained (in other words, that there are no significant transportation costs). It is also assumed that the associated training lacks OPFOR, controllers, or role players. Potential safety and environmental concerns could result in unknown additional costs, but it is expected that these would be small.

Module 16: Use of Public Facilities During Hours of Closure
There is no investment cost associated with this module. However, there may be risks that have not been assessed that could translate into additional costs.

---

[25] BG Omer C. Tooley, email to Russell W. Glenn, April 15, 2005.

Cost Analyses for Alternative/Other Training Concepts

Module 17: Use of Abandoned Domestic Urban Areas

We provide two cost estimates for this module, based on two exemplar facilities with distinctly different training potentials: (1) the abandoned town of Playas, NM, and (2) a recently closed school for the mentally retarded in Muscatatuck, IN. The New Mexico Institute of Mining and Technology now owns Playas; the Indiana State National Guard manages the Muscatatuck facility.

The small town of Playas was originally built to house workers and their families (about 1,000 people) associated with a nearby copper smelter plant. The town's occupancy began to dwindle when smelter operations were suspended in 1999. Its current population is approximately 20 families. The New Mexico Institute of Mining and Technology bought the town for $5 million in 2003 and is seeking clients (U.S. government and other) interested in leasing/renting the site for various training, exercise, and research events. The potential for military training exists, but live fire would not be allowed. Major facilities in Playas consist of

- The 640-acre town site and an additional surrounding area of 1,200 acres;
- 259 single-family homes (1,300 ft^2 to 3,800 ft^2);
- 25 apartment units;
- A community center;
- A fire station;
- A fully equipped medical clinic with ambulance;
- An airstrip (~5,000 ft);
- Wide streets with streetlights;
- Three water wells;
- An elevated water storage tank (200,000 gallons);
- A wastewater treatment plant.

Numerous other amenities (e.g., a convenience store, a bowling alley, and other buildings) are also part of the Playas complex. Figure 6.3 illustrates the layout of the Playas central area.

Figure 6.3
Central Area of Playas, NM

SOURCE: Photo courtesy Energetic Materials and
Testing Center.

The town is available for rental on a per-day basis: $10,000 for Department of Homeland Security or U.S. government events and $14,000 for other events. These rates can be prorated depending on how much of the town is needed. Using the government rental rate of $10,000 per day (and then discounting for present value), we calculated an aggregate cost for renting Playas 365 days each year through FY2011 of approximately $20 million. These costs are summarized in Table 6.9.[26]

The Muscatatuck facility currently consists of more than 69 buildings on 1,000 acres (at one time it supported a population of more than 2,500 people). This infrastructure is valued at $240 million and includes numerous multistory/multipurpose structures and an extensive underground tunnel system. The facility is being turned over to the Joint Force Headquarters, Indiana National Guard, for

---

[26] Unpublished JUT project site visit trip report, Playas, NM, November 2, 2004; briefing by New Mexico Institute of Technology President Daniel Lopez, "Playas, New Mexico . . . Imagine the Possibilities," November 2, 2004.

Table 6.9
Cost Estimates for Renting Playas, NM
($ thousands)

Type of Cost	FY2005	FY2006	FY2007	FY2008	FY2009	FY2010	FY2011	Total
Nonrecurring								
Recurring		3,523	3,440	3,359	3,281	3,204	3,128	19,935
Total		3,523	3,440	3,359	3,281	3,204	3,128	19,935

use as a MOUT facility that will support both the doctrinal CACTF and other, nondoctrinal training objectives up to brigade size. It will also provide a capability for urban CAS training in conjunction with the air-ground ranges at nearby Camp Atterbury Joint Maneuver Training Center and Jefferson Proving Ground. Indiana National Guard authorities have completed extensive coordination with local airfields in the immediate area and airports in Cincinnati, Louisville, and Indianapolis to minimize constraints on rotary- and fixed-wing training at Muscatatuck. Work is in progress to create robust support for aviation units, including no-drop engagements and rooftop land-ing zones.[27]

Nonrecurring and non–exercise-related recurring costs associ-ated with Muscatatuck are estimated at $103 million through FY2011 and are summarized in Table 6.10. Conversion of the exist-ing facility to one more suited for the support of urban operations training is planned in four consecutive-year phases with an estimated total cost of $54 million. Phase I ($11 million) will consist of selec-tive building demolition to provide a templated CACTF capability. Other minor modifications to ensure building safety will also occur in this phase. Phase II will include a MILCON element ($10 million) for power, utilities, and fiber-optic backbone. It will also include a large investment in targetry and instrumentation ($22.5 million). Phase III will focus on enhancing antiterrorism/force protection capabilities ($8.5 million) such as fencing with perimeter road and

---

[27] BG Omer C. Tooley, email to Russell W. Glenn, April 15, 2005.

Table 6.10
Cost Estimates for Muscatatuck
($ thousands)

Type of Cost	FY2005	FY2006	FY2007	FY2008	FY2009	FY2010	FY2011	Total
Nonrecurring	11,000	32,500	8,500	2,200				54,200
Recurring		8,686	8,482	8,283	8,089	7,899	7,714	49,153
Total	11,000	41,186	16,982	10,483	8,089	7,899	7,714	103,353

entry control points. Phase IV will consist of training enhancements of the interstate road section railroad mock-up and 727 fuselage and mobile MOUT site preparation ($2.2 million).[28]

Muscatatuck recurring costs are those of operations ($6 million) and sustainment ($3 million), the latter consistent with the *DoD Cost Facility Handbook's* rule-of-thumb for sustainment costs of 1.1 percent. The full existing Muscatatuck infrastructure is valued at $240 million.[29] Because the Indiana National Guard will have the benefit of free prisoner labor from a nearby prison facility for almost all of its maintenance and cleanup, these recurring costs may be understated for other abandoned urban sites. Based on a review of recurring labor costs at Ft. Polk and elsewhere, a total cost estimate for this module might have to be adjusted upward in the final analysis.

Module 18: BRAC'd Installations

George AFB was announced for BRAC closure in 1988 and was officially closed four years later, in 1992. Located within the unincorporated city limits of Victorville, CA, it covers more than 5,000 acres and includes two runways, more than 6 million ft^2 of ramp space and associated facilities, 1600+ housing units, 14 dormitory buildings, a hospital and dental clinic, and various offices and industrial structures.

---

[28] Survey response provided by LTC Ken McCallister, IN ARNG, Deputy J5/7, October 2004; unpublished JUT project site visit trip report, October 13, 2004.

[29] Operating costs are included in this module as a point of reference. The operating costs are excluded when we make comparisons across modules because they are common to all of the modules and were not included in the other module recurring costs.

A variety of redevelopment and reuse activities have occurred at George AFB since its closure, including the leasing of its facilities and infrastructure to support the Southern California International Airport (now called Southern California Logistics Airport (SCLA)), which opened in 1994. The U.S. Army is an important airport tenant, using SCLA as an airhead for transporting troops to the NTC at Ft. Irwin, CA.

SCLA (George AFB) has also been the site of major Marine Corps Warfighting Lab (MCWL) training exercises, providing insights into the potential costs of using a BRAC'd installation for JUT. For their training exercises, the marines used a 1-km^2 portion of the former base residential area. Structures included a family housing complex (700 buildings comprising 1,000 units), a theater, a commissary, and other miscellaneous buildings.

The total cost of conducting the USMC 57-day Millennium Dragon 02 exercise in this residential area was $1.23 million (FY2004 dollars); this included travel, MCWL personnel (but not the training audience) planning costs, facility rental, utilities, and installation of a T-1 line. The facility rental was the largest single cost at $550,000 (~$10,000 per day).[30] Rental cost is assumed to remain the same for future years, although it is expected that this particular facility will have a useful training life of only about five years and may require upgrades if effective training beyond this time period is desired.[31]

Total costs for renting the residential portion of SCLA (George AFB) through FY2011 are estimated at approximately $20 million (the same as the cost of renting the Playas facility) and are summarized in Table 6.11. This assumes that the facility is rented for 365

---

[30] Costs would obviously differ were the facility still the property of DoD. Maintenance and upkeep costs might well be passed along to user units, but the actual cost for DoD would depend on the condition of the installation at the time of transition (realignment) from previous use, among other factors.

[31] CNA study, February 2004; http://www.globalsecurity.org/military/facility/george.htm; http://www.afrpa.hq.af.mil/ols/george.htm; http://www.globalsecurity.org/military/facility/george.htm; http://www.eltoroairport.org/issues/george-afb.html.

Table 6.11
Cost Estimates for a BRAC'd Military Installation (George AFB)
($ thousands)

Type of Cost	FY2005	FY2006	FY2007	FY2008	FY2009	FY2010	FY2011	Total
Nonrecurring								
Recurring		3,523	3,440	3,359	3,281	3,204	3,128	19,935
Total		3,523	3,440	3,359	3,281	3,204	3,128	19,935

days each year and that no improvements are made during this time period, indicating that this option would probably not have a useful life beyond 2011.[32]

Thus far, we have assumed that this module could make a synergistic urban training opportunity available from a local redevelopment authority after a base is closed or realigned. An alternative way to examine the use of BRAC'd facilities is to estimate the cost to use a base for an urban training site if its status were changed from closed to realigned with DoD retaining control (or assuming control if the facility was initially the property of another department) of all or part of the site. The base or the urban-like portion of it would be realigned to become an urban training range. Several types of costs would be expected to be incurred, depending on the fidelity of the training opportunity to be provided.[33]

One implementation would be to keep the facilities as they are (put no improvements into the urban area to be used) and to annually fund only enough maintenance and caretaker support to keep them viable. Costs for caretaker status for some previously BRAC'd

---

[32] These costs also do not include other exercise-specific costs such as T-1 line installation and subsequent rental (the J8 study estimates that rental of a T-1 line could cost approximately $9,700 per month and installation could cost $1,500).

[33] Two costs are not included in this assessment, environmental restoration costs and an opportunity cost. If the base is not being closed, environmental restoration costs may not be incurred, but this is a determination outside the scope of this analysis. The opportunity cost means that if the base were closed and long-term savings would have accrued as a result, such savings would be forgone.

facilities range from $100,000 to $1.8 million annually.[34] Typically, these costs cover caretaker staff payroll, supplies, and limited maintenance contracts. The buildings would slowly decay over time unless additional sustainment and restoration funds were provided. Costs would obviously depend on the size of the site and whether the entire installation or just the urban portion of it was realigned. An estimate of $1.5 million in recurring annual costs for caretaker status is reasonable.

Another approach would be to improve facilities at the installation to make them more useful for training and/or to instrument them. This is similar to what is being done at Muscatatuck, and such improvements would require more than caretaker maintenance, as is the case at that location. Whether changes to physical facilities were needed would depend on the types of facilities in the existing urban area. The cost estimates for the Muscatatuck module (construction and/or instrumentation and maintenance) could be used to support estimates for a similar realignment of an urban base.

In either case, if the urban training site were to be permanent, a periodic restoration and modernization cost would be incurred as in the other real property modules. Also, we assume that in either case, there would not be a large permanent military population at the base. Therefore, all users would incur travel expenses. (As with the other module cost assessments, we do not directly attribute training operations costs (e.g., OPFOR, role players, controllers) to the BRAC'd installation module.)

Total costs for realigning a BRAC'd installation and maintaining it in caretaker status with periodic restoration through FY2011 are estimated at approximately $12 million and are summarized in Table 6.12.

One point merits further attention. We assume that a facility that undergoes a transition from its original use to serve as an urban

---

[34] Department of Defense Base Realignment and Closure Account IV, Army (BRAC 95) FY2004 Budget Estimate, justification data submitted to Congress, January 2003.

Table 6.12
Cost Estimates for a BRAC'd Realigned Installation
($ thousands)

Type of Cost	FY2005	FY2006	FY2007	FY2008	FY2009	FY2010	FY2011	Total
Nonrecurring								
Recurring		1,500	1,376	1,330	1,285	5,379	1,199	12,069
Total		1,500	1,376	1,330	1,285	5,379	1,199	12,069

training site would either remain the property of the original owning Service or be reassigned at no or negligible cost to another Service. An alternative possibility is that a joint organization (e.g., a combatant command) would assume ownership and management responsibilities. We do not consider this second alternative a desirable option given (1) the personnel and other costs that would be incurred in standing-up and maintaining a range-control operation (the Services already have such capabilities in place), and (2) the likely limited demand the joint community would have for a major urban training facility compared with potential Service use. Further, there may be federal legal restraints that would bar a joint entity from assuming ownership of a facility potentially scheduled for BRAC. Congressional action in this regard may be necessary should it be deemed desirable for a joint organization to obtain ownership.

Module 21: Abandoned Factories

There is no investment cost associated with this module, although there may be risks that have not been assessed that could translate into additional costs.

Module 22: Abandoned/Constructed Overseas Urban Areas

There is no investment cost associated with this module, although there may be risks that have not been assessed that could translate into additional costs.

Module 24: Classroom Instruction

We assume that instruction takes place in preexisting classrooms by resident instructors, with a formal program of instruction that includes training materials. Nonrecurring costs of providing a classroom and recurring base operations support costs are excluded, as are trainee costs. The cost of such training support is estimated at $100 per day per student.[35] Total costs for a given number of students would depend on the length of the training provided. The cost would be zero if the instruction were provided in the classroom by unit personnel without specialized training materials or school support. Also, many of the training expenditures included in this estimate are sunk costs. If joint urban subject matter were taught in lieu of other currently taught subject matter, the dollar cost would again be zero.

Module 25: Conduct of Combatant Command or JTF Headquarters, Large-Scale Schools, or Multi-Echelon/Interagency Exercises

Cost estimates for the JTF Echo and Foxtrot alternatives (both battalion-size training) as described in the J8 study are used as the basis for estimating costs for this module.[36] Important objectives of the J8 study were to examine how a JTF could be trained for operations in an urban environment and to provide a cost-benefit analysis of this training. We assessed training effectiveness by evaluating each alternative's potential for facilitating training of the Joint Mission Essential Tasks (JMETs) associated with a JTF involved in urban operations. The preferred JTF training alternatives were those that met the

---

[35] This figure is based on several estimates of daily cost for a student workload. For example, a rough estimate for a day of Army specialized skill training is $56; a day of Army Sergeant Major Academy training is $108; a day of Army intermediate school is $268. Unpublished RAND research indicates a considerable range of costs for different types of training, from $100 to $1,000 per student per day. The average across multiple courses is $400 per student per day. The average cost per graduate for an Air Force enlisted skills progression course is about $6,600 (Air Force Instruction 65-503).

[36] The J8 study concluded that the most cost-effective alternative for meeting the objectives was JTF Foxtrot, with the caveat that JTF Echo (option 1) could be comparably cost-effective but would first require identification (and subsequent earmarking for use) of a newly BRAC'd facility.

dual objectives of fulfilling as many of these training requirements as possible and doing so at the optimum cost.

The JTF Echo alternative is a training event in which JTF staff, service component staff, and O/Cs participate from a collocated training facility. Both simulated and live task-force components would operate from urban training field facilities. The joint MOUT training facilities required would be from one facility. Three facility options were considered: leasing a BRAC'd facility (George AFB), upgrading an existing facility (Ft. Polk), and building a new facility (Ft. Irwin). The calculations for this option (as shown in Table 6.13) assume use of the BRAC'd facility option due to the significant cost savings involved and the resultant likelihood that it would be selected if available. The actual choice of facility type will depend on decisions related to discussions presented in Chapters Seven and Eight.

The JTF Foxtrot alternative is a training event where JTF staff, component staff, O/Cs, and task force components would operate from actual urban locations in a distributed mode, using tactical equipment. Representative forces would comprise at least one battalion. No joint MOUT training facilities would be required. However, it would be necessary to lease training rights in an actual urban environment. Two civilian urban locations were assumed, San Diego, CA, and Savannah, GA. We assume that the nearby military installations would allow exercise managers to draw on them for air and ground support (e.g., refueling, storage, billeting, and messing) at less-than-commercial costs.

Cost estimates for these two JTF alternatives through 2011 are shown in Tables 6.13 and 6.14. Expenditures are expected to range

Table 6.13
Cost Estimates for J8 Alternative Echo-BRAC JTF Training
($ thousands)

Type of Cost	FY2005	FY2006	FY2007	FY2008	FY2009	FY2010	FY2011	Total
Nonrecurring								
Recurring	450	440	429	419	410	400	391	2,939
Total	450	440	429	419	410	400	391	2,939

Table 6.14
Cost Estimates for J8 Alternative Foxtrot JTF Training
($ thousands)

Type of Cost	FY2005	FY2006	FY2007	FY2008	FY2009	FY2010	FY2011	Total
Nonrecurring								
Recurring	1,091	1,065	1,040	1,016	992	969	946	7,119
Total	1,091	1,065	1,040	1,016	992	969	946	7,119

from $2.9 million (JTF Echo-BRAC) to $7.1 million (JTF Foxtrot using two civilian urban locations). As in all the module cost discussions above, these estimates are presented in constant 2004 dollars and are discounted to reflect present value.

Cost Analyses for Simulation Capabilities Modules

Costing simulation capabilities is difficult for several reasons. In the near term, the simulation modules could represent linkages of existing capabilities rather than new costs. New capabilities can be developed and implemented in the longer term. Some of this development is under way and would not represent new costs for urban use. Also, useful lives of software and hardware are much shorter than those of physical facilities. Use of live, virtual, and constructive simulation as a substitute for training ranges, ammunition, equipment wear and tear, and similar replacement functions could represent cost savings rather than outlays. (However, it is important to note that the joint training requirements the training modules address do not tend to be ammunition- or equipment-intensive.) This is discussed further under "Cost-Related Summary and Observations." Assuming these constraints, we estimated costs for three groups of the modules described above.

Modules 26 and 29: Tactical Behaviors in and Around Structures; Specialized-Technology Simulation

In the near term, these modules essentially represent a baseline fire-team capability replicated with six to eight personal computer stations, four to six staff, and limited infrastructure. (These items are not included in the individual cost estimates for these two modules.)[37] The nonrecurring costs were replicated on a three-year cycle for technology refreshment. A straightforward extrapolation can be made for forces up to platoon level. The costs shown in Table 6.15 are for one unit of capability. Duration of training, periodicity, and throughput needs drive total costs. For the longer term, special technology models (CAVE-type systems, network models, nonlethal-weapons-effects representations, and other enhancements) can be added as software improvements and as physical environment additions to the local-area, PC-based systems. The added hardware costs should be moderate and could be considered to be included in the technology refreshment costs in either case.[38] Software costs for developing games based on urban training tasks are estimated to be about $6 million in

Table 6.15
Cost Estimates for Fire-Team Simulation Modules
($ thousands)

Type of Cost	FY2005	FY2006	FY2007	FY2008	FY2009	FY2010	FY2011	Total
Nonrecurring	32			31			29	92
Recurring	346	338	330	322	315	307	300	2,258
Total	378	338	330	353	315	307	329	2,350

---

[37] Computers are typically Pentium 4 or better, with extensive RAM, and graphics cards (costing approximately $3,000 each). Specifically, for JCATS, the current single-station spec calls for a Dell Precision Workstation 670 with dual 3GHz processors, 1-GB SDRAM, 36-GB ultra 320 SCSI 10,000 rpm, Hi-Performance PERC320 SCSI RAID card, and a 19-in. flat panel monitor, costing about $4,500. FSW calls for an X-box platform but will likely be transitioned to a PC in the future. Most of the simulations have no license or software fees, although JCATS requires a license for Linux, and for non-DoD users, a $40,000 per year support fee. Informal polling of sites indicated a rough salary requirement of $70,000 per year, depending on experience and responsibilities,.

[38] Most of the coming generation of simulations can be run on PCs with consumer-level hardware cards instead of the previously dominant high-end SGI or Sun workstations. This holds down the costs significantly. Also, combinations of the simulators above should be maintained and operated by the same staff members. That is, a single analyst should be able to operate several different simulations, and one maintenance or database person should be able to work with multiple suites of equipment.

2004 dollars (after which the games would be available for use at minimal expense, barring upgrade purchases).[39]

We do not expect that units would link multiple simulation systems of this type in exercises (e.g., that a company of three platoons, each with three squads (= 9), each squad having two fire teams [(3)(3)(2) = 18], would employ all such teams). Systems are seen as standalone capabilities for very low-level training, with each system costing $378,000 per year, or $2.3 million over a seven-year period per system purchased. This does not rule out linking multiple systems, but doing so at the small-unit level (e.g., the above-noted example of every fire team in a company) would likely be prohibitively expensive given the per-system cost.

Modules 27, 28, and 32: Higher-Echelon Planning and Coordination; Joint, Multinational, and Interagency Operations; Geographically Distributed Joint Simulation

These modules all require large-scale composite simulations, such as JTLS/JCATS or JSAF/Urban Resolve. In the near term, existing systems could be linked across a high-speed network over many geographically distributed sites. In the longer term, with more accurate modeling of noncombatant behaviors, dynamic terrain, and detailed structures, supercomputer support, DREN connection, and large numbers of support personnel may be required.

Costs are estimated first for providing facilities and equipment designed around simulation capabilities existing in 2005; these costs are shown in Table 6.16. Costs were estimated by analogy from a FY2003 Army project for a squad through corps battle-simulation center. Adjusted for geographical construction and inflation, the non-recurring costs are about $22.3 million for the primary facility and infrastructure. Information systems cost another $5.5 million. The simulation exercise area consists of a central control facility for battalion through corps exercises (JSIMS), a brigade/battalion simulation

---

[39] Shawn Zeller, "Training Games," *Government Executive,* January 2005, pp. 45–49.

Table 6.16
Cost Estimates for a Simulation Center
($ thousands)

Type of Cost	FY2005	FY2006	FY2007	FY2008	FY2009	FY2010	FY2011	Total
Nonrecurring	27,472							27,472
Recurring		236	231	225	220	215	210	1,337
Total	27,472	236	231	225	220	215	210	28,809

(WARSIMS), and an exercise area capable of supporting two simultaneous exercises and other capabilities. The facility is designed to be able to accommodate a corps-level exercise without impacting the training schedule and support of lower-level exercises.

Modules 30, 31, and 33: Scenario-Variant Generation; Physiological and Other Stress Simulation; Environmental Degradation and Urban Biorhythm

These are specialized capabilities that are suited to such simulations as Full Spectrum Warrior, Full Spectrum Command, IUSS, IWARS, and MANA. As in the tactical simulations, these capabilities can be added to the suite of PC systems used in the baseline and should have minor cost impact. The cost estimates for these modules are the same as for Table 6.14 with the same assumptions. The stress representation expected from inclusion of IUSS code (or later development of this OOS functionality) makes this (and noncombatant behavior modeling) available only in the far term.

Summary of Simulation-Capabilities Costs

A range of costs exists for simulations capabilities, as represented by the three module groupings used above. At the lower end is the use of existing capabilities for small groups and purchasing sufficient hardware and software to make them available to those groups as needed. In a middle range is the development of new games or PC-based simulations focused on JUT tasks. Alternatively, this middle range could include modification of existing games or PC-based simulations to include JUT tasks. At the upper end are (1) the replication of physical facilities in which simulation training or exercises could take place, and (2) the development of hardware and software for such large-scale capability.

Cost Analyses for Training Support Elements

Module 34: Infrastructure Trappings

These "trappings" for the standard CACTF from the CAMTF study include electrical utilities ($8.26 million), mechanical utilities ($559,000), site improvements ($4.15 million), and roads, parking, and sidewalks ($2.21 million). The total cost of these items in FY2004 dollars is $15.5 million.

We estimated maneuver area site work to provide a much larger complex (access roadway, grading, excavation, storm water control) at $7.6 million; electrical service for buildings at $3.6 million; and utilities, including portable generators for power and light poles, at $7.4 million. This totals $18.6 million.

The Nevada training initiative estimated $13.1 million for infrastructure costs such as power, roads, and drainage.

Costs of the individual components could be estimated directly using the *DoD Cost Factors Handbook* and related documents (e.g., TM 5-800-4, May 1994). For example, sidewalks cost $2.60/ft^2 and street lighting costs $23/linear foot.

All these costs are geography-dependent and could be increased or decreased using standard area cost-factor indexes.

Targetry and instrumentation vary by size of facility. Representative costs in FY2004 dollars for a company-size, purpose-built facility range from $5.3 million to $10.4 million. The estimated cost in the CAMTF study for targetry and instrumentation in FY2004 dollars is about $5.7 million.

Module 35: OPFOR

The size and composition of the OPFOR will influence its cost, as will the source of its personnel. Various studies and interviews indicate four potential sources: volunteers, Guard and Reserve, active-duty military, and civilians.[40] Each source involves different costs. This subsection provides a per-person estimate for each source. While not exactly linear, the per-person cost could be scaled up depending on the size of the OPFOR desired and the extent of its use.

The CNA study used the Army FORCES cost model to estimate annual sustainment costs of an active light infantry battalion and a comparable Reserve and National Guard (ARNG) battalion.[41] It then employed assumptions to estimate the cost of a civilian OPFOR of the same size.[42] The authors also used travel costs to propose a traveling active-duty battalion that could cover more than one geographic site. Their annual cost estimate is shown in Table 6.17.

Module 36: Noncombatant Role Players

Noncombatant role players are a part of urban training that is increasingly being recognized as essential to quality preparation. Inter-

Table 6.17
Cost Estimates for OPFOR

OPFOR Source	Annual Cost ($ thousands)	Number of People	Annual, Per Person ($)
Single-site active-duty battalion	42,000	570	73,684
Single-site contractor battalion	27,000	570	47,368
Traveling active-duty battalion	44,000	570	77,193
Reserve battalion	54,500	570	95,614
ARNG battalion	58,800	570	103,160

SOURCE: CNA Study.

---

[40] This topic was raised in the RAND JRTC and Muscatatuck interviews; in the CNA and J8 studies; and on the FORCES (Army cost factors) website.

[41] There is an important embedded assumption here, namely, that the battalion is in existence and has no alternative demands. Thus, the only cost is the annual cost to sustain it; the estimate does not include life-cycle costs to structure and outfit it.

[42] A key assumption was that it would cost about 35 percent less because it is not necessary to provide all military unit functions (e.g., medical, food service) when civilianization of military functions takes place. Thus, the cost per civilian may or may not be as high as the cost per military participant, but there would be fewer civilians.

views at the various training sites indicate a wide range of possibilities for providing such players, including volunteers, prisoners, paid performers, Reserve military, and active-duty military. Estimates for paid service range from about $11 per hour to $30,000 annually per person. Requirements for particular expertise, especially language and cultural knowledge, influence costs. We use the $30,000 per year cost in our estimate. Costs would be expected to scale up proportionally with size of facility, with the largest facilities requiring the greatest numbers of role-player personnel.

It is difficult to accurately represent such staffing costs as part of a JUO training strategy, for several reasons. Many installation range and training personnel divide their time between various sites and responsibilities, making it extremely hard to determine what the urban-related personnel costs are. This fiscal analysis is intended to support development of an investment strategy rather than to determine total JUT-related expenses during the 2005–2011 period. However, for comparisons between live training and training dependent on simulations, such personnel costs should be determined.

Module 39: Joint Force Headquarters

For cost purposes, we assume a nominal standing JTF headquarters of 55 officers and an average annual per person rate of $135,000.[43] This is an annual cost of about $7.5 million, or approximately $20,300 per day, with the assumption that the unit exists (i.e., that we do not need to cost standing-up the JTF headquarters). One of two other assumptions must be made. If the standing JTF headquarters is providing only training support, its annual cost should be included, as was done with the OPFOR. Moreover, in this case, it need not be a real standing JTF headquarters but could be one composed of civilians or contractors. If, however, there is training value for an actual headquarters whenever it is used, then its annual cost should be omitted. In either case, there is a transportation/travel cost if the headquarters is to geographically collocate with the other units being

---

[43] FY2005 military composite standard rate for an O-5 (lieutenant colonel or commander).

trained rather than participate virtually through communications. Whether one or more of these headquarters would be needed to support all training depends on the frequency and duration of the training and the needed throughput.

Training Transportation

For some modules, training facilities and prospective trainees are not located at the same place; personnel must move to the training facilities. We used the cost to move an Army mechanized infantry battalion (without tracks) to the NTC from five different major installations as a starting point in calculating these estimated costs. This movement cost is divided by the distance traveled from each installation to derive a cost per mile for the battalion. This cost is then divided by a notional strength of 700 for the battalion to derive an average cost per soldier-mile for transportation to training. This cost is approximately $0.40. Thus, moving 100 soldiers for 1 mile costs $40 and moving them 100 miles costs $4,000.

Cost-Related Summary and Observations

Many of the training modules (primarily those eliminated from final consideration) are of marginal value for training large numbers of people because of capacity limits. Therefore, they should be considered as niche training opportunities that could be part of an annual training budget or a Service initiative rather than part of a long-term JUO training investment strategy.

The investment strategy presented in Chapter Seven is based on the approaches adopted in answering three major questions:

1. Is JUO training contextual and thus something that can be added into existing or planned Service urban operations training as an annual operations budget initiative, or does such joint training need to be separately planned, implemented, and funded?

2. Is it better to "make" training (i.e., build structures and facilities at local installations and bases) or to "buy" it (i.e., move people to where low- or no-cost training opportunities already exist)?

3. Is it better to have virtual and constructive training as an alternative to live training or as a supplement to it?

Joint Training: A Separate Entity or an Augmentation of Service Preparation?

Joint training takes place almost exclusively at Service structures and facilities and via simulations. The audience for the training requirements may be Service individuals or units or people staffing joint headquarters. These factors bear on how to cost the different modules for joint training. For example, if a new purpose-built facility is needed only to satisfy a joint training requirement, its associated costs could be determined as exclusively joint. However, the cost is incremental and possibly minimal if that requirement could be satisfied by adding it to an existing training regimen at a Service facility or occasionally using that facility for a joint-headquarters-controlled urban exercise. Alternatively, the joint requirement might add a day or more to an existing urban training regimen at an existing facility; this might ultimately require more facilities—or possibly not, depending on throughput needs. These approaches tend to imply that the primary training audience is in most cases a Service unit or individual and that the joint training requirement is contextual to their training. However, the training audience might also be an inherently joint organization, such as a joint force headquarters. Since much training of this character will involve primarily higher-echelon staffs rather than maneuver units, deployment to a live urban training facility might not be needed. Simulation or conduct of a joint headquarters exercise at some generic location could well be sufficient.[44] Ultimately, the

---

[44] This is similar to the state of affairs to which training in Germany evolved during the 1980s and 1990s. The costs in dollars and environmental problems (e.g., indigenous tolerance of life's disruptions, vehicle accidents, maneuver damage) resulted in reduced numbers of units participating at full strength. Headquarters alone deployed to the field, and simulations and models increasingly became fundamental to such undertakings. REFORGER (Return of Forces to Germany) exercises ceased. A good example of the resultant training is that

investment strategy must account for either the full cost of new JUT means or the incremental cost to existing training means. This issue is discussed further in Chapter Seven.

Build, Adapt, Rent, or Otherwise Acquire Training Capabilities?

There are two primary tradeoffs for an investment strategy. The first is between building training facilities and structures at installations where soldiers, sailors, airmen, and marines are located and moving them to existing facilities and structures.[45] In essence, is it more effective to build at facilities heavily populated with user units or to move those units to fewer sites used by organizations from multiple installations? The second tradeoff is between building battalion-size facilities and building smaller ones. Both of these options depend on troop density at installations, throughput requirements, availability of non–purpose-built facilities, and the distances to such field training capabilities. The basic factors inherent in these two types of tradeoffs are considered here. In Chapter Seven, assumptions about throughput are applied to assess ground rules for an investment strategy.

Figure 6.4 shows the results of an analysis of selected training modules that enables us to compare their costs on an annual cost-per-person basis. Calculations regarding the first three modules at the left side of Figure 6.4 assume that the facilities are built at installations where a substantial number of tactical units are home-based. The primary users are therefore organizations that do not have to travel other than in their immediate vicinity for urban operations training. For costing purposes, there are no transportation expenses associated with them. The fourth module involves moving half the personnel that use it to its location from remote sites (i.e., installations not in the immediate vicinity of the training capability). Such travel is not

---

conducted by the U.S. Army's Battle Command Training Program (BCTP), which is responsible for training higher-level service staffs.

[45] As part of the analysis of needs, we consider whether existing facilities/structures could be modified to incorporate the joint training requirement, whether simulations could be substituted, and whether existing facilities might increase operational hours or availability days. These are discussed below.

Figure 6.4
Average Annual Cost per Person (FY2005–FY2011) for Selected Training Modules, Based on 30-Year Life Cycle

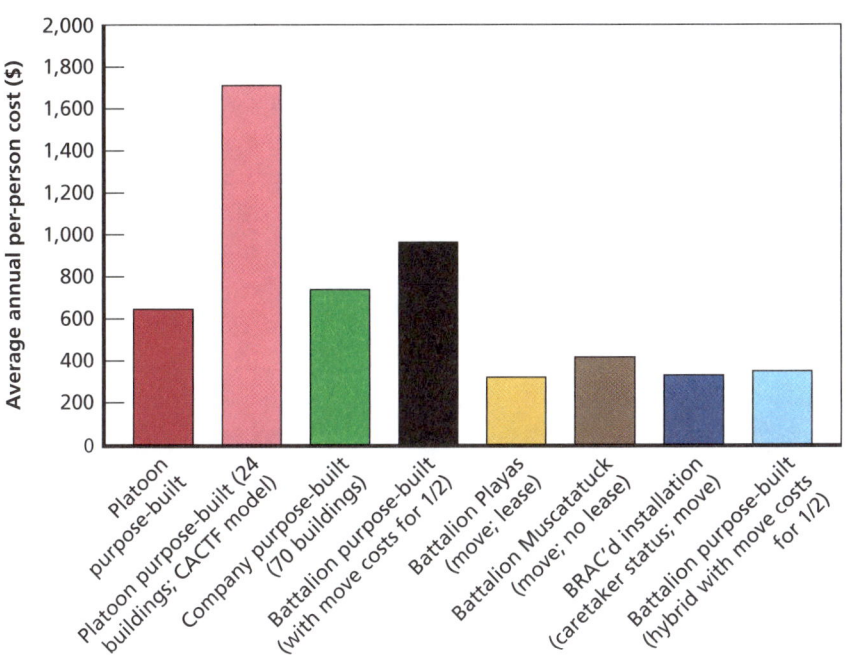

RAND MG439-6.4

an unrealistic expectation given that the facility offers the opportunity for an entire battalion to train simultaneously for urban operations.[46]

---

[46] As noted earlier, the Twentynine Palms facility as planned would be capable of supporting the training of the better part of a brigade, if not a full brigade. The cost per person trained in Figure 6.4 assumes that the facility supports a battalion at any given time. This underestimation results in higher per-person costs than would result were a brigade used as the standard. However, the following additional factors influence this decision:

- An assumption that 15 home-station units and 15 visiting units train on the facility per year would require high transportation costs and a very robust transportation support system. (The assumption of 30 annual rotations is consistent with similar assumptions made in calculating the annual per-trainee costs for platoon- and company-size sites.)
- Such an assumption regarding the number of home-station units is a poor one, as the largest ground component hosted at Twentynine Palms is the 7th Marine Regiment. While this unit could (and likely would) use the facility repeatedly in a given year, the

The next three modules require movement of all trainees to the sites from remote home stations. The final module is a hybrid facility as-described earlier, one that also hosts half of its trainees from remote locations. The first four modules have high initial (first-year) construction costs and substantial sustainment costs thereafter relative to the size of the unit they can host.[47] Muscatatuck and the hybrid module have lower costs of this type relative to size. The Playas module has lease and movement costs each year, while the BRAC'd installation module has recurring maintenance costs as well as movement costs. The last five modules represent facilities capable of supporting training for up to a battalion-size unit. (This assumes (1) augmentation of Playas with additional structures or the use of another abandoned-town facility with greater capacity leased at a similar cost, a not unrealistic assumption, and (2) that Indiana National Guard plans to increase the density of structures at Muscatatuck are acted upon.)

Several points stand out. First, the use of an abandoned urban area such as Playas and other more-creative modules, including the hybrid facility, are notably less costly than the other alternatives on a per-person-trained basis.

---

actual number of units that would have to travel to it to complete 30 annual brigade-equivalent rotations is virtually unachievable.

- Reducing the estimated number of rotations per year from 30 brigade-size events to 15 increases the per-person costs, because fewer individuals are using the capability.
- Such a reduction in the number of rotations makes further sense, as brigade-size rotations could well exceed the seven days assumed as the baseline for computations in this study.

[47] Computations for comparing the alternatives shown in Figure 6.4 are based on a 30-year life cycle with upgrades or overhauls every five years. This is considered realistic given the durability of extant urban training facilities such as the British Army Copehill Down Village complex and those in the United States. Though urban training does cause wear and tear, properly maintained facilities have retained virtually all of their training value (or improved same due to upgrades) given adequate range funding and quality management. This assumption regarding life span is significant. Using longer time periods (e.g., extending beyond the FY2011 endpoint of this study) for life-cycle costs means that the cost per person decreases as the initial nonrecurring costs are amortized over a longer period. A consideration we do not include in this analysis is whether there are sufficient MILCON appropriation dollars to fund the construction. We recognize that alternatives that make long-run economic sense may have significant up-front nonrecurring costs that make them impractical from a budget or appropriations standpoint.

Second, the all-movement modules (Playas, Muscatatuck, and BRAC'd installations) are economical for round-trip travel up to distances of about 2,500 miles and become more costly after that. Transporting soldiers to training involves significant movement costs, but not necessarily investment costs. The nonrecurring costs of a company-size purpose-built facility and those of a site such as Muscatatuck are not significantly different, but the latter has greater capacity, and thus cost per person is lower even with movement costs included. Sensitivity analysis shows that the modules requiring movement reach a breakeven point with the company purpose-built modules at a round-trip distance of about 2,500 to 2,800 miles and become more expensive thereafter. While not quantified in this study, the opportunity cost of time spent traveling instead of conducting other training should not be forgotten when considering where to develop urban training facilities and how many to develop.

Third, non-hybrid purpose-built facilities are costly; CACTF designs (which meet a standard of training only up to a platoon at a time) are extraordinarily so. Costs scale up faster than capacity for non-hybrid purpose-built facilities. Economies of scale do not appear to apply if increased capacity is offset by greater fidelity. This situation is exacerbated when units have to move to larger-capacity sites for training, which is very likely. The larger the site, the more expensive it is to build to a given level of fidelity, but larger sites are attractive because of the unparalleled training opportunities they offer. Units will want to use them, yet because of their cost, few will be built. It is unlikely that any one installation will be able to employ such a site at capacity via use by units assigned to it alone; thus, the per-person costs of the fourth and eighth alternatives shown in Figure 6.4 are very likely the more accurate representations of the true costs associated with such facilities.[48] It therefore appears that the "tradi-

---

[48] For example, we estimate that 30 battalion-size training events could be conducted annually at Module 1. If there are not 30 battalions in the geographical area, the cost of movement to the facility needs to be included in the cost of the modules, further increasing their expense. Local purpose-built facilities have greater long-term benefits if they are fully utilized by local units. Those benefits decline with lower usage rates.

tional" Shughart-Gordon-style platoon-size module provides the most effective training as measured by cost per person trained, at least among the non-hybrid, purpose-built options.[49]

Fourth, our calculations are based on U.S. Army live urban training usage rates of about 210 days per year. If more days of training are scheduled and used at module types similar to those discussed here, the efficiency of the purpose-built modules increases faster than does that of those requiring movement. But while there is improvement, the extent of savings is not sufficient to make them less costly than the movement-mandatory alternatives. Moreover, achieving such increases in use may not be feasible given the need for routine facility and staff downtime, instructor leaves and education, and periodic extended halts for overhaul or upgrades.

A fifth point to consider is that the comparison is not complete until costs associated with the *total* number of these types of facilities are factored into the calculations (so that total service and/or joint force throughput can be calculated).[50] Increasing the total number of purpose-built facilities does not impact *average* costs as long as the facilities are used to the capacity assumed for these computations. However, total costs do increase as facilities increase in number. For example, if five separate 70-building, company-size, purpose-built facilities are needed to train the force, the average cost per person for this solution option would still be about $750, but the total costs would quintuple.

Sensitivity analysis raises a sixth and final point in this consideration of whether it is more efficient and effective to build new fa-

---

[49] Other subjective factors should be assessed as well in making a final recommendation.

[50] Throughput is the number of individuals or units that train to the stated requirements, using a given facility or capability. Throughput capacity is the number of individuals or units that a facility can support during a specific period. Availability is affected by scheduled and unscheduled maintenance days, holidays, and inclement-weather days. A typical multipurpose training range has 85 total non-available days per year, which leaves 280 available days. The goal of the Army is to schedule for 80 percent of available days, or 224 days. Typical use is 90 percent of scheduled days, or 202 days. (The value of 210 used in this study is the result of assuming 30 seven-day rotations per year, which provides a usage value consistent with other assumptions used in cost calculations, e.g., that a rotation would be of seven days' duration) (HQDA, TC 25-8, April 5, 2004).

cilities, improve existing ones, rent from commercial enterprises, or otherwise acquire needed capabilities. Purpose-built facilities have long useful lives, assumed here to be 30 years. Their initial costs are therefore amortized over longer periods of analysis when the time frame is extended. For shorter life cycles, the relative costs of those modules with large nonrecurring costs (e.g., purpose-built facilities) will rise. The shorter the expected life of a facility, the more severe the consequences in this regard. Facilities designed with shorter expected life spans will therefore have higher per-person-trained costs than those shown here, other factors being equal.

Figure 6.4 is based on the recurring and nonrecurring costs shown previously in the module assessments; on unit sizes of 30 for a platoon, 100 for a company, and 500 for a battalion; on moving 500 miles one way (1,000 miles round trip) at 40 cents per person per mile; on 210 days of facility availability; and on training-event duration of seven days. Sensitivity analysis was done on movement distance and event duration.

Figure 6.5 shows the effect of increasing movement distances to up to 5,000 miles round trip. Increasing movement distance increases the per-person costs of those options involving movement. (Thus, the platoon- and company-size facilities remain constant along the x (distance) axis, since units using them undergo only intra-installation movement, which is not measured here.) The costs of the three all-move modules ramp up quickly, while the two half-move modules ramp up more slowly, as would be expected. Differences remain because of different levels of recurring and nonrecurring costs. Options with the most other costs change least with movement distance. Tripling movement distance to 3,000 miles round trip increases the per-person costs of most of the move modules to higher than that of some of the platoon- and company-size purpose-built modules. The hybrid facility is the least expensive on a per-person basis. At a round trip distance of 5,000 miles, the all-move modules approach the per-person costs of the battalion-size purpose-built module. This analysis suggests that regional movement options are more cost effective than

Figure 6.5
Average Annual Cost per Person, Based on 30-Year Life Cycle and
Movement Distance (Platoon- and Company-Size Purpose-Built Facility
Costs Are Constant, and 100 Percent Home-Station Usage Is Assumed)

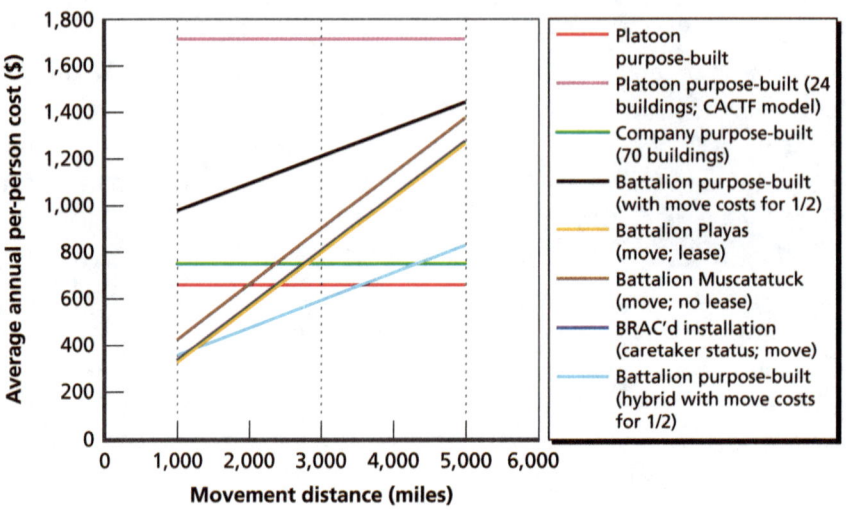

RAND *MG439-6.5*

national ones. If movement distances triple, (non-CACTF) platoon-
and company-size purpose-built facilities are competitive on a cost-
per-person basis.

Figure 6.6 shows the effect of decreasing event duration from
the seven days used as the standard in this analysis thus far. Decreasing duration to less than seven days decreases per-person cost proportionally for the purpose-built facilities (halving duration halves per-
person cost) and does not change per-person cost as significantly for
the move options. It costs the same to move a person regardless
of training duration; opposite to the effect above, the recurring
and nonrecurring costs are now the costs that go down on a per-
person basis.[51] The platoon-size purpose-built and the company-size

_____

[51] This occurs because the per-person cost results from two calculations. First, annualized
recurring and nonrecurring (fixed) costs are divided by the number of trainees; second,
movement cost (a variable) is multiplied by the number of trainees. The two functions oper-

Figure 6.6
Average Annual Cost per Person, Based on 30-Year Life Cycle and Event Duration

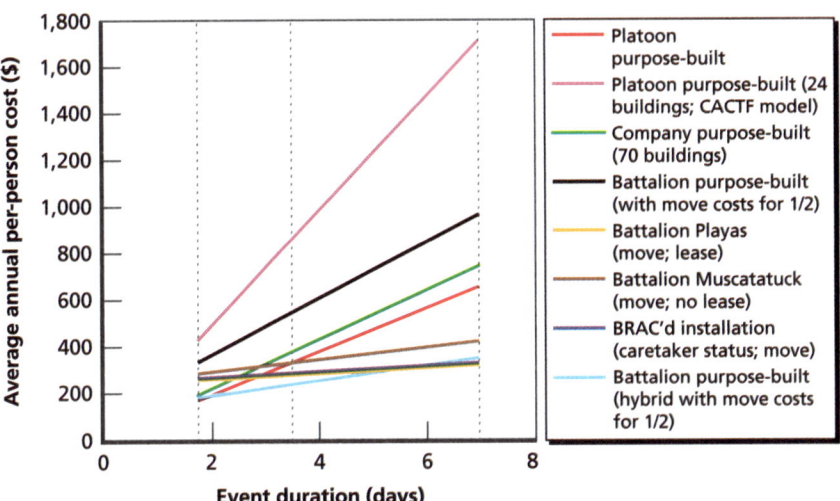

purpose-built are now competitive with the move options. The hybrid option is the least expensive over the broadest range of durations. Also, there is interplay between reduction in event duration and increases in movement distance that are not captured in this analysis. Decreases in event duration would lead to more than one-half of personnel moving to fixed facilities because facility capacity would be higher than before, compared to local population density.

Logically, the strategy would be to procure or invest in the lowest-cost training that first met JUT requirements and then move up the cost ladder to the point where all training requirements were satisfied. This assessment is made in the next chapter.

---

ate in different ways. As training duration goes down, the number of trainees goes up, so those modules with proportionally more fixed costs go down the most. As movement distance goes up, variable costs go up, so those with the most movement costs go up the most.

Additional Comments About the Proposed Twentynine Palms Urban Training Facility

In the cost estimate for the "battalion and larger purpose-built facility" module, it was assumed that the planned Twentynine Palms training site would host a battalion per event. The 900 "full-up" buildings (capable of supporting internal and external training) and 600 additional façade or container-type structures could very likely support an entire brigade. It is therefore necessary to reconsider the high per-person training costs to provide a fair evaluation, for if three times the number of personnel train there annually, it would seem that these costs should decrease.

Given the initial assumption that this site could have a full battalion in the facility at one time, previous standard assumptions would be that each battalion would have seven days "in the box," and 30 battalions (or 10 brigades) would move through the site in a year. This is comparable to JRTC rotations. Fifteen battalions (five brigades) would be local (some combination of active-duty and Reserve components), and 15 would need to travel to the facility. However, the site could be large enough to accommodate two battalions or a full brigade at one time. Costs would change as discussed below.

Another operational model could allow for a brigade to occupy the geographic area with two battalions simultaneously in the facility. For example, if the brigade had the facility for 14 days total, it could have the first and second battalion on site for five days; the second and third battalion on site for four days; and the first and third battalion on site for five days. Thus, one battalion would have 10 days of training, and two battalions would have nine days of training. Fifteen brigades (45 battalions) would use the facility each year. Five of the brigades would still be local, and the other 10 would move to use the facility.

A third operational model could allow for a complete brigade to occupy the facility. In theory, 30 brigades could be accommodated in a year. In reality, this is probably too many brigades to effectively move into and out of the area without overburdening local transportation, billeting, and related capabilities. It is more likely that training duration would expand to 10 days, which would allow for a total of

21 brigades each year. The cost estimates for this alternative assume that five of these brigades would be local, and 16 would move to the facility for training.

The costs of the four module alternatives are compared in Figure 6.7. Perhaps surprisingly, adding more throughput to the facility does not significantly influence per-person cost. While per-person cost for construction and maintenance does decrease as these costs are amortized over greater throughput, the decrease is offset by the added movement costs. Moreover, these costs were assessed for a round trip distance of 1,000 miles, but as the number of units increases, it is more likely that they would come from greater distances. Scale does not always benefit on a per-person basis if there is not enough local strength to fully use the capacity of the facility.

Figure 6.7
Alternative Costs per Individual Trained for the Brigade-Size Purpose-Built Facility at Twentynine Palms

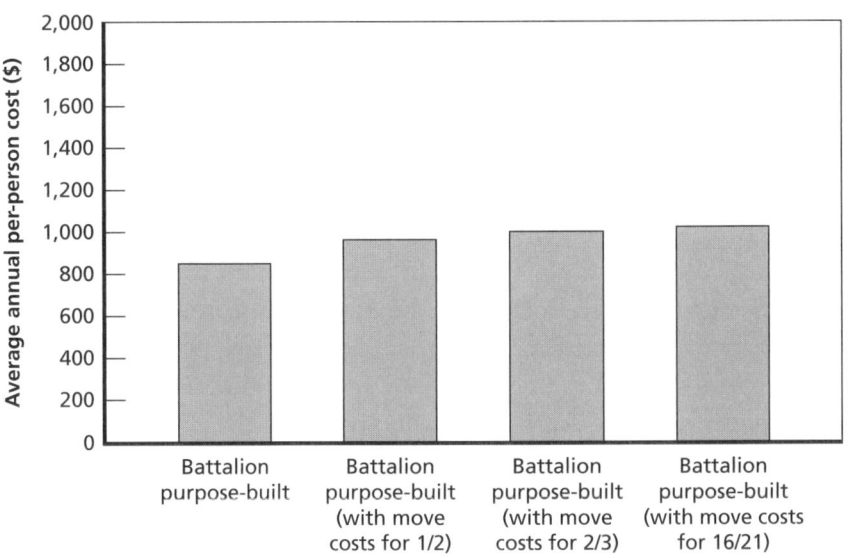

The end result of these various computations reflects the fact that other issues impacting the per-person cost of training at a planned facility result in little difference between the initial and various alternative values.

Virtual and Constructive Training: Alternatives or Supplements?

A previous RAND study discussed the issues associated with tradeoffs between live and either virtual or constructive simulation.[52] These issues are summarized here in the context of JUT.

Given the goal of trained (proficient) individuals and units, a greater or lesser number of budget dollars could be allocated to JUT versus other training or resource needs, and within those budget dollars spent on urban training, a greater or lesser amount could be allocated to live, virtual, or constructive JUT. These issues seem straightforward, but other considerations also affect the tradeoffs.

First, time is a limited commodity. Frequently, it is suggested that virtual or constructive training can be added to the existing training program as if time were infinitely available. Second, proficiency (benefit derived per unit of training time spent) is not likely to be equivalent across the three types of instruction. Military personnel tend to put a premium on live training. This is especially true for culminating events, which many JUT exercises are likely to be. A complete tradeoff of one for the other is probably not possible. Third, periodicity (how often the training task must be repeated to retain requisite skill levels) and duration of training events affect proficiency and efficiency. Fourth, it may not be possible to conduct all training needed by the desired means because resources are not infinite. The number of live training sites is limited, as is the number of virtual and constructive training seats. While one might prefer more repetitions of a training event, such repetitions are likely to be constrained by this lack of resources. (Note, however, that as simulations are more routinely PC-based, accessibility to them for training could increase

---

[52] John Schank, Harry Thie, Clifford Graf, Joseph Beel, and Jerry M. Sollinger, *Finding the Right Balance: Simulator and Live Training for Navy Units*, Santa Monica, CA: RAND Corporation, MR-1441-NAVY, 2002.

dramatically. While such an option is already available for some low-level tactical urban instruction, this is primarily of value to Service training. Significant advances in a manner that would substantially help JUT are not foreseen in the 2005–2011 time frame.) Fifth, availability and fidelity matter for both live and virtual/constructive training. Availability (influenced by capacity, accessibility, and number of a required capability), discussed immediately above, generally refers to the facility and simulator time available to units and individuals. Fidelity relates to the quality of the facilities or simulators (i.e., how well they replicate the real world in addressing training objectives). Simulators might not need to train the whole task in order to be considered useful; lower-fidelity simulators could focus on partial tasks. Finally, experience is a factor in simulator value. Rehearsal in a simulator before an actual live event has more value for inexperienced personnel than it has for experienced personnel.[53]

Some training events (e.g., those that are high-risk and otherwise infeasible to conduct live) will be best done with simulation. Others could be accomplished at lower cost for equal proficiency through a mix of live and simulation training. For example, a skill might be introduced or practiced on a simulator, and then exercises employing the skill could be conducted during live training. Or, for multiple iterations of an event, some live iterations could be replaced with simulators. Measuring the effectiveness of simulator training is difficult, and there is a general preference for conducting at least one live certifying event with a formal AAR.[54]

In sum, the basic tradeoffs of using simulation and live training at their current levels of availability and fidelity appear to be between substituting simulation hours for live hours and adding simulator hours to live hours. A much more complex tradeoff for a longer-term investment strategy includes considering changing the availability and

---

[53] Ibid., p. 54.

[54] This is an area worthy of more research, as is recognized in the modeling and simulation community. See, for example, "Playing to Win," *The Economist*, Vol. 373, No. 8404, December 4, 2004, pp. 24–25.

fidelity of both live and simulation-supported training by investing in them disproportionately in relation to further development of live capabilities.

Our analysis of training modules does not include an assessment of trading off simulation events for live training events. For the 2005–2011 period, the state of availability and fidelity of simulators make such an assessment moot. Virtual and constructive simulation will have the greatest value in a complementary role when it comes to virtual and constructive training. Investments could be made in better software and hardware capabilities in the longer term, but the availability of such simulations is unlikely to preclude the need for a substantial proportion of live training during JUO skill development for many years to come. In essence, the use of constructive and virtual simulations is valuable, but it will continue to serve in a complementary rather than a primary role in any but exceptional circumstances during the 2005–2011 period.

Investment Versus Annual Training Budgets

Many of the modules discussed in this report do not require investment per se, but they do have added costs that must be included in annual training budgets. Only the purpose-built facilities for live training, the physical facilities for simulation centers, and software development of new simulations require up-front investment to meet nonrecurring costs.

These investments are driven not necessarily by the joint training tasks but rather by Service training needs that are the precursor to joint training. However, because of the building-block nature of JUT in particular and military training in general, the JUO training need will not be met if the primary training investments are not made. Incremental joint training inherently requires additions to training regimens, which has a net effect of extending the duration of each training event. The extended duration reduces throughput and, in the net, requires additional training resources than would have been needed without the JUT tasks.

CHAPTER SEVEN

Developing a DoD-Wide Joint Urban Operations Training Strategy

The target was a white house with a red roof. . . . He described it to the pilot. . . . The leading A-7 came in very low, under 200 feet, for no fewer than three passes. . . . As they drew near, the pilot was heard to say over his radio that he could see people near the house. This did not fit with the actual target. . . . Too late. . . . The pilot had fired into a gray building . . . causing chaos and seventeen casualties [in the U.S. brigade command post].

Mark Adkin, *Urgent Fury,*
Describing an urban air-ground engagement in Grenada, 1983

We write a lot of the lessons we've learned, but we don't learn from the lessons we've written.

LTC Kevin Murphy,
Deputy Commander, Combined and Joint
Special Operations Task Force, Afghanistan
Bagram, Afghanistan, February 2004

Varied and unique range safety requirements limit cross-Service use of aviation ranges during joint exercises. In order to facilitate joint training and co-use of Service aviation ranges, a collaboratively developed and agreed-to range safety policy and to some degree, procedures, is required.

Cross Service Range Use Standardization
Working Group Aviation Range Safety
Sub-Group Charter (draft, 2005)

This chapter draws on all the information presented thus far to develop a JUO training strategy, which we present in several phases. The first phase provides some final groundwork before the building of the strategy itself. The elements presented in this initial phase are not so much components of the strategy as they are observations offered to heighten the value of the ultimate product. The second phase delves into strategy construction itself. We then look at factors influencing the execution and maintenance of the strategy and address some hard implementation challenges before closing with a roadmap for strategy implementation.

Some Principles for Joint Urban Training

The modules we developed are the components of a JUO training strategy for the 2005–2011 period. They sit like bricks stacked by the side of a building site, ready for use. Before beginning construction, however, we provide several observations gleaned from prior research efforts as well as from interviews conducted, in the hope of making the final outcome of that assembly optimally valuable. These observations, which might be considered as principles for JUO training, are the following:

- The training strategy must be comprehensive.
- The training strategy must be dynamic.
- Much improvement is needed in lower tactical-level JUO training, but the greatest shortfalls are at the highest echelons.
- U.S. trainers must remain in "receive mode."
- Joint training modules are only some of a training strategy's building blocks.
- Systems of effective capabilities underpin successful training.
- Even the best training and the most effective training strategy can sometimes not fully prepare a force.
- Size has a quality all its own. Corollary 1: Size can be cheated. Corollary 2: If the consequences of an action in an urban area are not reflected during training, the instruction is flawed.

- Bigger is better. Bigger and denser is better yet.
- If a capability exists in the field, find a way to replicate it for training.
- The total training audience in, around, or over an urban training site may not equate to the number of personnel actually receiving substantive urban training on relevant requirements.
- Simulations, virtual and constructive training, and synthetic environments will not be capable of fully replacing live training during the 2005–2011 period.
- It is important to promote innovation and reconsider proven methods.

The Training Strategy Must Be Comprehensive

A JUO training strategy must address the full spectrum of operations, from those involving support missions devoid of human adversaries to scenarios in which WMD might be, are, or have been used. It must recognize that many types of challenges occur simultaneously rather than sequentially—that offensive, defensive, stability, and support missions involving several or all of the nation's Services have occurred and will again occur side-by-side at the same moment in time. Recent operations in Central Asia and the Middle East also make it evident that interagency preparation is in many instances as important to success as is joint training. Further, training in governing, fiscal management, and similar skills is now given little time even at the war-college level. Military leaders, Service and joint, cannot afford to limit their expertise to the "management of violence."[1] Training should prepare them to better understand urban social infrastructures, to identify and interact with influential representatives of various demographic groups, and to use the resources at hand (funds distribution included) to best serve coalition objectives.

---

[1] The "management of violence" is how Samuel P. Huntington (quoting Harold Lasswell) described the special expertise of the military professional (Samuel P. Huntington, *The Soldier and the State: The Theory and Politics of Civil-Military Relations*, New York: Vintage, 1957, p. 11).

The Training Strategy Must Be Dynamic

An urban environment is an especially vibrant one, all the more so during combat operations, when survival depends on rapid and effective adaptation. While the major components of a JUO training strategy will remain fairly stable for some time, those components and the relative emphasis given to particular aspects of training should be reviewed frequently to ensure that training maintains pace with change. Constant monitoring of observations, insights, lessons, and analysis involving ongoing operations should be an essential part of any training. Rotation of personnel who have demonstrated notable ability in the field into training assignments will help to ensure relevance, as will the assignment of O/Cs and selected other faculty to operational theaters. Requirements for periodic "lessons learned" submittals by deployed units and maintenance of a system to collect, analyze, and appropriately disseminate the results in a timely manner constitute a further positive step. Yet even such commendable initiatives to secure insights into recent operations may be insufficient. Trainers must also consider how the lessons from recent operations will pertain to future contingencies, both domestic and international, and they must adjust pertinent aspects of instruction.

Much Improvement Is Needed in Lower Tactical-Level JUO Training, but the Greatest Shortfalls Are at the Highest Echelons

Urban operations present unparalleled complexities to pilots, section leaders, and generals alike, but the level of complexity increases at higher echelons. The increased recognition of the importance of urban operations means that tactical leaders are granting more time for training in them. The same is less true at the highest joint echelons. Exercises that prepare those who most need to understand the interrelationships between urban actors, systems, and surrounding rural or other nearby built-up areas are rare. Classroom preparation about urban governing or urban operations in general is all but lacking except for occasional electives.[2] Events that substantively bring military deci-

---

[2] Discussions with representatives at the National Defense University, National War College, Army War College, and Naval War College revealed that there is little if any urban-

sionmakers in contact with other-agency representatives, indigenous authorities, or commercial enterprises of relevance to operations are extremely rare. Problems confronting coalition members in Iraq are less attributable to failures at the tactical level than to those at the operational and strategic level. Not a few are attributable to decisions by military leaders and those in other parts of federal governments. Currently, military training does little to address the shortfalls.

U.S. Trainers Must Remain in "Receive Mode"
Correctly updating JUT will rely considerably on input from lessons learned and other sources. Other nations' expertise and experiences have much to offer if U.S. trainers are willing to listen, and American military personnel should establish and maintain multinational contacts to gain benefit from alliance and coalition member savvy. U.S. representatives should at the same time not be chary of diplomatically passing along information to those from other nations who might make good use of it.

An interesting phenomenon was taking place in Iraq at the end of 2004. British ground forces, some of whom at times found U.S. methods too aggressive, and Americans, who at times did not agree with their UK colleagues' "softly, softly" approach, found themselves learning from each other and moving toward a common middle ground. Two officers at the British Army's OPTAG training complex, once a center for preparing units for Northern Ireland tours and now an organization with a greater Middle Eastern operations focus, found that U.S. and UK

> methods of operation [in Iraq] are starting to come together. Danger is causing us to close the gap. . . . We [the British] are

---

specific instruction during coursework at higher-level joint and Service schools. That which does exist is generally an elective, the continued existence of which is largely dependent on a specific individual assigned to the faculty at the time. Such instruction is far more commonplace in tactical-level instruction, e.g., the USAF FAC-13 "Urban CAS" block of instruction; the USMC officer basic course, "Urban Patrolling"; and the aforementioned training at the U.S. Army Infantry School involving terrain walks in Columbus, GA.

coming to the higher threat operations. . . . We've almost converged on the same TTPs [tactics, techniques, and procedures].[3]

Joint Training Modules Are Only Some of a Training Strategy's Building Blocks

We have argued that the traditional and proven building-block approach to training had pertinence in constructing a JUO training strategy. Even the best courses of instruction, highest-quality facilities, and most realistic simulations cannot overcome inadequate student preparation. Excellent urban training by Service components is essential. That training must include preparation of individuals and units for their participation in joint, multinational, and interagency urban operations. Similarly, the facility, simulation, and other capabilities that dominate the analysis in the following pages must be employed in the same manner; more-complex and demanding training should follow only proven achievement of subordinate skills.

Systems of Effective Capabilities Underpin Successful Training

Depending on the architect and contractors, the same building materials can result in an effective structure that promotes worker cooperation or one that frustrates interaction and quality work. A JUO training strategy is likewise more than simply the sum of the capabilities it comprises. The capabilities themselves must be available in sufficient *quantity* to provide for training all joint forces requiring preparation in a timely fashion. Those capabilities must individually and collectively be of a *quality* that ensures that the training produces operational readiness. The *size* of capabilities has to allow units to train as they will operate after deployment, i.e., as functional combined arms, joint, and interagency teams. Sufficient *throughput capacity* (e.g., not having to create and maintain more of a given type of capability than is necessary, while allowing periodic refresher visits to maintain perishable individual and collective skills) is fundamental to

---

[3] Interview with Lieutenant Colonel J. N. Nick Watt, Chief Instructor, and Major Paul Dutton, British Army OPTAG, by Russell W. Glenn, Folkstone, UK, December 17, 2004.

efficient and effective training. Efficiency will similarly and increasingly demand *linkage* so that live, virtual, and constructive capabilities can be blended to maximize training value for dollar spent and to prepare personnel to use new or future technologies that are not yet available in the training environment.[4] Cybernetic linkage of live, virtual, and constructive training involving systems that will have synergistic effects when actually brought together in operational areas can preclude having to physically collocate the separate parts during exercises. Increasing development in this aspect of training has potentially very significant dividends. Money can be saved, and deployment times can be compressed.

Even the Best Training and the Most Effective Training Strategy Can Sometimes Not Fully Prepare a Force

Based on the analysis described in Chapter Five and summarized in Appendix G, five of the 34 JUO training requirements cannot be trained to a "run" level of readiness in the 2005–2011 period despite the implementation of the strategy developed in this study:

- Conduct urban HUMINT operations
- Provide urban fire support
- Consolidate success in the urban environment
- Shape the urban environment
- Transition to civilian control

The reason for "not being able to get there from here" has to do with the inherent complexity of urban areas, a complexity born of an extraordinary density of challenges and their interactions. Because of

---

[4] Some of these technologies might be fielded in the operational theater, where leaders believe best use can be made of them, before they are introduced into the training base in sufficient numbers. Virtual training will potentially help to avoid the difficult decision of whether to allocate low-density systems to operational units or to the training base. The problem is a recurring one. Soldiers assigned to Vietnam in the mid-1960s frequently trained with M14 rifles, receiving M16s only after boarding troop ships or arriving in theater. The same situation confronted the British Army in 2004 in Iraq, where units were issued the hand-carried "Minimi" automatic grenade launcher on arrival, but few soldiers had seen or been trained on it previously.

the sheer number and type of actions inherent in four of the five requirements above ("provide urban fire support" is the exception), the best that trainers can hope to accomplish is to provide trainees with the skills needed to adapt instruction to meet real-world demands. In the case of "provide urban fire support," safety restrictions, inadequate numbers of aircraft simulator linkups to other urban training capabilities, and less-than-fully-effective means of replicating indirect fires mean that training sufficient numbers of FSCs, pilots, and others vital to these missions by 2011 will be impossible.

This situation can be significantly redressed by conducting training in such a manner that those trained expect to confront the unexpected, that the occurrence of surprise does not surprise. Preparation that addresses these five requirements must focus less on specific skills than on intellectual and emotional preparation to manage new challenges under crisis conditions. Such training can provide readiness at a "walk" level of expertise; only field experience and on-the-job training will allow individuals and organizations to achieve a "run" level.

Size Has a Quality All Its Own

With notable exceptions (e.g., snipers, patrols, reconnaissance elements, and some Special Operations teams), most units train and operate as part of a platoon or larger-size organization. That platoon, company, battalion, or even higher-echelon organization is itself a team. Good training demands that it prepare as it will fight or accomplish other missions in the field. A company conducting live training in an urban operations facility or a platoon using a simulation that focuses only on light infantry maneuver will get far less out of its sessions than it would if it were to train as it would function under operational conditions. Training facilities of inadequate size can in fact teach the wrong lessons. As its repeated appearance in the discussion of joint urban training shortfalls (in Chapter Four) made very evident, lack of sufficient size—one of the critical factors noted under "Systems of Effective Capabilities Underpin Successful Training" above—ranks among the major shortcomings in American urban training capabilities today. Simulations can help, by artificially putting units on the flanks of a live training complex, for example, but those replications fall far short of reality. Nor does it appear that simulations of sufficient sophistication in this regard will be available until after 2011.

A caveat is essential. While "bigger is better" is generally true for units participating in joint or Service urban training events, the cost of developing and maintaining large sites with high resolution may be excessive if sufficient usage rates cannot be maintained. Facilities capable of providing realistic training for up to a battalion task force or equivalent are desirable and much needed. But it is questionable whether a facility that can provide training for a brigade-size unit is needed. While possession of such an asset would potentially allow superb training for smaller units and the occasional full-up brigade event, the number of opportunities to conduct the latter, in either a Service or joint context, is probably too small to justify creation of such a facility in any but the most exceptional of circumstances.

Corollary 1: Size Can Be Cheated

The need for size can be addressed in nontraditional ways. Those managing the urban ranges at Ft. Polk have made theirs the Cadillac of urban training facilities through the clever use of augmenting strands of buildings. As highlighted previously, shipping-container-type and façade structures, with and without interior rooms as appropriate, have been placed along entry roads and trails surrounding more permanent urban training elements to reproduce the effects of shanties, suburbs, and other nearby adjacent urban nodes. These add to the quality of training for units not actually in an urban facility "box" proper and permit simultaneous training on such essential skills as "react to convoy ambush in an urban area." While not a replacement for size, these and similar augmentations help to mitigate the consequences of insufficient size.

Similarly, infantry officer coursework at Ft. Benning, GA, uses terrain walks or overflights of actual urban areas to give students insights into the character of villages, towns, and cities. No purpose-built urban training site and no simulation for many years to come will be able to present the heterogeneity and complexity of a modern megalopolis. Terrain walks in these and even much smaller urban conglomerations can augment other live instruction or training that employs synthetic environments.

Corollary 2: If the Consequences of an Action in an Urban Area Are Not Reflected During Training, the Instruction Is Flawed

Although military leaders often talk of "strategic corporals" (i.e., decisions made by a leader at the lowest tactical levels can have strategic consequences), the potential repercussions of good or bad decisions and actions are rarely incorporated in urban training. Urban areas are tightly knit social and physical systems. Because actions almost inevitably have secondary and higher-order effects, training should demonstrate the sensitivities of such environments. For example, throwing candy from convoys can result in children being run over and killed.[5] The death of a child can undo weeks or months of building positive community relations. Actions within urban areas will often also affect the areas surrounding and dependent on a village, town, or city. Training should also replicate these interrelationships.

Bigger Is Better. Bigger and Denser Is Better Yet

Size alone will not present leaders with the challenges they are sure to find when deployed. A large training site, real or synthetic, that is only sparsely populated with structures and noncombatant role players does not present the problems a pilot is likely to find as he attempts to distinguish which of the buildings "with white walls and a red roof" below him is the one his forward observer (FO) wants him to strike. Ground unit leaders who can look right and left to see the limits of the training site will not encounter the dangers of OPFOR infiltration to the extent that they will in an actual theater of operations. Density *within* structures is important also. Unrealistically large rooms lacking closets, hallways, storage nooks, and other compartments normally found in buildings fail to fully meet training demands.

---

[5] Vehicles in U.S. convoys in Safwon in 1991 and Al Kut, Iraq, in 2003 ran over and killed Iraqi children who dashed out to collect food thrown to them.

If a Capability Exists in the Field, Find a Way to Replicate It for Training

Some "on-the-job training" is always part of deploying to a theater. If possible, however, no training that could more effectively take place in a safer, less time-constrained environment should be left to last-minute preparation in a theater. History proves that the cost of learning in combat is too high under the best of conditions; failure to properly train personnel magnifies the risk to human lives.

The Size of an Organization with Elements In, Around, or Over an Urban Training Site May Not Equate to the Organization Being Trained

The battalion commander whose attackers free Ft. Polk's Shughart-Gordon training site of OPFOR and the wing commander whose A10s occasionally overfly an urban complex in support of urban training at Camp Lejeune have not trained their units for urban operations unless every component of the command has participated in all relevant aspects of training. The platoons that clear the village and the pilots that make mock training runs might approach a satisfactory level of preparation. Those on the outskirts isolating the built-up area, whether on the ground or in the air, have not been sufficiently challenged.

Simulations, Virtual and Constructive Training, and Synthetic Environments Will Not Be Capable of Fully Replacing Live Training During the 2005–2011 Period

Advances in simulation-supported training have been considerable, but they have yet to meet the challenge of adequately representing the complexities of urban operations. Aircraft simulators and some replications for practicing basic ground maneuver skills and lower-echelon decisionmaking adequately address many aspects of urban operations for relevant individuals. In other cases, virtual and constructive training are at best supporting elements for field or classroom instruction.

It Is Important to Promote Innovation and Reconsider Proven Methods

Some of the best existing urban training capabilities are the products of innovative thinking. The Yodaville air-ground urban training site was the brainchild of then-Major Floyd "Yoda" Usry, a marine helicopter pilot and instructor at Marine Aviation Weapons and Tactics Squadron 1 (MAWTS-1). Economically but very effectively fabricated out of discarded shipping and munitions containers, the facility allows rotary- and fixed-wing pilots to drop inert weapons on targets that appear surprisingly realistic from the air (see Figure 3.1).

The U.S. Army contracted for the creation of a small, three-building replica of an Afghan leaders' compound on the Bagram base currently used by American and other coalition forces. The replica is small, but it provides squads and Special Operations units needed practice in immediate proximity to actual operational areas.

The Australian Army was once considering using tarpaper or plastic "walls" at a training site near Darwin in the north of the country. In addition to dividends in cost savings, use of such penetrable materials would allow units to recognize the penalties for firing into walls (and, by extension, ceilings) that do not stop rounds and thereby bring death or injury to comrades. (The same principle has potential value in more robustly constructed facilities. Replacing some solid walls with penetrable material painted to match firmer surfaces would have similar training benefits if penetrating Simunitions training rounds struck comrades.)

The USMC has similarly shown willingness to enhance urban training in other-than-ordinary ways, using the abandoned base housing area at George AFB, a closed military hospital in Oakland, CA, and actual cities (e.g., Little Rock, AR; Boise, ID) for training, after coordinating with community leaders. The Indiana National Guard has many innovative ideas for combining commercial, military, and state enterprises to mutual benefit as it continues to work the transition of the former Muscatatuck asylum facility into an urban training complex.

At the same time, both the aforementioned urban classroom instruction and supporting professional-development programs are sorely lacking. No U.S. staff or war-college curriculum approaches adequate consideration of governing in urban areas (or military governing in general). Urban courses are few and not included in required core curriculums. Instruction and exercises involving regular force combat scenarios and those dominated by stability or support missions are needed at higher-level joint and Service institutions. Professional-development offerings such as correspondence courses and topical reading lists would also provide value. Classroom and other educational efforts should include instruction on the capabilities of other federal agencies, NGOs, and PVOs and material on means of effectively supporting indigenous citizens and governments in times of crisis.

Designing a JUO Training Strategy

It is now time to take the above observations and the work summarized in the previous chapters to develop both an immediate- and a longer-term JUO training strategy.

A JUO Training Strategy for the Immediate Term (2005–2007)

Strategy development in this study has two primary components. This, the first, addresses those actions that would have immediate or near-immediate effects if taken. There is, fortunately, much that can be done in this regard that could conceivably bring much of the U.S. armed forces to a "run" (operational readiness) level of proficiency for a considerable number of JUT requirements. That these actions can be taken fairly quickly and generally at limited expense does not diminish their potential impact. They address requirements largely ignored to date and that segment of the military hierarchy most in need of attention (the highest-level Service and joint commands), in addition to potentially enhancing readiness at the tactical level. Though applicable to the "engage" component of the USECT concept, they also tend to influence the understand, shape, consolidate, and transition phases that heretofore have received too little attention, as recent operations have clearly demonstrated.

Notably few higher-echelon joint training events have incorporated significant and realistic urban environments. There is similarly very little education that prepares military or other personnel to feel comfortable dealing with such events, much less readying them for actual operations. However, this adverse news has a counterpart. Commanders employing Module 25 (conduct of combatant command or JTF headquarters, large-scale schools, or multiechelon/interagency exercises) would find that they can provide a "run" level of preparedness for 12 of the 34 previously identified JUT requirements. Additionally, Module 24 (classroom instruction) training can provide a "walk" level of preparedness for eight of these 12 requirements, and a "crawl" level for another three, an excellent example of how building-block training can prepare units and individuals for both more-complex training events and real-world undertakings (as classroom instruction could provide the training needed for maximizing the benefit derived from higher-echelon exercises).[6] There is further good news: High-level headquarters and similar exercises are fairly economical, costing an estimated $2.9 million to $7.1 million per event if the headquarters deploys and the event includes links to units in the field (see Tables 6.12 and 6.13). Such deployment and outside-organization participation are often not necessary, allowing such events to be conducted at lower cost. That these events could address more than one-third of the requirements of concern means that the return for dollar invested would be excellent. Further, classroom JUO training outlays are negligible; they should be measured more in terms of opportunity costs than of dollar expenditures (the loss of course instruction on topics that would be covered was the time not being spent on urban training).

The 12 requirements for which exercise training would permit organizations to achieve a "run" level of preparedness are listed below. The extent to which classroom training can address each is shown in parentheses for each entry:

---

[6] Classroom instruction also permits achievement of a "run" level for "identify critical infrastructure nodes and system relationships," as does the conduct of major training exercises.

- Conduct stability operations in the urban environment (C)
- Conduct support operations in the urban environment (C)
- Conduct urban operations exercises (C)
- Integrate urban operations with other relevant environments (W)
- Coordinate multinational and interagency resources (W)
- Govern in the urban environment (W)
- Identify critical infrastructure nodes and system relationships (R)
- Plan urban operations (W)
- Orchestrate resources during urban operations (W)
- Understand the urban environment (W)
- Achieve simultaneity in meeting requirements (W)
- Conduct training across multiple levels of war (W)

Joint and high-level Service headquarters would benefit from the immediate introduction and sustained inclusion of full-spectrum, realistic urban challenges into annual combatant command exercises such as Cobra Gold and JFCOM Mission Readiness Exercises (MRXs).[7] It is essential that such urban contingency training avoid devolution into exercises dominated by regular-force-on-regular-force combat to the effective exclusion of the interagency, diplomatic, governing, stability, support, and other urban operations demands that are at least equally important for operational and strategic commanders.

Modules involving exercises and classroom instruction will find a valuable complement in professional reading programs and other professional-development initiatives, such as asking veterans with appropriate expertise to address unit members in professional-development sessions. Including guests from other Services would enhance the joint training value of these initiatives.

---

[7] JFCOM MRXs include support by a JTF headquarters comprising representatives from army corps, marine expeditionary forces (MEFs), and numbered fleets and air forces (comment to authors from Lt Gen (USMC, ret.) G. R. Christmas in his review of this study, received February 14, 2005).

Another requirement that is readily within "run" training status for units in the immediate term is "navigate in an urban environment." Module 10 (terrain walks in actual urban areas) provides this opportunity, again at little cost. Nearby towns or cities offer the training environment necessary, although commanders wanting to challenge their personnel with unfamiliar terrain may choose to go farther afield.

The immediate term also offers the potential to develop the assets addressed by three of the six training support element modules, capabilities that could have dramatic benefits for future joint urban training. Participation by an expert OPFOR (Module 35) and noncombatant role players (Module 36) will lend much-needed realism to live training at urban facilities, headquarters or classroom training involving virtual training support, and many other scenarios. School faculty and exercise support should include individuals with the requisite expertise in this regard; it is desirable for subject-matter experts to become a fixture during joint and Service training events.

Training at purpose-built or other field venues should have backing of equal quality when the event merits it. It may not be necessary to have an expert OPFOR and noncombatant actors at some lower-tactical-level training. Many such training events involve tasks (e.g., air-ground training) in which live OPFOR and others are simply not needed or can be replicated by inanimate objects (in shoot houses or during other training involving live ammunition). During higher-level collective training, however, expert participation will be essential to quality instruction. Facilities with high usage rates (e.g., those similar to Shughart-Gordon) will continue to need their own OPFOR and noncombatant role players. It may be necessary to import such experts for other sites and for events at the pinnacle of the training pyramid.

Creating such capabilities is not an overnight enterprise. OPFOR and noncombatant actors require training themselves and time to prepare for exercises. Further, OPFOR and noncombatant role-player group membership will have to change for different scenarios and theaters for which units are preparing. There is an outstanding need for formal joint training doctrine addressing how to develop and manage OPFOR and noncombatant capabilities. This doctrine should cover the size of representation appropriate for given training scenarios; how to organize, train, equip, and otherwise prepare participants; and the many other considerations required to develop these capabilities.

Similarly, joint force headquarters representation is essential at major exercises. These include the higher-echelon command post, school, and interagency exercises discussed above and field exercises with sufficiently broad and sizable participation to merit such an asset's support. A standing joint force headquarters (core element) (SJFHQ(CE)) is appropriate for this role, especially during training events involving theater headquarters of which it could become a part during actual deployment, or units that could similarly deploy with the exercise headquarters. The SJFHQ(CE) would both provide and receive benefit from training during joint urban events in which it participates. Augmentation from combatant commands or other sources may be necessary in cases where the SJFHQ(CE) lacks the manpower or full range of expertise needed to adequately support a training event.[8] It may be necessary to create a permanent traveling JTF headquarters element organized exclusively for training support if the number of exercises requiring such support exceeds field capabilities to provide it. Its funding and management could conceivably fall to JFCOM's JNTC. Regardless of the source of the headquarters support, its personnel should participate in all relevant aspects of JUT events, from initial planning through rehearsals and other preparation, execution, and after-action activities, including completion of final reports.

In short, 13 of the outstanding 34 JUT requirements can be addressed to a "run" level of readiness in the immediate term, and significant steps toward enhancing the training in support of these and other requirements is readily within grasp given efforts to address the

---

[8] For more discussion of SJFHQ(CE), see Charles W. Cosenza, "Standing Joint Force Headquarters (Core Element): Its Origin, Implementation and Prospects for the Future," *Joint Center for Lessons Learned Quarterly Bulletin*, No. 6, June 2004, pp. 3–8.

aforementioned training support elements of OPFOR and noncombatant actor development and joint force headquarters participation. Commanders with extraordinary funding or the good luck to be located near premier urban training facilities can achieve a like standard for other requirements in the near term, but insufficient resources exist for the entire force to do so. It is necessary to look at addressing those needs in the more distant future of 2008–2011.

A JUO Training Strategy for the Longer Term (2008–2011)
The JUO training strategy for the longer term builds on the elements discussed in the previous section. Maintenance of a robust schoolhouse, headquarters training events, education for those making decisions influencing the operational and strategic levels of war, and multinational/interagency exercises supported by qualified OPFOR, noncombatant, and joint force headquarters elements will be essential to U.S. readiness to conduct urban operations. Those preparatory events should include domestic as well as international contingencies. Domestic exercises provide double benefits, permitting maturation of interagency understanding while better readying the nation for homeland security operations. Interagency participation is equally essential during training to prepare for international contingencies.

Although those training in the field may not feel the impact for several years, the time to make critical decisions about urban simulations funding is now. Current simulation and modeling efforts tend to be virtually independent. Organizations conduct expensive work using suboptimal terrain databases, only to later learn that far better resources were available from another organization. Similar simulation development efforts are allowed to continue, each consuming millions of dollars annually, because of a combination of bureaucratic self-interest and the lack of an overarching, responsible agency that can objectively evaluate costs and benefits and then exercise the authority to terminate efforts despite sunk costs that might run into the tens of millions of dollars. Virtual and constructive simulations and urban modeling will dramatically influence training in future years. That influence will have greater positive impact and will likely be felt sooner if hard decisions are made as quickly as is feasible after the completion of hard-nosed analysis of which programs should survive and which should not.

Virtual and constructive training complement live training in the 2008–2011 time frame and will increasingly continue to do so. Their benefits are especially notable in instances where safety, technological development (testing not-yet-fielded systems), environmental impact, or expense makes purely live training infeasible. Yet the resolution essential to fully replicating the complete spectrum of urban challenges is many years off, ensuring that virtual and constructive training will continue to support live training to and beyond 2011 in all but exceptional cases. That does not mean that continued investment in this area is not advisable. On the contrary, further development promises considerable return on investment.

Unfortunately, it is difficult to determine how much should be spent on urban simulations. First, establishing which programs merit funding requires the hard decisions called for above. Those decisions themselves necessitate preliminary research. Second, most simulations of interest to the armed services are not exclusively urban in character. Rather, urban conditions, environments, challenges, and their interactions with other scenario elements are inseparable components of larger virtual and constructive capabilities. The same is true of the systems that link those entities to live and other virtual or constructive training, linkages that themselves have to be part of capability development and merit funding.

The 16 JUT requirements that remain outstanding—those in addition to the 13 for which "run" status is attainable in the near term—require other means to meet a "run" level of operational readiness. Based on our analysis and as shown in Appendix G, five modules have particular potential for addressing the remaining requirements at this level:

- Module 1: Battalion and larger purpose-built facility (addresses 10 of the 16)
- Module 7: Hybrid facility (addresses nine of the 16)
- Module 17: Use of abandoned domestic urban areas (addresses 11 of the 16)

- Module 18: BRAC'd installations (addresses 13 of the 16)
- Module 22: Abandoned/constructed overseas urban areas (addresses 9 of the 16)

The last of these, Module 22, is of limited value to this analysis. Abandoned or constructed overseas facilities should not be relied upon as primary urban training capabilities (aside from the occasional instance in which U.S. units are located in close proximity to such assets). They are instead complements to preparation conducted prior to deployment. They will serve as skill-refresher assets or provide practice for dealing with evolving conditions rather than being resources on which units should have to rely for first-time achievement of "run" status. A JUO training strategy should support the development of such capabilities as specific theater needs arise (whether they involve the conversion of existing built-up areas or building sites from scratch), but they will not receive further consideration here.

That reduces the number of modules requiring further consideration to the remaining four. Module 1 is very attractive from an availability perspective in that DoD will own battalion and larger purpose-built facility sites, but the considerable expense this option involves, especially when movement is necessary for use, is a serious drawback (see Figure 6.4). Module 7, battalion purpose-built hybrid facility, is competitive from the perspectives of cost and (potentially) availability, but it falls short, primarily because it addresses only nine of the outstanding 16 training requirements at the "run" level of proficiency.

The remaining two are particularly attractive in that they can achieve R status for a larger number of the 16 remaining requirements: Modules 17 (11 of the 16) and 18 (13 of 16). They are also among the more economical solutions in terms of cost per person trained (see Figure 6.4). A drawback to Module 17 (use of abandoned domestic urban areas) is that users may have to travel considerable distances to use such facilities. Further, availability is questionable unless DoD leases civilian-owned facilities on a long-term basis, a key concern during unpredictable times when surge training is essential, as was the case for preparation to support operations in Iraq and

Afghanistan throughout 2004 and into 2005. Restrictions on live fire and environmentally related issues are also likely at these facilities. Further, it has been estimated that the realistic life expectancy of these facilities is only five years, barring considerable upkeep, meaning that DoD would inherently be relying on civilian entities to find and develop such sites repeatedly, and to do so in a manner conducive to sophisticated military exercises. The risk in that regard seems significant.

Module 18, BRAC'd facilities, is immediately attractive because it addresses all but three of the outstanding requirements, more than any other option. A reconsideration of how such facilities might be managed potentially enhances this attractiveness. Original BRAC'd facility cost estimates assumed use on a lease basis, with user payments being a fairly economical $10,000 per day (average), and annual expenses being a rather low $3 million to $3.6 million.[9] However, the situation would be more akin to that of Muscatatuck if DoD were to retain ownership of one or more closed military installations, and cost estimates would share many of the characteristics of the armed forces using an abandoned civilian area, but with the added benefit of retaining complete control. This would comprise less a BRAC than transition of an installation from one set of functions to another (e.g., from housing a headquarters to support of urban training). Benefits are numerous. For example, this option would have less negative impact on the local civilian community (e.g., retention of jobs), a social/political benefit that is potentially quite significant. The residents in the vicinity of Muscatatuck are very supportive of the Indiana National Guard assuming responsibility for the facility because of the economic benefits continued use promises for the local community.

Returning to Figure 6.4, it is evident that the transition of several well-located installations would reduce the travel burden and therefore the overall costs of their use, making BRAC transition sites an attractive alternative from several perspectives. While their costs

---

[9] See Table 6.11.

are potentially higher than those of leasing, expenses would very much depend on the specific case at hand.[10] For example, increasing the density of structures at Muscatatuck largely accounts for the non-recurring costs that comprise more than half of its $103 million estimated costs for fiscal years 2005–2011. The estimates shown in Table 6.12 indicate that this "transition" or "realignment of purpose" option may well be more cost-effective than DoD leasing such capabilities from another authority. It is also notable that realigning a base for use as an urban training site could contribute to two of the eight selection criteria that DoD plans to use in choosing bases for BRAC: Criterion 3, "ability to accommodate contingency, mobilization, and future total force requirements at both the existing and potential receiving locations to support operations and training," and Criterion 6, "economic impact on existing communities in the vicinity of military installations."[11]

BRAC review procedures and other considerations would need to be revised for those facilities thought to have potential use as urban training sites if the BRAC-transition option is adopted. Reviews of potential BRAC installations should include a procedure that measures the suitability of locations for transition to JUT facility status. This evaluation would permit determination of (1) whether a site has transition potential, (2) the costs involved in developing that potential, and (3) the longer-term expenses involved in maintaining the site. Suitability factors included in any such evaluation should include measures of local residents' acceptance of the change in status, other environmental considerations, and on-hand assets of value to urban training. For example, Aberdeen Proving Ground, MD, has often been mentioned as a possible BRAC candidate. Its proximity to water and rail transportation would favor its selection for transition. Distance from potential user units is a mixed issue; there are few nearby active tactical units but many Reserve and National Guard organiza-

---

[10] See Table 6.10.

[11] David E. Lockwood, *Military Base Closures: Implementing the 2005 Round*, Congressional Research Service Report, January 4, 2005, p. 3.

tions. Proximity to Baltimore-Washington International airspace and possible site cleanup requirements are potential downside issues. These are only a very small sample of the types of factors that a full transition evaluation should include.

U.S. armed forces readiness would benefit from the DoD community's early identification of lucrative BRAC candidates and requests to hasten their transition to meet the already real needs for better Service and joint urban training. The reception for such an initiative on the part of residents in the vicinity of these sites and of politicians with interest in the area would probably be mixed. Transition to urban training status would likely entail loss of some jobs at the installation, but it would in many ways be preferable to complete closure. Regional predispositions toward military activities will also impact acceptance of such transitions. (The above arguments also pertain to DoD assumption of properties owned by other federal agencies that have been closed or are under consideration for closure.)

An added attraction of the BRAC-transition/BRAC'd facility lease option is the potential to select future and already BRAC'd facilities from locations that minimize travel times for potential user units. An offshoot of this option would be to use parts of active installations that have been abandoned or are underutilized, as was done at the recently opened urban training site at Ft. Lewis, WA.

As attractive as this alternative is, it suffers from the major shortfall that selection and development of potential BRAC-transition facilities are largely out of DoD control. Base closures—in truth, any significant alteration of a military facility's status—is an inherently political issue. Practical implications and military necessity will have only limited impact on decisions. This should not be an argument for abandoning initiatives to develop such resources, especially in cases of already-closed locations, but common sense dictates consideration of other options.

Returning to the three other modules that meet a considerable number of the outstanding joint urban training requirements— Modules 1, 17, and 7—two merit further consideration for the role of backup to BRAC facilities. Module 1—battalion purpose-built facilities—fails to make the cut because of its high cost, especially when considerable movement is necessary. Module 17—use of abandoned towns (e.g., Playas-type sites)—is very attractive from a cost perspective. Occasional exercises employing facilities such as these could be both valuable and not overly costly. However, caution would be necessary to ensure that site restrictions do not constrain military training to the extent that too few JUT requirements are achievable. Ultimately, and as previously noted, it is the likelihood of such constraints and the reliance on civilian organizations to find, develop, and maintain facilities in sufficient number and quality that argue against adopting this course of action as the primary backup to BRAC-transition installations. Such reliance would be trading political risk for a commercial-enterprise gamble, a situation unacceptable when the ultimate cost might be American lives.

The remaining option, Module 7—purpose-built hybrid facility—has a fundamental drawback in that such facilities support only nine of the outstanding 16 JUT requirements at a "run" level of readiness.[12] Table 7.1 summarizes the requirements that would still be unfulfilled if a strategy relying on the BRAC-transition or hybrid module were adopted.

Two of the shortfalls are common to both options. Only a battalion-size or greater purpose-built facility and abandoned domestic urban areas permit meeting the requisite standard for "conduct urban SIGINT, IMINT, MASINT, COMINT, ELINT, and other intelligence efforts." Such facilities routinely have the large numbers of enclosures of different types, variation in densities of building materials, heterogeneity of structural types, and differences in building heights necessary to meet this requirement. (Some BRAC and hybrid facilities would also meet these standards.) Fortunately, most organi-

---

[12] As shown in Figure 6.6, hybrid facility costs per individual trained drop below those for BRAC facilities for training events of five days or less. However, the costs in the vicinity of the five-day break point are not dramatically different. Further, given that the focus is on larger units (battalion and greater), exercises of less than five days' duration would probably be of limited value, especially when total time on site (which would include preparation, reconnaissance, other missions, AARs, recovery, and other miscellaneous activities) is taken into account.

Table 7.1
Requirements Not Met at a "Run" Level Using a Strategy That Combines Modules 18 and 7

Module	Unmet Requirements
18: BRAC'd installations	• Communicate in an urban environment • Conduct urban SIGINT, IMINT, MASINT, COMINT, ELINT, and other intelligence efforts • Conduct urban operations during and after a WMD event
7: Hybrid facility	• Conduct airspace coordination • Conduct urban SIGINT, IMINT, MASINT, COMINT, ELINT, and other intelligence efforts • Provide security during urban transition operations • Conduct urban noncombatant evacuation operations (NEOs) • Conduct U.S. domestic urban operations • Conduct urban combat search and rescue (CSAR) • Conduct urban operations during and after a WMD event

zations do not need to train to the level of sophistication that would constitute a "run" status in this regard. Intelligence organizations can gain much of what is needed at BRAC-transition and hybrid facilities (a "walk" level of proficiency). Training for many intelligence units (e.g., those at lower echelons that depend on receiving much of their SIGINT, IMINT, MASINT, or other-than-HUMINT from higher headquarters) will not require a facility capable of attaining an R level. The exceptions can use actual urban areas or deploy to a BRAC, hybrid, or other site that can meet such needs (should they exist). (The Playas site, though an abandoned domestic urban area, would not meet R levels of readiness for this requirement in its current state.)

The second common unmet requirement, "conduct urban operations during and after a WMD event," is exceptional and is met by only one of the modules developed in this study. The contamination and levels of destruction that R-level training demands for this requirement present serious problems. Replicating the many types of biological and chemical agents in a manner that meets operational conditions without raising safety or decontamination concerns is difficult enough. Reproducing the extent of destruction a nuclear strike would impose and allowing for the extensive recovery efforts that training would entail further magnify the difficulties. Fortunately, there is once again a silver lining of sorts. While such training for the entire force is desirable, only specialty units *must* practice at the levels of realism needed to attain "run" readiness. Thus, such training can take place at selected locations on a case-by-case basis without notable negative effects on the force as a whole. A final point: Both this and the "conduct urban SIGINT, IMINT, MASINT, COMINT, ELINT, and other intelligence efforts" requirement could benefit from high-quality simulations. Virtual training might be able to provide for R levels of training accomplishment were such capabilities further refined.

The remaining shortfall for BRAC-transition facilities is in attaining a "run" level for the "communicate in the urban environment" requirement. Here again, a W level of achievement is attainable in a BRAC location. "Run" would be feasible were a site to have a density of structures, sufficiently tall buildings, and enough underground facilities to put communications, GPS, Blue Force Tracker, and similar systems through their paces. Some BRAC'd facilities have such a capability, but those portions used for training thus far lack some or all of the above requirements (e.g., the portion of George AFB used for urban training has underground drainage systems, but building density is marginal, and there are no sufficiently tall structures within the maneuver area). Unfortunately, this is not a requirement that affects only specialty units; virtually all organizations need to experience the communications difficulties an urban area can pose so that they can develop solutions and workarounds. Hybrid and battalion-size and larger purpose-built facilities are the only facilities that routinely possess the size and density to meet these demands (and only in some cases). Barring the availability of either, augmenting BRAC-transition capabilities to increase structural density (as is being planned for Muscatatuck) and using parts of actual cities provide the best alternatives. The second alternative (represented in part by Module 16, "use of public facilities during hours of closure") fails to attain an R rating in Appendix G because of the difficulty of replicating tactical operations in such an environment, but demonstrations of system limitations in subways, urban canyons, and elsewhere, combined with instructions on solutions to the resultant challenges, could provide a "strong walk" level of proficiency suitable for all but the extraordinary operational challenge.

The remaining shortfalls of the hybrid-facility option would have to be met through similar innovations or uses of other training capabilities. Solutions are easier for some skills than for others (e.g., for "conduct airspace coordination" and "conduct CSAR," aircrews should be able to find active military installations or other facilities over which to practice airspace coordination, and the number of ground personnel that would have to move to support such training is small compared with entire units). This is not completely satisfactory given the desire to train entire tactical units, not just those with the special skills needed to qualify fully as air controllers, but it indicates the types of measures that will be necessary if sufficient BRAC-transition facilities are not developed or without dramatic expenditures on very large purpose-built sites. In the future, as in the past, new training requirements will challenge old approaches; innovation during instruction will ever be in demand for joint force training.

Balance will be a crucial component in attaining long-term JUT success. Ensuring that all of the Service and joint building blocks are properly prepared, regardless of echelon, means that leaders from squad level to Chairman of the Joint Chiefs of Staff and combatant commander will have to adapt training for an ever-evolving environment while constantly assessing readiness to confront existing and emerging urban operations challenges.

Key Considerations for a JUO Training Investment Strategy

In this section, we consider some of the key considerations in developing a JUO training investment strategy, including what to build; how many facilities to build; the best locations for battalion- and larger-capable BRAC, hybrid, or other types of urban training facilities; and what current capabilities should be upgraded.

What to Build

The traditional building-block approach to training requires that Service organizations have regular access to urban training facilities. There is a need for major installations (those housing a brigade or larger maneuver unit) to have urban training facilities capable of supporting at least platoon-size operations. Such installations should also have additional facilities capable of supporting basic urban skills training such as isolating a building (to deny access or egress that might allow reinforcement, resupply, or escape), room-clearing, and building-clearing to support the small-unit training that is at the base of the JUO training pyramid. These facilities can be located adjacent to an installation's platoon-size sites, but the positioning should be such that the basic skills training does not preclude simultaneous use of the larger site.

This call for urban training facilities does not constitute a proposal to build expensive CACTF-standard facilities throughout the Services. Many installations already have urban sites suitable for foundation task training such as room-clearing and building isolation. Abandoned buildings will in many cases be available to support some or all such instruction, and they may provide complexes suitable for training platoons. Conversion of the structures/areas to urban training sites should be fairly economical.

The same approach is necessary for fixed- and rotary-wing aircrew training. Many skills (e.g., target identification, approach tactics, attack angles) can be practiced adjacent to or over actual urban areas, active installations, or abandoned portions of posts or bases. Simulators support such training to a greater or lesser extent depending on the simulator itself, the software available, and the tasks for which training is sought. As in the case of ground unit training facilities, air and aviation units should have the live, virtual, and/or constructive capabilities on their installations or within a short distance so that basic urban skills training is readily achievable. While there is a need to set standards for such Service-oriented training, initiatives such as those that led to the construction of Yodaville and motivated urban training construction at Nellis AFB are notable for recognizing the need. Similar sites in the Midwest and the eastern United States are called for.

It is obvious that these smaller-unit Service facilities can also support joint training at lower echelons. Other tactical capabilities (such as shoot houses and urban convoy ranges) will also be desirable or necessary for some Service and joint commands. Funding should be provided for Service development of facilities to train on foundation skills. The expenditures needed will be a function of existing capabilities, the availability of suitable structures for conversion, missions of units at a given installation, and likely use for joint exercises.

If units are adequately prepared before they pursue training at larger-capacity sites (e.g., those capable of supporting a company team or battalion task force), the amount of training time needed at larger facilities will be less than it would be if arrivals were not already urban savvy. Leaders who combine use of platoon-capacity sites with terrain walks in actual urban areas and other innovative training methods will be better prepared yet. Still, it will ultimately be both desirable and necessary to participate in larger-scale events.

The components of the JUT investment strategy proposed here follow directly from the previous discussion. The investments will provide capabilities that serve both larger-unit Service instruction and JUT events. A JUO training strategy should include the following:

- Service air and ground facilities for base-level tactical training, such as air-to-ground engagement, room- and building-clearing, and platoon-size training. Approval for further projects should follow an investigation of onsite capabilities, including abandoned or low-use areas that could be converted into urban training facilities.
- Proposals for air and ground training facilities that include consideration of likely usage rates. Company- and larger-size sites should be located only where home-station or visiting units will take reasonable advantage of their capacity. A complex suitable for training at least a platoon at every home-station installation with the minimum of a maneuver brigade is desirable. (This constitutes a site with at least 25 buildings, the actual number

depending on the underlying terrain, vegetation, building density and character, and other characteristics.)

- Continued funding of urban-training-related modeling and simulation after a comprehensive review of ongoing initiatives and subsequent decisions about consolidation of ongoing and proposed efforts. Such management should be a continuing process rather than a one-time event. Priority for funding should go to efforts that encompass full-spectrum conditions in models and simulations (i.e., offensive, defensive, stability, and support operations; sophisticated human factors and social-infrastructure replication involving noncombatants and OPFOR; and physical terrain) that are conducive to (and compatible with) supporting live or other virtual and constructive training and that provide the means of completing the links essential for that support.

- Conduct of urban operations exercises, especially those with higher-echelon orientation and participation by interagency counterparts to DoD personnel. Exercises should emphasize the same areas noted for simulations in the previous paragraph (i.e., full-spectrum operations and social interactions in lieu of exclusively Cold War–type, combat-dominated events).

- Classroom and professional-development programs that support urban operations readiness.

- Design, training, and maintenance of expert OPFOR, noncombatant role players, and joint force headquarters entities to support joint and major Service urban operations training. This should encompass a combination of local OPFOR and noncombatant assets (for facilities that can support permanent capabilities in this regard) and traveling capabilities (likely necessary for joint headquarters participation in virtually any exercise, selectively for OPFOR and noncombatant assets).

- Analysis of BRAC'd and pending-BRAC facilities for potential transition to urban training facilities capable of hosting realistic training for a battalion task force or larger unit, as well as the provision of funds to modify, maintain, and manage those installations or parts of installations selected for this purpose. Ideally, such sites will have numbers of buildings with similarities sufficient to confuse both those navigating on the ground and those attempting to perform target identification from the air. Analyses should include installations owned by other than DoD agencies.

- Construction of hybrid facilities, either purpose-built or as expansions of existing purpose-built and abandoned installation/town properties in lieu of BRAC-transition capabilities where BRAC facilities are unavailable. The above comment about number of buildings and similarities of structures applies equally here.
- Provision of support for Service construction of shoot houses and other specialized capabilities for units requiring such niche training, e.g., SOF and USAF base security forces. As in the case of providing a platoon-size facility at each installation acting as a home station for a maneuver brigade or larger unit, use of existing facilities or alteration of extant structures may offer economical means for meeting requirements.
- Purchase, lease, and/or modification of sites for special-function urban training (e.g., WMD cleanup and complex intelligence-process preparation).
- Integration of interagency considerations in live, virtual, and constructive training.

Those responsible for providing funding for such capabilities should consider the following:

- Similar urban training resources within reasonable travel distance of the proposed location.
- Intelligence use of existing urban training facilities or portions of installations, ranges, or other built-up areas that could complement or serve as the basis for an urban facility.
- Home-station unit use of the facility, with accessibility and likely use by visiting units, including Reserve component and other-Service organizations.
- Evolutions in relevant tactics, techniques, and procedures.

- Potential for joint usage that provides realistic training for multiple Services and joint headquarters.
- Replication of the full spectrum of urban contingencies, not only those limited to or dominated by combat.
- Durability, maintainability, and adaptability to changing equipment, training resources, and field conditions.
- Level of detail (realism in replication) sufficient for quality training but not excessive (e.g., the appearance of Yodaville from the air).
- Decisions about tradeoffs such as size of facility versus extent of instrumentation, live-fire capability versus the inherent restrictions in maintaining such a capability, higher-echelon versus tactical training.
- Integration of virtual and constructive training capabilities with live training capabilities, as appropriate.
- Character and number of structures, building density, height of structures, inclusion of infrastructure (streets, curbs, utilities, services).
- Plans for OPFOR and noncombatant role-player participation in realistic numbers and after sufficient pre-event preparation.
- Integration of the facility with surrounding rural and other urban training capabilities.
- Installation transportation, billeting, and other capacities that will influence training-site throughput.

How Many Facilities?

We strongly encourage a rigorous study of how frequently larger-unit collective training is needed to maintain requisite skill levels (i.e., how often collective urban training is required by various unit types at different echelons). Pending such an undertaking, on the basis of Dutch Army training policies for dismounted and mechanized units and the rotation tempo at Ft. Polk's urban facilities, we believe it is desirable for light infantry organizations to conduct a battalion-size urban training exercise every nine to 15 months, with mechanized and armored units undergoing a rotation involving significant urban operations every 18 to 24 months.[13] The duration of a training event at the Dutch Army's Marnehuizen is 10 days (the same as a typical rotation at Ft. Polk's urban training facilities). Improved urban training at lower echelons (e.g., that possible for platoons if every major installation had a 25-building or larger training facility) could reduce the time needed at greater-capacity sites. Field observation and study should be undertaken to determine whether installations require modification once these lower-echelon units routinely train at greater frequency than is currently the case. Distance to training sites for using units, operational tempo and rotation schedules, and additional factors should also be incorporated in decisions about developing these facilities. Duration of unit deployments and policies that will influence the rate of personnel turnover in units are further points that require inclusion in any such analysis.

Determining how many units would need to train annually on urban training sites capable of hosting a battalion-size task force is a back-of-the-envelope calculation given the current high tempo of operations overseas and the historical lack of such training. (Such inexperience and the resultant lack of data make Reserve component training requirements especially hard to estimate.) Remaining with the previous assumption of seven-day rotation cycles at such sites and assuming that 120 battalions (60 heavy and 60 light) will require urban training per year, the requirement for heavy units is $(60)(7)/2 =$ 210 days/year if a conservative estimate of training once every 24 months is used. The equivalent number for light units is $(60)(7) =$

---

[13] These values are for the larger Marnehuizen facility alone. The training is in addition to the current five days spent at Oostdorp by light units for squad training. The Dutch further recommend that units with heavy equipment return to Marnehuizen for at least five days of refresher training at least once within that 24-month period (interview with Lieutenant Colonels Henk Oerlemans and Johan von Houten, Dutch Army, by Russell W. Glenn, Oostdorp and Marnehuizen, the Netherlands, December 7–8, 2004; and Ft. Polk, LA, trip notes regarding site visit and interviews by Christopher Paul, Brian Nichiporuk, and Barbara Raymond, October 26–27, 2004). Ft. Polk plans on sixteen 10-day rotations through its Shughart-Gordon facility per year.

420 days/year given a 12-month cycle.[14] Retaining the earlier assumption that such facilities would be able to support 210 days of training annually, it is apparent that there should be a sustained demand for three such facilities in the CONUS. None currently exist.

This estimate is not all-inclusive; there is also a need to provide such training during Combat Training Center (CTC) or equivalent rotations. The JRTC and NTC (or a nearby facility) should logically have urban training complexes available to commanders who wish to emphasize this aspect of preparedness. Designating these locations as two of the three for which there is sufficient demand means that most units would experience such instruction only during CTC rotations, an undesirable state of affairs. We therefore suggest that a total of four facilities be developed, two to support CTC-type rotations (both supporting U.S. Army and USMC as well as joint events) and one each to allow for the conduct of joint and Service events without interfering with or relying on CTC rotation schedules. The latter two would be located in the eastern and south-central United States for the reasons explained below. We also recommend that USMC units be incorporated in CTC rotations, just as U.S. Army units currently are.

Where Is It Best to Locate Battalion- and Larger-Capable BRAC, Hybrid, or Other Types of Urban Training Facilities?

To a considerable extent, the question of "where" is a function of unit locations: It is desirable to minimize travel distances for training, but relying on this variable alone is overly simplistic.

The organizations that need such large facilities include maneuver units of battalion and larger size and their supporting elements for ground forces (U.S. Army and USMC) and any aviation unit that is

---

[14] These values should be considered no more than rough approximations used to gauge the magnitude of demand. In addition to the influence of operational rotations and the difficulty of estimating Reserve component demands, the influence of increased urban training for staffs, students at Service and joint schools, more platoon- and company-level training, and simulation enhancements make any estimates general approximations. We are calling for a study on retention of urban skills after individual, and especially collective, training, since such information is currently all but nonexistent.

likely to conduct operations over built-up areas. Ground maneuver units of this size are by and large concentrated in divisions and separate brigades.[15] Aviation units include helicopter-based organizations, most of which are associated and/or collocated with large ground maneuver elements (with the exception of some U.S. Army corps aviation units). Fixed-wing installations hosting wings provide a good basis for gauging concentrations of Service aviation units.

Maps showing the locations of military installations nationwide indicate that the greatest density of Service bases is in a band that runs from the northeastern United States near Boston and swings in an arc into the southeast and across the southern states into California.[16] Alaska also has several installations, and Hawaii (notably Oahu) is especially dense with them. There are also individual nodes of particular force concentrations (e.g., Ft. Lewis, WA). However, focusing on the greatest concentrations of candidate organizations (separate brigades, divisions, and aviation wings) and recalling that the concern is not only higher-level Service but also joint urban training concentrates attention on that swath across the CONUS.[17] Under more

---

[15] Unfortunately, this assumption is not without its drawbacks. National Guard units can have subordinate units dispersed over a large geographical area. Some active divisions have affiliations with Guard brigades that can be in completely different states. However, the level of resolution needed to make initial decisions regarding the location of urban training facilities is such that the ultimate choices should not be dramatically influenced using the procedure that follows.

[16] Return of (primarily U.S. Army) units to the CONUS from Korea, Germany, and elsewhere could significantly impact the number and types of units at military installations nationwide. Army restructuring from legacy to unit-of-action organizations will also influence post capacities and character. However, existing active-duty installations with the capacity to accept additional units also by and large fall within this band, and some currently have excess capacity (e.g., Ft. Bliss, TX). Decisions regarding where to build urban training capabilities should unquestionably include consideration of any such realignments. The recommendations in this study should continue to be valid given the information available in this regard at the time of our writing. For a discussion of potential unit movements, see Sean D. Naylor, "The Coming Brigade Shuffle: How Adding New Combat Units Will Radically Alter the Army's U.S. Footprint," *Army Times*, January 31, 2005.

[17] This should by no means rule out seizing the opportunity to realign an installation in these non-CONUS states, but distances, water gaps, and/or significant environmental issues may mitigate against capitalizing on this option even if it presents itself. Further, the total number of units and personnel that would be able to regularly take advantage of such a site would be less than that in many CONUS locations.

careful scrutiny, the concentration in the northeast is light. Both USMC divisions, all but one CONUS-based U.S. Army division and separate brigade, and most USAF organizations of concern are south of the Mason-Dixon line in the east. (Ft. Drum, NY, is the U.S. Army exception.) Seventy-four percent of the USAF's airframes are home-based in the south, and ranges there account for 80 percent of its available CONUS training hours.[18] Including pertinent Reserve and National Guard units does not preclude a conclusion that shortening the arc to start in North Carolina rather than New England is appropriate.[19]

Looking yet more closely at CONUS-based maneuver and larger aviation unit basing, we find that three regions within the arc are notable:

- Kentucky (home to the 101st Airborne Division (Air Assault)), North Carolina (82nd Airborne Division and 2nd Marine Division), and Georgia (3rd Infantry Division) encompass roughly half of the USMC's and U.S. Army's CONUS-based division strength.[20]
- Texas' Ft. Hood is home to both the 1st Cavalry Division and the 4th Infantry Division. Ft. Polk, LA, is home to a regimental/brigade-size unit as well as its urban complexes. Ft. Riley and

---

[18] "A/G Ranges and Units," Department of the Air Force briefing, undated. This briefing divides the United States into three focus regions. The percentages cited here pertain to the combined totals for the southeastern region (Kansas, Missouri, Kentucky, and Virginia on its northern boundary; Kansas, Oklahoma, and Texas on its western) and the southwestern region (west of the southeastern region, with its northernmost states being Colorado, Utah, Nevada, and California).

[19] Basic training installations are not a notable concern here. Urban training at that level should retain an individual and small-unit center of mass. Use of single structures or small complexes of buildings is sufficient for attaining the requisite level of training needed in most instances. Addressing those skill areas and tasks requiring larger areas (e.g., base security operations) via use of these and still-active parts of installations will meet the vast majority of requirements for such training.

[20] Of the U.S. Army's 10 active force divisions, three are internationally based (two in Germany, one in Korea) and one is located in Hawaii, leaving six in the CONUS.

Ft. Carson have the better part of a division equivalent between them.

- The USMC 1st Marine Division is based in Southern California. USN, USAF, and USMC flight training and exercise capabilities are especially focused on the southwestern U.S. hub of Nevada, Arizona, and California. The Army's National Training Center is also in this region.[21]

The same three areas contain the bulk of the most promising existing urban training sites: Ft. Knox's Zussman site (KY), Yodaville (AZ), the McKenna MOUT complex (Ft. Benning, GA), Camp Pendleton, California's MOUT facility, and the planned USMC mega-MOUT facility at Twentynine Palms, CA. (Similar sites at Ft. Bragg, NC, and Camp Lejeune, NC, are also within the easternmost of the three unit-concentration areas.)[22] The JRTC system of urban training capabilities that includes Shughart-Gordon falls "between stools" in a positive sense, being a part of the Texas and central Midwest installations, while not overly distant from the Kentucky–North Carolina–Georgia node.

There is another point of importance in considering where best to develop the nation's largest and most capable urban training facilities. U.S. Army units periodically rotate through major exercises at CTCs. Those in Europe deploy to the Combat Maneuver Training Center (CMTC) at Hohenfels, Germany.[23] Mechanized and armored

---

[21] For a more comprehensive discussion of U.S. military installation locations, see *2005 Guide to Military Installations Worldwide*, Springfield, VA: Military Times Media Group, November 2004.

[22] The Muscatatuck site is attractive in part because it *does not* fall within the three areas. It is centrally located, from a northern Midwest perspective. National Guard units in the region will therefore potentially have a facility for training that is not in great demand by large active force units yet is still reasonably accessible to SOF forces of all components.

[23] Units in Germany will be challenged to achieve the same level of urban proficiency as those in the United States that can train on facilities capable of handling a battalion or larger. CMTC is already very constrained for space, and chances of securing space for developing a separate urban facility are virtually nil. There is discussion of building an urban training capability in Korea. Information on this topic was limited at the time of writing, as the concept is apparently in its early stages.

units go to the NTC at Ft. Irwin, CA, and more lightly equipped maneuver organizations use the JRTC. USMC units go to equivalent-level training at Twentynine Palms, CA. As noted, combat fixed-wing pilots train next door to the NTC outside of Las Vegas, NV. USMC rotary-wing pilots use Yodaville, in Yuma, AZ. Ground force units from the 1st Marine Division have also been routinely training at the former George AFB in Victorville, CA, to hone urban proficiencies. This argues for having premier urban training facilities at or near these locations, since (1) many major maneuver units based in the CONUS already deploy to them periodically for exercises, and (2) the southwestern U.S. hub readily provides opportunities to integrate urban aspects into all four Services' tertiary training events and joint-sponsored events conducted in the area.

While any battalion and larger urban training capability is likely to be Service-owned and operated, use for major joint-sponsored urban operations exercises and by other-Service component head-quarters should be a condition of funding. Reserve component organizations should be allocated usage time when unit-training and pending-deployment status supports such a decision.

The choice of locations for battalion- and larger-size facility de-velopment seems fairly straightforward given the above discussion. Air and ground unit home stations favor putting a large facility in or near each of the three major areas of concentration (in addition to those at the two CONUS-based CTCs). That each should ultimately include capabilities to train the full range of fixed- and rotary-wing aviation as well as ground components is apparent. This would argue for a total of five such facilities. As will be explained below, however, having two facilities in the southwestern United States (one to sup-port CTC training and one to support units based in the vicinity) would likely not result in sufficient use to justify the maintenance of both. Four such training sites should meet Service and joint training demands:

- One in the Kentucky–North Carolina–Georgia region
- One at Ft. Hood, TX
- One at the JRTC, Ft. Polk, LA

- One in the southwestern United States in the vicinity of Twentynine Palms and the NTC.

First priority should be given to the location that has the quickest payoff in terms of timely readiness to support this larger-scope joint and Service training: the JRTC at Ft. Polk. Already the premier ground maneuver urban training site in the country, the JRTC demonstrates insightful use of hybrid construction and is home to a training staff dedicated to including joint elements in training events. That the early 2005 JNTC urban training event took place there reflects this collocation of capabilities. Facility expansion would require preliminary evaluation that goes beyond what we have conducted here. The feasibility of expanding the space dedicated to urban instruction is one key question; the JRTC conducts much training that does not specifically involve urban scenarios. Further, plans for stationing a 10th Infantry Division (Mountain) brigade at the installation and Ft. Polk's role as a CTC mean that such a facility could experience tremendous demand. High usage is desirable, but neither JRTC rotations nor the quality of training for its home-station units should be allowed to suffer undue degradation. Upgrading one of the urban training capabilities at Ft. Polk (other than Shughart-Gordon) to provide a venue for platoon and smaller-unit instruction could help to address this situation.

Second in priority for large urban site development is the southwestern U.S. area of unit concentration. Unless a near-immediate opportunity to secure an appropriate BRAC facility occurs, continued use and upgrade of the former George AFB and development of a hybrid facility in the Twentynine Palms–Ft. Irwin–Nellis AFB area appear to be the logical choices. Which of these makes the most sense will be a function of unit accessibility, supporting infrastructure, and other factors. There are strong arguments favoring continued use and upgrade (or possible repurchase of portions) of George AFB. It is readily available, and the local population has been supportive of USMC use (though the extent to which that support would be affected by significantly increased use is unknown). It is conveniently located between prospective users at Camp Pendleton, Twentynine Palms, and the NTC. It would be necessary to expand the maneuver area to increase the number of structures, to expand their variety, and to provide the opportunity for combined arms training. (The facility is currently adequate only for training by light infantry and its supporting elements.)

There are also strong arguments for developing the capability at either the NTC or Twentynine Palms. The NTC, like the JRTC, hosts much of the U.S. Army's CONUS-based active and Reserve component capability every year. It does not require significantly more travel for local USMC unit home stations. Alternatively, well-developed plans for the Twentynine Palms mega-MOUT complex provide a foundation from which to adapt if reducing costs is thought necessary. Twentynine Palms has the added advantage of a home-station unit to make use of the facility (the 7th Marine Regiment); the installation is also well positioned to host marine units from Camp Pendleton or army organizations from Ft. Irwin. It is important to note that both the NTC and Twentynine Palms are heavily used due to unit training rotations every year.[24] This training encompasses a wide spectrum of FoF and live fire on a variety of terrain types. Urban training would constitute only a portion of selected rotations. It therefore seems that with careful scheduling and inter-Service cooperation, a single large site could serve (1) local USMC unit training, (2) USMC CTC-type rotations, and (3) U.S. Army organizations undergoing CTC rotations.[25]

---

[24] For example, Twentynine Palms hosts 10 MAGTF rotations (eight active and two Reserve) annually, each consisting of approximately 22 training days. This 220-day average usage for the facility is in keeping with our planning figure of 210 days. However, urban training at March AFB accounts for only eight of the days, or an average of 80 days annually (though other urban training on smaller sites does or will occur in preparation for the ultimate urban training event). March AFB is no longer available to the USMC, as of August 2005. This is a significant motivation for the USMC to develop a quality site at Twentynine Palms.

[25] Assuming continued interest in building urban operations proficiency and the possibility that the percentage of training time spent on it during CTC-type rotations will increase, it could be advisable to build a fifth battalion- or greater-size urban training facility in the southwest U.S. region. A decision about this possibility should at a minimum include consideration of (1) USMC and U.S. Army usage and demand of existing facilities in the region,

However, whether such a site should be along the lines of the USMC's proposed complex of some 900 fully capable and 600 less-capable buildings should receive further scrutiny. While very high-quality training would be provided by this robust capability, the cost of such a site seems prohibitive despite the USMC's desire to conduct urban training at the MAGTF (brigade-plus) level.[26] While we applaud the desire to train a full MAGTF in a combined arms environment, the marginal value of training more than a battalion or MEU-size element simultaneously and the frequency of opportunities to truly do so require further consideration. Should it be determined that construction of a battalion-size or larger urban training site similar to that in the current Twentynine Palms plans be envisioned, it is likely that costs can be reduced through a combination of one or more of the following actions:

- A reduction in the number of structures from the current 900 "full-up" interior/exterior training buildings and 600 other structures.
- Use of more-economical approaches to structure construction (e.g., use of corrugated metal "walls" for some building interiors).
- Import of unused buildings from other locations (if economically feasible).

These and other ways of reducing costs merit consideration given the high cost per individual trained for the Twentynine Palms facility as it is currently planned. Those costs could decrease, and greater expenditures could be thought better justified were the USMC and U.S. Army able to share use of the resultant facility for

---

and (2) any increased understanding regarding the desirable frequency of collective urban training.

[26] We were informed on the eve of publishing this report that the likely cost of such a capability is now estimated at $450,000 to $500,000 rather than the less-than $350,000 originally expected and used in the calculations in Chapter Six (interview with Maj Gen Thomas S. Jones, Commanding General , Training and Education Command, by Russell W. Glenn, Boswell, PA, February 21, 2005).

CTC-type rotations and for local unit use. Such a large facility would also be valuable for supporting major joint urban training events.

It is worth including the possibility of expanding Camp Pendleton's urban training facility as a dark-horse candidate. The currently rather bare-bones capability is attractive in part because of its valley location. Expansion would demand putting structures on hillsides, adding a level of underlying terrain three-dimensionality rarely found on the seemingly inevitably flat MOUT training sites. The hillsides are potentially locations for notional slum communities akin to those found in many developing-nation urban areas. That there is a sizable home-station population ready to make use of such a site is a further benefit. The effect of Southern California's traffic volume on the ability to move units in and out of Camp Pendleton is a potential concern; rail could be a much-preferred option for the final leg of any training deployment journey. Further, the Camp Pendleton asset suffers the same fatal shortcoming as does the former George AFB site: Combined arms training is at present infeasible. Any upgrade would have to redress this situation.

The final potential southwest U.S. location, the NTC, currently has no plans at the same stage of development as the Twentynine Palms plans. Were one available, an existing abandoned or under-utilized building complex could serve as a possible starting point for construction of a facility at Ft. Irwin. Plans to build from scratch would merit scrutiny similar to that required for a capability at Twentynine Palms.

Two additional battalion-size or larger urban training sites are recommended, one each for the Kentucky–North Carolina–Georgia concentration of units and another in the south central United States. Ft. Hood, TX, is the obvious choice for the latter. It is home station for both the 1st Cavalry Division and the 4th Infantry Division, as well as for numerous other corps air and ground assets. Texas also provides a home for a number of USAF units. Absent the availability of an abandoned or underutilized building complex on the installation or a fortuitous nearby BRAC event, we recommend constructing a hybrid facility of sufficient capacity.

The situation is somewhat more complicated in the eastern U.S. region of interest. That region seems to have no existing facility that is immediately attractive for expansion. Here, as in the other three areas (and nationwide, for that matter), joint and Service trainers should be constantly on the lookout for BRAC conversion opportunities.[27] If no appropriate BRAC facility can be found in the area, the somewhat central locations of Ft. Bragg/Pope AFB, NC, Ft. Knox, KY, and Camp Lejeune, NC, make them attractive candidates. Ft. Bragg has a site from which to expand, though more research is needed to determine the feasibility of that option. Use of this North Carolina post (or those at Camp Lejeune or Ft. Stewart, GA) is additionally attractive from the perspective of a home-station unit that could make frequent use of a nearby capability. Pope Air Force Base's size and proximity to army and marine divisions means that air accessibility is excellent; it might also lend itself as a possible location for an urban training site if expanding the one at Ft. Bragg is found impractical. Since all urban training sites should support fixed- and rotary-wing aircraft, local airspace restrictions and the impact of potentially increased air traffic on community relations are vital factors to be studied. Here, too, Pope AFB could offer definite advantages.

The Ft. Knox site, while possessing an excellent base facility from which to expand, suffers the disadvantage of not being a home station of a regular user. The Camp Lejeune area has several of the

---

[27] Recall that the term *BRAC* as used here includes other-than- DoD installations. As with the case of the Indiana National Guard recognizing the value offered by Muscatatuck, DoD trainers should be ever vigilant in watching for prospective candidates for JUT facilities. For further details on procedures and guidance regarding the 2005 BRAC, see "Memorandum for Infrastructure Executive Council Members Infrastructure Steering Groups Members Joint Cross-Service Group Chairman," Subject: 2005 Base Closure and Realignment Selection Criteria, Washington, DC: The Under Secretary of Defense Acquisition, Technology and Logistics, January 4, 2005; Memorandum, Subject: Internal Control Plan for Management of the Department of the Navy 2005 Base Realignment and Closure (BRAC) Process Policy Advisory Two, Washington, DC: Department of the Navy, June 27, 2003; "BRAC 2005," briefing, Washington, DC: Headquarters U.S. Air Force SAF/IEB, dated as of 18 June 2003; Memorandum, Subject: Transformation Through Base Realignment and Closure, Washington, DC: The Secretary of Defense, November 15, 2002; and Memorandum, Subject: Transformation Through Base Realignment and Closure, Washington, DC: Secretary of the Army, December 12, 2002.

advantages noted for the Ft. Bragg/Pope AFB complex. Although access is somewhat less convenient, it nonetheless should be considered as a prospective candidate site. Other locations, such as Ft. Jackson, SC, are less attractive because they do not serve as home stations for large tactical units.

Assessing the Upgrade Candidates: Creating Facilities Capable of Supporting Major JUO Training Events

Tables 7.2 through 7.7 list the facilities we believe have the greatest potential for adaptation to provide a facility capable of supporting Level 4 JUO training.[28] The six tables compare capabilities across facility, accessibility, forces supported, infrastructure, architecture, and scenarios supported, respectively. The features listed in the first column are those we believe are vital to achieving the desired level of training. They need not be met through live training (e.g., the factor is satisfactorily addressed if virtual or constructive real-time links with a facility provide adequate realism for airframe or indirect fire support of units in the urban complex). As noted in the previous section, Shughart-Gordon is the most developed site with respect to supporting Level 4 JUO training; the planned Twentynine Palms facility addresses many relevant aspects as well. Sites in the Kentucky–North Carolina–Georgia region would need to be greatly enhanced to become capable of such a level of support, as would those elsewhere in the United States.

---

[28] Level 4 is the ultimate level of JUO training, defined previously as:

- Urban training events in which two or more separate Service components orchestrate their activities to achieve common objectives at one or more levels of war (strategic, operational, or tactical).
- Events that might involve greater reliance on one or more Services than others during some phases; virtually continuous interoperability of multiple Services' capabilities in multiple primary environments (i.e., land, air, sea, or space) within and between phases dominates training and is essential to accomplishing assigned missions or attaining specified objectives in the training scenario.
- Events where significant interoperability occurs horizontally between command echelons across Service boundaries and vertically between Services and joint echelons.
- A joint headquarters, joint effects cell (JEC), or other joint synchronization element orchestrates joint organization and inter-Service capabilities.

Table 7.2
Facility Features That Would Enable Level 4 JUO Training at Selected U.S. Facilities

Feature	Selected Facilities[a]									
	S-G	Y	P M	M	29 M	FL	Z M	M M	FB M	CL M
25 buildings (platoon size)	X		X	X	X	X	X	X	X	X
75 buildings (company size)				X	X	X				
300 buildings (battalion or larger size)				—	X					
"Sprawl" on arteries to central complex	X				X				?	?
AAR shelter	X		X		X		X	X	?	?
Resident OPFOR, size equal to formation to be trained	X								?	?
Permanent O/C manning	X	X							?	?
Noncombatant role players, larger-than-formation training	X			—					?	?
Connectivity instrumentation (DIS)	X				X			X	?	?
Effects simulators	X						X	X	?	?
Threat emitters									?	?
Indiv. engagement replication of at least MILES quality (Simunition-type preferred)	X		X	X	X				X	X
Arial range (either inert drop or NDS)		X			X				?	?
Role-player trappings (animals, phones and radios, trade goods)	X								?	?
Costume and sign shop	X								?	?
Vehicles (ambulance, school bus, police, utility, various civilian)	X			X			X		?	?

[a]S-G = Shughart-Gordon; Y = Yodaville; P M = Pendleton MOUT; M = Muscatatuck; 29 M = Planned Twentynine Palms MegaMOUT; FL = Ft. Lewis, WA; Z M = Zussman MOUT, Ft. Knox, KY; M M = McKenna MOUT, Ft. Benning, GA; FB M = Ft. Bragg, NC MOUT; CL M = Camp Lejeune MOUT.

The characteristics shown in Tables 7.2 through 7.7 are representative of those a facility should have to support high-resolution live urban training, whether Service- or joint-centric. Characteristics marked with an X exist at the facility noted. An X in a size row (e.g., 25 buildings) means that the facility meets or exceeds the size requirements for the stated echelon (25 buildings = platoon, 75 = company team, 300 = battalion task force). A dash implies that the site approaches but falls short of meeting the need. A question mark indicates that the information was unavailable to us at the time of writing. Blank entries indicate a lack of the capability.

Table 7.3
Accessibility Features That Would Enable Level 4 JUO Training at Selected U.S. Facilities

Feature	Selected Facilities[a]									
	S-G	Y	P M	M	29 M	FL	Z M	M M	FB M	CL M
Highway	X		X	X	X	X	X	X	?	?
Rail	X		X		?				?	?
Airport	X		X		?				?	?

Table 7.4
Forces-Supported Features That Would Enable Level 4 JUO Training at Selected U.S. Facilities

Feature	Selected Facilities[a]									
	S-G	Y	P M	M	29 M	FL	Z M	M M	FB M	CL M
Combined arms	X		X	X	X		X		?	?
Rotary-wing	X	X	X	X	X				?	?
Fixed-wing	X	X	X	X	X				?	?
Amphibious	X		X	X					?	?

[a]S-G = Shughart-Gordon; Y = Yodaville; P M = Pendleton MOUT; M = Muscatatuck; 29 M = Planned Twentynine Palms MegaMOUT; FL = Ft. Lewis, WA; Z M = Zussman MOUT, Ft. Knox, KY; M M = McKenna MOUT, Ft. Benning, GA; FB M = Ft. Bragg, NC MOUT; CL M = Camp Lejeune MOUT.

Table 7.5
Infrastructure Features That Would Enable Level 4 JUO Training at Selected U.S. Facilities

	Selected Facilities[a]									
Feature	S-G	Y	P M	M	29 M	FL	Z M	M M	FB M	CL M
Limited/no "no-go" areas	X		X	X	X				?	?
Clutter/debris/filth	—				X				?	?
Water/plumbing				X					?	?
Electricity	X		X	X					?	?
Lights	X			X	X				?	?
Telephone				X					?	?
Furniture			X	X	X				?	?

Table 7.6
Architectural Features That Would Enable Level 4 JUO Training at Selected U.S. Facilities

	Selected Facilities[a]									
Feature	S-G	Y	P M	M	29 M	FL	Z M	M M	FB M	CL M
Changeable features (moveable walls and/or buildings)					X				?	?
Heterogeneity (slums, shanty town, walled compounds)					X				?	?
Similarity (buildings, areas of sufficient likeness to confuse ground/airborne navigating/ targeting)									?	?
Real urban contents (government, hospital, prison, asylum, religious structures, HAZMAT)	—			X	X				?	?
Urban canyon		X			X					
Subterranean complex throughout; basements	X		X	X	X				?	?
Adequate density of structures; varying density	X	X	X	—	X			X	?	?

[a]S-G = Shughart-Gordon; Y = Yodaville; P M = Pendleton MOUT; M = Muscatatuck; 29 M = Planned Twentynine Palms MegaMOUT; FL = Ft. Lewis, WA; Z M = Zussman MOUT, Ft. Knox, KY; M M = McKenna MOUT, Ft. Benning, GA; FB M = Ft. Bragg, NC MOUT; CL M = Camp Lejeune MOUT.

Table 7.7
Scenarios Supported That Would Enable Level 4 JUO Training at Selected U.S. Facilities

Feature	S-G	Y	P M	M	29 M	FL	Z M	M M	FB M	CL M
					Selected Facilities[a]					
Force on force	X		X	X	X				?	?
Convoy					X				?	?
VCP			X	X	X				?	?
Demonstration/riot	X		X	X	X				?	?
WMD					—				?	?
HA	X		X	X	X				?	?
NEO	X		X	X	X				?	?

[a] S-G = Shughart-Gordon; Y = Yodaville; P M = Pendleton MOUT; M = Muscatatuck; 29 M = Planned 29 Palms MegaMOUT; FL = Fort Lewis, WA; Z M = Zussman MOUT, Fort Knox, KY; M M = McKenna MOUT, Fort Benning, GA; FB M = Fort Bragg, NC MOUT; CL M = Camp Lejeune MOUT.

Major JUT events will generally be focused more on the operational level of war and joint headquarters participation than on tactical unit activities. While the latter will offer (and receive) valuable complementary training to joint and upper-echelon Service participants, most of the training requirements for individuals and organizations at higher echelons can be met with minimal, if any, live participation by tactical units at urban training facilities. Major JUO training events are therefore likely to only rarely involve participation by large numbers of tactical units.

The factors listed in Tables 7.2 through 7.7 are not comprehensive. They broadly represent critical elements that merit highlighting as representative of those needed for any facility to be capable of supporting battalion- and larger-size training at Level 4. Full evaluation of each location that takes into consideration the 34 identified JUT requirements and supporting modules is needed to more fully identify all shortfalls. Monitoring-type instrumentation and live fire are not generally considered necessary for the support of joint training in this regard but may be desirable from a Service perspective. (See Chapter Eight for further discussion of these two areas.)

Challenges for JUO Training Strategy Implementation

In this section, we address three challenges that will impact implementation of a JUO training strategy.

Whether to Build, Adapt, Rent, or Otherwise Acquire Training Capabilities

Of the three issues raised in Chapter Six, this challenge relates to the first, "Joint Training: A Separate Entity or an Augmentation of Service Preparation?" The other two were "Build, Adapt, Rent, or Otherwise Acquire Training Capabilities?" and "Virtual and Constructive Training: Alternatives or Supplements?" All but a few of the U.S. live urban training sites and simulations are Service-owned, operated, and controlled. The creation of a JNTC to provide oversight and design for joint training across functions and environments is potentially a significant step toward enhancing joint training effectiveness. Thus far, effectiveness relies on Service capabilities for conducting exercises (with notable exceptions such as the JSAF simulation).

This is by no means inherently negative. First, the bulk of training at tactical echelons is Service training. Therefore, it makes sense that urban facilities and simulations be designed to meet Service requirements at this level. Second, much of the physical and bureaucratic infrastructure has long been in place; construction, maintenance, and staff development are in many cases largely if not completely paid for. Third, facilities are at least to some extent where they are needed, albeit of insufficient size.[29] Many are on installations that house their primary users, though demands by such units can conflict with other needs. There are fears, for example, that the displacement of the 4th Brigade, 10th Mountain Division, to Ft. Polk will overtask

---

[29] Though by no means always. An Office of the Secretary of Defense training report notes that the distance between USAF ranges and installations "exceeded the established flying limitation—19 percent of the total air-to-ground training requirements for fighter jets" ("Military Training: DoD Report on Training Ranges Does Not Fully Address Congressional Reporting Requirements," General Accounting Office Report to Congressional Committees, GAO-04-608, June 2004, p. 15).

the installation urban facilities that are already operating at a high tempo in support of units readying for action overseas.

There are a number of drawbacks to Service ownership and control as well. While tactical-level training may be served by existing facilities, the situation is quite different at the operational and strategic levels of war, and it is here that the joint urban readiness deficiencies are the greatest. Purpose-built and other facilities designed to support tactical preparation are by and large of limited value in this regard. As noted, classroom instruction, seminars, war games, and exercises of other sorts are the general order of the day when the trainees are members of JTF or higher-echelon headquarters and those with whom they will work. High cost and the recognition that lower-echelon units are often underutilized in larger exercises in part precipitated late-1980s reductions in such manpower-intensive events as REFORGER. Further, the level of resolution needed to exercise headquarters personnel can now frequently be met, at least in part, by using computer simulations. There is an occasional demand for tactical joint urban training as well, and Service-specific designs, scheduling priorities, and billing methods are among the factors that can degrade or preclude joint exercises. This is natural and expected; Service training priorities differ from those of joint force commanders. Still, the state of affairs is hardly conducive to meeting JUO training requirements.

This does not necessarily argue for facilities built exclusively for or run by joint headquarters. The fact that live joint field training is generally either functional in character (e.g., air-ground targeting) or at the pinnacle of the training building-block process means that it is logical for joint training to rely on Service capabilities. Joint events can be conducted either during periods reserved for use by joint entities or as part of Service-run exercises. The bottom line: There appear to be few convincing arguments for joint ownership and management of large-scale urban training facilities, whether those capabilities are already in existence or are part of plans for the future.

Joint Range Coordination

Joint and inter-Service uses of Service training capabilities are currently negotiated either at periodic joint range conferences or on a case-by-case basis between the parties interested in conducting training. This is true whether the facility to be used is under the control of an active component or Reserve or National Guard organizations. DoD guidance advises that components "to the maximum extent possible, shall share training resources, ranges, maneuver areas, and other facilities and devices that have training or test potential."[30] Further, "the cost, scheduling, and priority of such dual use shall be in accordance with the policies and directives in effect that govern their management and oversight."[31] The cooperative meetings have proven fairly effective, but the result is that Service needs generally take precedence on ranges owned by the Services.[32]

The issue is complicated by differences in Service and range regulations. The latter will always be problematic to some extent; specific locations inevitably have local conditions that impact aviation operations, fires, or other aspects of facility usage. However, some variations in allowable procedures are attributable primarily to Service perspectives or other factors. Service guidance regarding aircraft overflight of forces on the ground during training is an example cited during several interviews we conducted during this study. Other examples were similarly noted and likewise were sources of frustration. Varied Service approaches to charging for range usage were notable in this regard. Some of these issues will be in part addressed via the

---

[30] Department of Defense Directive Number 1322.18, Subject: Military Training, September 3, 2004, p. 4.

[31] Ibid.

[32] There have been notable exceptions, primarily those involving U.S. Army and SOFs conducting training together at installations such as Zussman at Ft. Knox or the facilities at Ft. Polk. Similarly, U.S. Army–USAF urban training in Louisiana and that conducted by units at the Ft. Benning McKenna MOUT site with links to Hurlburt Field, FL, demonstrate that effective training involving the sharing of facilities can benefit multiple Services simultaneously. The Indiana Army and Air Force National Guard was coordinating with active-duty SOF elements for use of the Muscatatuck urban training site at the time of writing, but the extent to which this would incorporate Guard and active forces is unknown.

establishment of JUT standards; many, perhaps most, will not. Understanding that some differences should remain, safety and optimal use of training time and other resources would be well served by reducing unnecessary variations.

Assumption of responsibility for assigning training and range priorities to a central authority could result in increased efficiency, greater variety in training challenges, and better joint urban training. Such responsibility and authority logically ought to be vested in a joint entity, probably the same entity that has overarching responsibility for joint training of the nation's armed forces: the Joint Forces Command. There appears to be no legal barrier to such an assignment. Title 10, sometimes used as a foil to joint headquarters' assumption of responsibilities that traditionally lie with the Services, seems to pose no such obstacles. DoD Directive 1322.18 tasks the Under Secretary of Defense for Personnel and Readiness with overseeing and providing policy for military personnel training programs, the collective training of military units and staffs, and programs of instruction for all individuals, units, staffs, and organizations in the tasks that support the operational needs of the combatant commanders.[33]

The benefits of centralized range coordination have already been recognized by the Services themselves. The Army's Deputy Chief of Staff for Training "is responsible for establishing range priorities and requirements and managing . . . range modernization and maintenance."[34] The USAF Director of Operations and Training, Ranges and Airspace, "acts as the executive agent for range management for the Air Force." Further, the Air Force "has an integrated approach to range management, to include range planning, operations, construction, and maintenance."[35] The USMC has similarly "established an

---

[33] Department of Defense Directive Number 1322.18, Subject: Military Training, September 3, 2004, p. 4.

[34] "Military Training: DOD Report on Training Ranges Does Not Fully Address Congressional Reporting Requirements" Government Accounting Office Report GAO-04-608, June 2004, pp. 7–8.

[35] Ibid., p. 9.

executive agent for range and training area management to implement its vision for mission capable ranges. . . . In addition to its own ranges, the Marine Corps engages in extensive cross-service utilization by depending on extensive and extended access to non–Marine Corps training ranges."[36] The USN has gone further yet, centralizing "its range management functions, to include training and testing ranges, target development and procurement, and test and evaluation facilities, into the Navy Range Office, Navy Ranges and Fleet Training Branch," which provides "a single voice for range policy and management oversight, and provide[s] a single source sponsor."[37] Perhaps most encouraging are inter-Service and joint efforts to centralize range coordination, particularly recent initiatives such as the June 2003 "Memorandum of Agreement on Overarching Range Cooperative Guidelines Between the Department of the Air Force and the Department of Navy" and "Cross Service Range Use Standardization Working Group Aviation Range Safety Sub-Group Charter."[38] Both lend recognition to a need for greater cooperation and commonality in procedures to enhance Service and joint training.

Creation of a single joint range coordination and management organization should result in significant efficiencies. Joint and Service exercises could be allocated to support preparation for more-sophisticated joint training and Service needs in a manner that provides all participants the time required to meet event standards. A uniform means of charging for range time could be developed across the force to simplify use by joint or other-Service organizations. Ownership would remain with the Services, but initial scheduling, range standards, safety regulations, cost accounting, and other standardization—responsibilities in keeping with the directive issued by

---

[36] Ibid., pp. 8–9.

[37] Ibid., p. 8.

[38] "Memorandum of Agreement on Overarching Range Cooperative Guidelines Between the Department of the Air Force and the Department of Navy," signed by USAF General John P. Jumper (June 16, 2003), Admiral V. E. Clark (June 22, 2003), and USMC General M. W. Hagee (June 3, 2003); and "Cross Service Range Use Standardization Working Group Aviation Range Safety Sub-Group Charter," undated draft, Office of the Secretary of Defense, Personnel and Readiness.

the Chairman of the Joint Chiefs of Staff to "formulate joint training policies and coordinate the military training of the members of the armed forces"—would be the responsibility of the joint entity.[39]

This joint oversight would facilitate the assignment of range time to units scheduled to deploy to theaters during the same time period, e.g., SOF and regular force units that otherwise might have difficulty training together before operations overseas. Such training could serve as a rehearsal for actual operations.[40] A fuller investigation of this possibility should also include consideration of better coordinating range, proving ground, and other experimental test-facility usage. Training at Eglin AFB, FL, for example, frequently is a second priority to testing. Closer coordination with a joint range coordination capability at such installations could provide additional training venues now overlooked due to the inability to synchronize experimentation and training events.

Joint urban range coordination should result in more efficient use of active, Reserve, and National Guard component facilities nationwide. Allocating responsibility for all urban sites to a centralized management capability would potentially improve JUO training across departments and components, allowing prioritization of requests based on deployment schedules and other critical variables. However, the joint community's touch should be a light one, its direct influence on range usage based on a "by exception" approach. The objective is to ensure sufficient joint access to Service capabilities while also serving Service needs to enhance the joint quality of their training. Scheduling responsibilities would return to the owning Service after it was ensured that these goals (e.g., providing time for major joint urban training exercises and exceptional other-Service uses of another component's facilities) were met. Further involvement of the joint coordinating entity would be called for only in exceptional circumstances, such as preparation for short-notice deploy-

---

[39] Department of Defense Directive Number 1322.18, Subject: Military Training, September 3, 2004, p. 5.

[40] Ft. Polk interview and trip report based on visit to Ft. Polk by Christopher Paul, Brian Nichiporuk, and Barbara Raymond, October 26–27, 2004.

ments. In short, neither this recommendation nor those below imply impinging on Title 10 responsibilities. Rather, they emphasize responsibilities already allocated to joint headquarters and seek to assist both the individual Services and the U.S. military at large to fully capitalize on funds allocated for preparing the nation's armed forces for operations in urban areas.

The Authority to Manage Ranges Effectively: Joint Oversight of Range Funding

Authority without the means to influence is of limited value. Providing the joint community with funding allocation authority over the development of urban training capabilities offers a number of benefits. It would give the joint community the opportunity to preclude development of multiple projects serving similar purposes in the same geographic area. Combined with the authority to schedule ranges, this would mean that ground and air ranges would be built at the most logical locations, and access would be equally available to all users. It would promote allocation of range maintenance funding in a manner best suited to the needs of the common community. Interagency trainers would have a single point of contact with which to coordinate fiscal and scheduling matters.

While the joint community would thereby maintain oversight responsibility, it would not have to develop range managers, a skill area already well honed by the Services. Actual oversight of construction and subsequent on-the-ground management (and ownership) would remain with the Service owning the host installation. The joint entity would allocate the funds; the Service would manage their expenditure. Requests for training funds from DoD would be centrally managed by the same joint authority, providing for central monitoring of live, virtual, and constructive requirements and subordinate requirements within those larger categories, such as providing standing joint force headquarters training capabilities to support exercises throughout the force. The central authority would also coordinate unit participation in joint exercises, providing combatant commanders with a single point of contact. The joint entity would be responsible for establishing the guidelines and conducting training and evaluation for OPFOR, noncombatant role players, and other training support elements. Such centralization is in keeping with Service realization that central management of ranges is effective and efficient, while also supporting the spirit of joint cooperation encouraged by the Goldwater-Nichols Act.

Three Additional Considerations for Joint Training Strategy Implementation

Joint urban training is, of course, but one component of the massive effort put forth in support of preparing the nation's armed forces for military operations. It therefore stands to both benefit from and lend benefit to other parts of the training system. Recent research on urban operations in Afghanistan and Iraq points to three areas in particular that are worthy of consideration in implementing a JUT strategy:

- Joint urban live, virtual, and constructive training standards
- Better linkage of lessons from the field and joint force urban training
- Other training in support of the JUT strategy

Joint Urban Live, Virtual, and Constructive Training Standards

The strategy developed provides a structure for conducting JUO training, but it is only a starting point. Adoption will require additional guidance. Joint urban operations tasks, conditions, and standards will be needed to ensure quality training in support of the strategy, just as is the case with any well-conducted military training. Such an effort will encompass a number of components.

First, it has been mentioned that what qualifies as an example of a given module can differ considerably from case to case. The size, character of buildings or synthetic representation, structure density and quality, availability and quality of OPFOR and noncombatant role players, and many other characteristics will differ among facilities. Variety is good. Pilots overflying and engaging targets at Nellis

AFB, Lackland AFB, or Yodaville will ideally confront very different challenges, as will marines and soldiers training at urban sites in Camp Lejeune, Ft. Knox, or Muscatatuck. There is nevertheless a need to define in detail what minimum qualities each module should possess so that JUO training requirements will be met to the extent possible. Aviation ranges should replicate the dangers and signatures likely to be confronted in the airspace over actual urban areas. The challenges of coordinating an air strike through a ground controller should be represented in their many and varying complexities.

Sometimes, even the best of facilities lack apparently obvious characteristics. A simple tactical example demonstrates this: American soldiers were wounded on at least two occasions during Operation Just Cause when they opened an interior door and hurled a grenade through the opening to clear the enclosure. On both occasions, the "room" was a closet, and they were wounded by their own rebounding grenade. Yet too few purpose-built urban facilities include closets with doors at all, much less in the numbers and locations likely to be found in actual situations.

Second, joint and Service standards should not only be compatible, they should be identical when covering the same tasks and conditions. This demands that there be a coordinating entity. U.S. Joint Forces Command is a likely candidate given the presence of the Joint Urban Operations Office in its J9 staff section and J7, JFCOM's responsibilities regarding joint training. Training standards, like JUT requirements, need to support combatant command joint mission-essential tasks. That combatant commanders are required to submit their task lists to JFCOM further supports designation of an entity in that command as the agency responsible for standards definition and coordination. (There is obviously a strong argument for assigning the development and maintenance of JUO training standards to the same organization that is responsible for urban range scheduling and related fiscal management.)

Third, such standards apply with equal importance to live, virtual, and constructive training capabilities. Poorly designed simulations teach the joint force the wrong lessons, as do inadequate live training facilities. Currently, development of urban terrain models and the simulations that employ them lack well-defined standards. Service biases, not surprisingly (and perhaps inadvertently), tend to underplay their counterparts' capabilities. Some components are given short shrift even within Service-sponsored simulations. (Ground force information operations and stability-operations-oriented missions compare poorly to coverage given direct-fire engagements, for example. There is an equivalent weighting in favor of air-to-ground and air-to-air tactical engagements over urban intelligence collection tasks in simulations supporting aviators.) There is just as much call for standards that will support proper training involving virtual or constructive elements as there is for live-training standards. The authors of the standards should not limit the scope of specific tasks and conditions to joint concerns alone. Studies demonstrate that multinational and interagency challenges are both notable and much in need of training focus.[41] All joint urban training for military organizations—live, virtual, and constructive—requires additional standards in this regard. Those standards would ideally address training system compatibility; currently, virtual and constructive training capabilities range from fully compatible to not-at-all compatible with each other. Full compatibility can offer much in synergistic benefits, and the lack of compatibility should be a matter of conscious choice (or at least knowing acceptance) rather than the result of a dispersed and therefore not fully effective management system.

Fourth, JUT standards should be complementary and all-encompassing in scope. Classroom training at joint and Service staff colleges, at war colleges, and at other echelons should, like events conducted in the field, be mutually supporting. Standards should specify what tasks classroom and professional-development initiatives are to meet in the service of operational readiness. They should obvi-

---

[41] For example, see Russell W. Glenn, Christopher Paul, and Todd Helmus, *Men Make the City: Joint Urban Operations Observations and Insights from Afghanistan and Iraq.* Santa Monica, CA: RAND Corporation, to be published in 2006; and Russell W. Glenn, and Todd Helmus, *Men Make the City 2: More Joint Urban Operations Observations and Insights from Afghanistan and Iraq,* Santa Monica, CA: RAND Corporation, to be published in 2006.

ously also apply to related course exercises and those conducted in the field by joint headquarters and organizations during all types of events.[42] Exercises at all echelons, including those at the highest joint and interagency levels, should have guidelines to help in developing scenarios and covering essential material.

Responsibility for establishing such standards seems at least in part to lie with the Under Secretary of Defense for Personnel and Readiness (P&R). DoD Directive 1322.18 specifies that the Under Secretary is to "develop policy for and oversee joint architectures and standards for integrating live, virtual, and constructive environments to support training, including prioritization of capabilities, implementation, sustainment, and compliance adjudication."[43] Indeed, taken in conjunction with the Under Secretary's previously mentioned responsibility to "oversee and provide policy for individual [and] collective training [that supports] the operational needs of the Combatant Commanders," the directive would appear to require that authority to set policy for and ensure adherence to all aspects of U.S. military training.[44] However, the JNTC is responsible for standards in support of joint training. This delineation of responsibilities implies OSD P&R oversight of a U.S. JFCOM effort to develop JUT standards. Extending this allocation of authority to encompass live as well as virtual and constructive training would seem to be an effective way of taking advantage of the JNTC, the larger JWC, J7 JFCOM, and JUO, J9 JFCOM base of expertise, while meeting the specifications laid out in DoD Directive 1322.18.

---

[42] Establishing such standards would significantly assist both the Services and joint entities such as the JNTC and Joint Warfare Center (JWC) during the design of training and mechanisms to support preparation for urban contingencies.

[43] Department of Defense Directive Number 1322.18, Subject: Military Training, September 3, 2004, p. 4.

[44] Ibid. That OSD P&R has assumed this mantle is evident in "Cross Service Range Use Standardization Working Group Aviation Range Safety Sub-Group Charter," undated draft, OSD, P&R, and its charter of the Project Kingpin study, "Overarching Instrumentation Policy and Standards Development to Support Joint Force Training and Joint Force Testing: Scope/Framework/Process Definition," SRI International, draft dated January 21, 2005.

A potential additional benefit of consolidating responsibility for range coordination, fiscal management, and standards establishment is improved uniformity across Service training guidance. Elimination of current "inconsistencies," such as the at-times-different USMC and U.S. Army training standards regarding aircraft overflight of ground personnel, would ease training management and potentially improve safety via reductions in misunderstandings. Variations in training constraints due to individual range differences are problem enough without unnecessary inconsistencies.

Better Linkage of Lessons from the Field and Joint Force Urban Training

As the results of Desert One in 1979 Iraq demonstrated, the quality of "jointness" is not an inherent good during field operations. The orchestration of multiple Services has value when it benefits mission accomplishment and greater strategic aims. A joint operation is a means to those ends rather than an end in itself. The same is not true in the world of military training. There, joint operations can be ends in themselves. Practicing "jointness"—building understanding, familiarity, and capabilities—can and should be an objective during one or more stages of JUO training. The strategy outlined in Chapter Six seeks to help in that regard.

In earlier chapters, we focused on how dynamic the urban environment is. The need to design a strategy that is inherently adaptable followed in natural progression. The Services conducting urban operations also change, adding yet further demands for constant monitoring of training to ensure that it best prepares Service personnel for current and future undertakings in densely populated built-up areas.

Keeping pace with the virtually constant evolution demands close contact with events in operational environments. It is essential to consistently provide input to joint and Service schools, urban facility staffs, simulations developers, and others in the training community from lessons-learned capabilities such as the U.S. Army Center for Army Lessons Learned (CALL) and Joint Forces Command Joint Lessons Learned Center. The observations and insights from such sources should not suffer the long delays of months, and even years, that sometimes currently plague release. Training effectiveness will be directly linked to the degree of confidence shown in instructors and students using those inputs appropriately, just as the proficiency of a unit is tied to the confidence shown in its leaders. Both groups must be trusted to properly apply lessons from the field as they prepare individuals and units for operations.

However, the dissemination of experience gained from ongoing training or operations cannot be left exclusively to formal lessons-learned organizations. Assigning recently returned veterans to instructor and O/C positions, constant study by the assigned individuals, and visits to operational theaters by instructors and curriculum developers should be standard procedure. These recommendations seem obvious, but they should not be taken for granted. We mentioned earlier that the British Army's OPTAG frequently sends its instructors to Iraq to observe urban operations and evolving TTP. While the staff members of U.S. lessons-learned organizations frequently make such journeys, few instructors do so. Combining such trips with a policy of assigning veterans to schools would provide the best of several worlds: Individuals with teaching and recent field experience could update their knowledge in the service of better instruction. Their minds are honed by previous operations, providing fertile ground for detecting changes. Such visits cannot replace the need to assign recent veterans to instructor positions, but they can extend the currency of those who have left a theater months before.

The prospects for assignment of veterans to training base positions are more encouraging, although there will always be instances in which those best qualified to teach are sent to other jobs that also demand their abilities. Army Lieutenant Colonel Hal Moore described his frustration at not being assigned to an instructor position after his tour in Vietnam, one that included command of an infantry battalion during the now famous action on LZ (Landing Zone) X-ray described in *We Were Soldiers Once . . . and Young*:

> I had hoped that my next assignment would be the Infantry School at Fort Benning, where I could pass along what I had learned in Vietnam to the young officers who were headed for

combat. It was not to be. In fact, only one of the hundreds of officers who had gone throughout airmobile training and a year in the field with the 1st Cavalry Division was assigned to the Infantry School. I was sent instead to Washington, D.C., where I was told my next job would be on the Latin American desk at the U.S. State Department. Great job for someone whose only foreign languages were French and Norwegian.[45]

Things have changed today. The U.S. Army Aviation School at Ft. Rucker, AL, is already bringing experts back who will allow students to benefit from their experiences over urban areas in Iraq and Afghanistan.[46]

Getting expertise into the training base is only an initial step. Most military personnel attend schools every several years. Instilling knowledge only in students passing through formal training institutions or members of units rotating through major training centers is not enough. Information must also go directly to those deploying, likely to deploy, or simply interested in personal professional development. Lessons-learned organizations have to make their products readily available to anyone with a legitimate interest. The U.S. military is a professional force; trusting its members to properly apply field experience is a risk worth taking, one in which the benefits outweigh the possible shortcomings. Training facilities, schools, and proving grounds conducting technical testing are also sources of material that should go to units. The U.S. Navy's recent centralization of training ranges with test and evaluation facilities is a significant step in this direction, one that should facilitate such exchanges.[47] The Yuma Proving Ground, conducting testing on IED detection and related procedures, similarly provides its findings to those responsible for training members of the joint forces.

---

[45] Harold G. Moore and Joseph L. Galloway, *We Were Soldiers Once . . . and Young*, New York: Random House, 1992, p. 344.

[46] Roxana Tiron, "Pilots Spurring Training, Tactics Revolution," *National Defense*, June 2004, available online at http://www.nationaldefensemagazine.org/issues/2004/Jun/Pilots. htm (accessed on January 11, 2005).

[47] "Military Training," GAO-04-608, June 2004, p. 8.

While training generally requires that students go to schools or major training facilities, the U.S. military has proven that it is also capable of sending instructors to students instead. The Army's Battle Command Training Program (BCTP), the USMC MAGTF Staff Training Program (MSTP), and the Air Force Senior Mentor Program are all tasked with training higher-echelon headquarters. The first two programs have long sent teams to unit home installations or active theaters. Instruction may be broad in content, addressing whatever range of topics a commander selects for his unit. It can also be very specifically focused, as was the case with the creation of Team Foxtrot in late 2002 and early 2003, a BCTP element manned by subject-matter experts who traveled to CONUS installations, Europe, and Kuwait with the mission of helping U.S. Army divisions, corps, and army headquarters prepare for urban operations in Iraq. Several members of Team Foxtrot also addressed USMC audiences. MSTP members have returned the favor, providing training to students at the Army's School of Advanced Military Studies in Ft. Leavenworth, KS. The USAF Senior Mentor Program supports all exercises in which a joint force air component commander (JFACC) participates.

Flexibility and responsiveness are keys to success for these and other premier Service training programs. More than 300 personnel from the NTC recently deployed over the Rocky Mountains to assist the 3rd Armored Cavalry Regiment (ACR) in preparing for operations in Iraq. NTC team members replicated both noncombatants and insurgents. A short timeline, one too limited to allow the 3ACR to both deploy to the NTC and ship its equipment to the Middle East in time to meet mission demands, motivated the movement of the trainers to the trainees.[48]

These efforts to disseminate information to individuals and units in the period immediate prior to deployment lack an effective longer-term counterpart. Kevin Murphy's observation (cited at the beginning of this chapter) about not learning from the lessons we

---

[48] Elaine M. Grossman, "No Time to Spare Before Iraq Deployment? Trainers Make House Calls," *Inside the Army*, January 17, 2005.

have written has been validated again and again in history. It is essential not only that the military make relevant lessons available to those in immediate need, but also that leaders and trainers preparing for urban contingencies five, ten, or more years in the future be able to access those lessons most pertinent to future challenges. There is, and has long been, a call for a JUO database that provides a single point for conducting information searches. A commander notified that his squadron is to deploy in 60 days or an instructor tasked with helping a unit prepare for urban contingencies should be able to go to such a source and quickly determine what documents and other material are available on topics of interest. Ideally, many, if not all, of these sources (e.g., books, reports, interviews, videotapes) would be immediately accessible through the research tool. CALL and other organizations have taken steps toward this highly desirable goal, but joint funding to create and maintain a truly comprehensive database and supporting system would be money wisely spent.[49]

Much of what any Service or joint headquarters learns is also pertinent to others. SOF, USMC, and USA urban TTP would benefit from the availability of a centralized, well-managed, and well-organized lessons-learned database and dissemination system. Pilots in all four Services require such a system; fixed-wing and rotary-wing pilots, whether from the Army, Navy, Marine Corps, or Air Force, would benefit from inputs by those in their common community. There are arguably even lessons that would cross the fixed- and rotary-wing divide. There should be a single, overarching lessons-learned entity responsible for management. There should be a single format and a central hub for submitting lessons learned. A centralized organization could provide training centers and units with guidance about how to submit observations. The organization should be sufficiently staffed to rapidly and wisely disseminate incoming material to the units and training entities that need it. Again, there is also a need to tear down the bureaucratic barriers that inordinately delay the release of materials of value to those deployed or preparing to deploy.

---

[49] Email from George Mordica, CALL, to Russell W. Glenn, November 16, 2004.

America's First Battles was published in 1986. Its authors decried the nation's habitual lack of readiness for war, a shortfall reflected by poor performance in the first major engagements of virtually every one of the country's wars. Operation Just Cause in Panama, Operation Desert Storm, and Operation Iraqi Freedom provide evidence that the U.S. armed forces have done much to reverse that dubious record. Mogadishu and ongoing challenges in Iraq demonstrate that there is nevertheless no room for complacency. Taking lessons from operations and training and passing them to those who can benefit is a key to the successful implementation of a JUO training strategy. It is both an area of demonstrated success and one requiring further enhancement. Success or failure is, as always, measured in men's and women's lives.

Other Training in Support of the JUT Strategy

Do Not Conduct JUO Training in Isolation

Both the joint community and the Service community will be well served by supporting specific individual training in preparation for as-yet-unidentified contingencies and to complement predeployment instruction after arrival in a theater of operations. The need to govern and fight and the need to meet the demands for better interagency cooperation are but two examples of instruction that overlaps urban and other instruction. Military officers and senior noncommissioned officers (NCOs) should receive formal education about the responsibilities associated with each. Much of that instruction will involve urban considerations. Rare is the military man or woman with the expertise needed to manage civil electrical infrastructure, reestablish education systems, or revitalize regional economies. However, Service and joint schools can provide enough knowledge so that commanders victorious in urban combat can accurately determine the types and quantities of support needed to rebuild urban areas. They can help in readying leaders for negotiations with urban leaders and in determining how to best use funds to make an immediate and, ideally, lasting positive impact on towns and cities in the aftermath of war. Instruction can prepare them not only for reporting theater needs in terms other agencies understand, but also for effectively coordinating the handover of civil governance and rebuilding responsibilities to other agencies' representatives or members of a developing indigenous government.

The Services currently fund training for medical doctors, lawyers, and those destined to teach at academies or assume other academic or technical responsibilities. That training too rarely encompasses graduate-level studies in nation-building fields such as infrastructure planning and management, financial administration, or health services or education systems design. Joint and Service sponsorship of such education, followed in at least some cases by a release to the civilian world with a Reserve rather than an active-duty obligation, would provide a resource of educated, real-world-experienced men and women on which the military community could draw during operations similar to those in Panama City in 1989, Mogadishu in 1993, or Afghanistan and Iraq.

Preliminary training that fully readies the force is the preferred capability, but five of the identified JUT strategy requirements cannot be met at a "run" level of preparedness. Steps to bring individuals and organizations to that level include formal training after arrival in-theater. Those with past experience pertinent to the five requirements are the preferred candidates for initial assignment to such areas. They will be best able to adapt their knowledge and instruct others, thereby ensuring a constant flow of mission-ready personnel through operational area urban centers. Instruction can be on-the-job in nature or provided in formal schools established in-theater to prepare new arrivals before they assume their responsibilities.[50]

---

[50] Such schools were established "out of hide" (i.e., without special manning allotments for their conduct) by units in Vietnam. The schools focused on passing lessons learned to newcomers, helping them to adapt U.S.-based instruction, and instilling unit pride. Joint headquarters or another command element remaining in-theater during unit rotations could conduct similar programs to address one or more of the seven requirements and thereby move trainees closer to R status. For a further discussion of these in-theater schools, see Russell W. Glenn, *Reading Athena's Dance Card: Men Against Fire in Vietnam,* Annapolis, MD: Naval Institute Press, 2000.

Train for the Generic and the Specific

Not surprisingly, the urban training sites, simulations, and case studies traditionally used in instruction have had a northwest European orientation. The Cold War was *the* war, Vietnam, Grenada, the Falklands, and other contingencies notwithstanding. More than a decade after the demise of the Warsaw Pact, the Berlin Wall, and the Soviet Union, too many urban training sites, simulations, and case studies still remind us of the Cold War rather than Mogadishu, Iraqi towns and cities, or Afghan villages. Change is occurring, but the past is proving too comforting a touchstone to abandon. Expenditures for future joint and Service urban training capabilities should be questioned if they fail to break free of the past's paradigms.

Because of funding and other resource constraints, the military cannot afford to build new capabilities every time the threat profile changes character, but proper development of training standards and the employment of inherently adaptable training modules can help to address the problem. Every urban area is unique, but all have common characteristics. Structures share more in common than the people who inhabit them; training in buildings that look somewhat different from those in an operational theater can still be very effective. In short, from the standpoint of building design (whether in live, virtual, or constructive capabilities), sites that possess a variety of construction types, building materials, traveled ways, infrastructure (e.g., open sewers versus enclosed), and other elements, either within a given site or between various sites, will serve trainees well. Trainers should seek to design their capabilities so that they can be tailored to specific environments at minimal cost in time and funds (e.g., changing signs into regional languages and altering the nature of refuse, animals, furniture, stairwell and door locations and design, rooftop profiles, and the like). Designing "generic" training sites and adapting existing sites so that they can better reproduce conditions similar to those of current and near-term likely threat conditions will provide frames on which regionally specific details can be draped. This is true of synthetic terrain as well. (However, advances in the speed and reductions in the cost of designing synthetic terrain based on actual theaters are such that calls for generic designs in synthetic terrain may diminish in the longer term.)

Populations share many characteristics as well, but social differences in urban areas generally have a greater impact than terrain features. Language, community mores, factions and their interfaces, religions, governmental and authority structures, and the like differ significantly even within a region or country. These differences mean that demands on permanent OPFOR and noncombatant role players, whether permanently assigned to a given site or traveling between sites (or, figuratively, between simulations), have to adapt to replicate specific environments. This will influence (1) the number of individuals employed in these roles (since time will have to be allocated for study and both individual and collective training), (2) the composition of role-playing groups (to incorporate language speakers and those with requisite cultural knowledge), and (3) scenario design (so that behaviors reflect local conditions).

A Roadmap for Application of the JUO Training Strategy

Figure 7.1 provides a concise overview of how the primary actions in the recommended JUO training strategy should be undertaken during the 2005–2011 period. The modules associated with each component activity appear in parentheses after the descriptions in the first column.

JUT strategy components generally fall into one of three broad groups: (1) actions on which other efforts depend, (2) actions dependent on these for initiation, and (3) actions independent of other events. It is desirable to move forward with simulations development and construction as soon as possible. This is especially critical given the ongoing challenges the United States and its coalition partners are confronting in Iraqi and Afghan urban areas. It is nonetheless essential to develop JUO training standards beforehand. This development will permit joint and Service sponsors to proceed with the desperately needed construction of platoon and larger training sites, air-ground

Figure 7.1
Overview of How Primary Actions in the JUO Training Strategy Should Be Undertaken During the2005–2011 Period

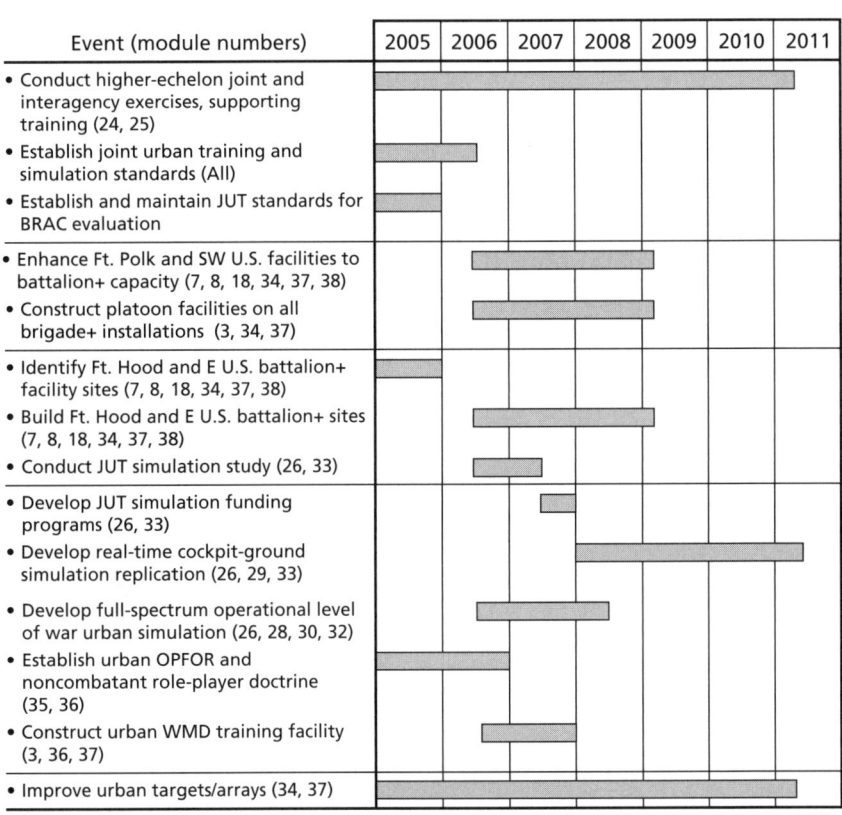

RAND *MG439-7.1*

facilities, and simulations that will complement the live training at these locations and elsewhere. Critical decisions about which simulation programs are to continue will likewise need to await standards definition (although the research to support the difficult decisions about simulation program selection can certainly begin prior to completing the standards development). Such delays are unfortunate, but they need not be times of inactivity. Site selection for urban training venues should begin. Installation personnel should initiate environmental, transportation, logistic, and other studies as appropriate to minimize delays once standards are released.

Training that is under way should continue, and new training should begin. Preparing forces by joint headquarters exercises, urban terrain walks and overflights, and the use of barracks and other existing resources does not require anything other than committed leaders, coordination, and limited (if any) funding. Fortunately, the JUT needed most—preparation of higher-echelon headquarters and the other-agency representatives with whom they will work—can begin immediately. (Some such events are already slated.) Several support module activities, including OPFOR and noncombatant role-player development and target array improvement, similarly do not need to wait on other action. Early development of these assets will add greatly to the quality of JUT events even before "ideal" sites or simulations are brought into play.

The road ahead is not a straight one. There will be twists and turns, dips and climbs, precipitated by changing objectives, adaptive adversaries, clever interest groups, and many other factors. The strategy, like the modules that are its building blocks, must be constantly reviewed for validity just as are the assumptions underlying a plan. It is a tool constantly under construction rather than a finished product to put on the shelf.

Implementing the JUT Strategy: A Cost Estimate

We conclude this discussion with an estimate of the prospective costs of our proposed JUT strategy for the 2005–2011 period. This estimate is derived by applying the values presented in Chapter Six to the strategy. It is representative only. A number of factors will impact the actual cost of implementation, including the number of BRAC (versus hybrid) facilities used, the ultimate disposition of units returning to the United States from Germany and elsewhere, combatant commander decisions about whether to conduct urban-related exercises, choices about the types and numbers of OPFOR and noncombatant role-player capabilities, and other variables. The estimate should nonetheless provide an idea of costs and the types of decisions that will influence them.

The following assumptions and decisions underlie this estimate:

- Combatant commands and the joint staff will each sponsor one headquarters or JTF urban training exercise annually. While each exercise may not adhere to the specifications in the JTF Echo or JTF Foxtrot training event, variations in numbers and exercise design use the dollar costs associated with these as a reasonable approximation of expected outlays. The higher JTF Foxtrot values are the basis for headquarters-related estimates given the expectation that several exercises will involve significantly greater numbers of player commands than are represented by single JTF participation.
- Costs associated with joint classroom or field navigation training for urban operations are zero for the reasons stated in Chapter Six and given the expectation that much of this instruction will be an inherent part of training involving more than urban environments.
- Given the recommendation that a minimum of one facility capable of supporting platoon training be placed at every installation with a brigade or larger ground force, it is assumed that 22 such facilities will be necessary.[51] As noted above, these sites will include upgrades of existing capabilities, the use of underutilized portions of active installations, and the use of other options that will preclude having to build from-scratch complexes. The platoon-size facility cost estimates in Chapter Six therefore provide a conservative (i.e., likely higher than actual) total.

---

[51] The number 22 is itself an estimate given that the final number of installations needing such facilities will depend on basing realignments as units return from overseas stations. Two of the 22 locations are themselves overseas (Okinawa and Korea); it is possible that they will not be needed, while other sites in the United States may be deemed appropriate. The 22 locations are Pendleton, Lejeune, Twentynine Palms, Irwin, Polk, Schofield, Kaneohe, Richardson, Wainwright, Lewis, Carson, Riley, Hood, Bliss, Campbell, Stewart, Benning, Knox, Drum, Bragg, Okinawa, and Korea.

- Each of the four ground force urban training sites that are battalion size or greater will have an accompanying air-ground training capability.
- Costs of joint force headquarters support of selected urban training events are mitigated by the fact that such entities will also support other-than-urban exercises.
- Expenses related to WMD, communications-specific, and CSAR urban training events are not included because of the wide variation in potential approaches to such training and the likelihood that commanders will employ actual urban areas as occasional venues.
- It is assumed that one fire-team urban simulations trainer will be allocated to each installation that serves as home station to a brigade or larger unit. It is assumed that aircraft simulator urban training will employ systems serving a broad spectrum of training; their costs are therefore not included in the estimate.
- Urban training simulations costs for larger units and those involving development of scenario variants are assumed to be inherent in the annual budget provided for operation of the Suffolk, VA, Joint Warfighting Center capability.
- Infrastructure trapping costs are not totaled separately. We assume that initial outlays are incorporated in construction/ upgrade expenditures. Annual O&M will vary greatly depending on location, frequency of use, and other factors.
- The estimate does not include transportation costs associated with urban training.

Other assumptions associated with individual line items are shown in Table 7.8. Costs assume that expenditures are initiated in 2005. Allocation of year-end funding could address selected expenditures. Adjustments to account for later outlays will be necessary for strategy elements begun in years after 2005.

Table 7.8
Estimated Costs for Sample JUT Strategy Implementation
($ millions)

	2005	2006	2007	2008	2009	2010	2011	Total
Headquarters/JTF exercises[a] (7/year)	7.6	7.5	7.3	7.1	6.9	6.8	6.6	49.8
Classroom training	0	0	0	0	0	0	0	0
Navigation training	0	0	0	0	0	0	0	0
OPFOR support for headquarters/JTF exercises[b]	27.0	27.0	27.0	27.0	27.0	27.0	27.0	189.0
Noncombatant role-player support for headquarters/JTF exercises[c]	5.3	5.3	5.3	5.3	5.3	5.3	5.3	37.1
BRAC (realigned/ transition)[d]	0	1.5	1.4	1.3	1.3	5.4	1.2	12.1
Hybrid facility[d]	189.0	2.1	2.0	1.9	1.9	30.1	1.8	228.8
Platoon facility[e]	227.9	2.4	2.4	2.3	2.2	40.8	21.6	299.6
Air-ground facilities[f]	133.0	1.6	1.6	1.5	1.5	1.4	1.4	142.0
Joint headquarters support[e]	7.5	7.5	7.5	7.5	7.5	7.5	7.5	52.5
Fire-team simulations[e]	8.3	7.4	7.3	7.8	6.9	6.6	7.2	51.5
OPFOR support for BRAC/hybrid facilities[g]	35.4	35.4	35.4	35.4	35.4	35.4	35.4	247.8
Noncombatant role-player support for BRAC/hybrid facilities[h]	28.8	28.8	28.8	28.8	28.8	28.8	28.8	201.6
Totals	669.8	126.5	126.0	125.9	124.7	195.1	143.8	1,511.8

[a] Cost of headquarters exercises is (Annual cost for JTF Foxtrot alternative)(7 exercises/year).

[b] OPFOR support for headquarters/JTF exercises is assumed at 50 personnel (one-tenth of an OPFOR battalion)(7 exercises/year).

[c] Noncombatant role-player support is assumed at (25 personnel)(7 exercises/year).

[d] This estimate assumes one of the four battalion or larger and air-ground facilities is a BRAC (transition to urban training use) site. The remaining three are assumed to be hybrid facilities. Costs here are for the ground facility only.

[e] See discussion immediately preceding this table.

[f] One per battalion or larger ground facility. For estimate purposes, two of the four are assumed to be close to parts (thus reducing shipping costs of materials).

[g] It is assumed that units conducting training at platoon-size facilities provide their own OPFOR. It is also assumed that each battalion or larger training facility has a permanent active duty OPFOR company of 120 personnel. The costs are therefore ($42M/site)(120/570)(4 sites) = $35.4M.

[h] It is assumed that units conducting training at platoon-size facilities provide their own noncombatant role players. It is also assumed that each battalion or larger training facility has a permanent noncombatant role player cadre of 10 individuals at $30,000/year, 20 specialty individuals (e.g., representing specific cultures or with particular language skills), and 300 others paid $100/day for 210 days of training per annum. Given four sites, the annual cost is [$0.300M + $0.600M + (300)($100)(210)](4) = $28.8M.

CHAPTER EIGHT

Concluding Thoughts

The streets were "shimmering in the sun like gold from all the expended brass lying on the ground."

> Private First Class David Turner,
> Baghdad, April 2003

Civilian role player whose goat has been declared killed by exercise observer/controller: "Sir, my goat is dead. It left the herd and was killed in one of your minefields. You will pay me, yes?"

Commander, exasperated: "Well, lady, wasn't the minefield marked?"

Role player: "Sir, my goats . . . they cannot read."

> Reconstruction from an interview with a Ft. Polk O/C

Joint and service doctrine state that in order to be effective, military training should be based on pre-established measurable standards, be realistic and challenging, and be conducted as part of a joint and combined arms team. At present, the training of conventional U.S. forces in urban operations falls short of meeting these accepted principles. Efforts are underway within the services and Joint Staff to improve urban training. However, a joint training strategy to guide and coordinate these efforts has not been developed.

> GAO/NSIAD-00-63NI, *Military Capabilities*,
> February 2000

Having laid the foundations for a JUO training strategy, we now present some concluding thoughts on its development.

Instrumentation

Significant divisions exist within the urban training community on the issue of instrumenting facilities to monitor performance and provide feedback during AARs. The Bagram, Afghanistan, urban site, Shughart-Gordon, and others include extensive instrumentation of this kind. The Dutch Army has instead chosen to forgo such capabilities, relying on O/Cs to both monitor ongoing events and provide post-exercise feedback. Some individuals we interviewed at Ft. Polk believe that money should be spent on more buildings to increase the size of an urban training site rather than on additional instrumentation, such as cameras. While we were uncommitted at the beginning of this study, we now tend to support a greater reliance on a human-in-the-loop for monitoring training. Cameras and supporting equipment are expensive both at the time of purchase and during ongoing maintenance and replacement because of upgrade demands and wear-and-tear. Instrumentation for the larger facilities called for in this study would likely be extremely expensive. Funding to provide increased realism seems to offer a greater return on the training investment.

Moving away from a reliance on instrumentation will incur additional manpower costs. The Dutch Army prefers to have an O/C for every squad or platoon participating in training at its Oostdorp and Marnehuizen sites (depending on the location and scenario). Tasking expert OPFOR and noncombatant role players with the responsibility for providing generic and specific functional feedback is one way to mitigate the extent of additional expenditures (e.g., employing a military medicine expert as a noncombatant role player and also having that expert observe and report on how well trainees handle casualties). The desirability of more-selective instrumentation and the alternative of relying exclusively on O/Cs and greater use of OPFOR and civilian role players as evaluators should be more fully investigated. (OPFORs frequently provide some feedback during current AARs; the recommendation here is for an expanded role in this regard.) It is essential to distinguish between the instrumentation addressed here, which is used primarily for monitoring and providing feedback, and other capabilities that enhance training realism or facilitate linking separate sites and simulations to add realism to urban instruction. The latter are and will remain essential.

Urban Live-Fire Training

Urban live-fire training using actual ball and/or tracer ammunition, fragmentation weapons, or other lethal munitions is vital to properly prepare some units and personnel, but such training will seldom impact JUT above the lowest echelons. While refining virtual and constructive simulations to better replicate urban engagements and interpersonal interactions is highly desirable, use of inert ammunition (as is done at the USMC's Yodaville facility) or Simunitions will meet many Service lower-echelon joint training requirements. Given the extraordinary safety precautions necessary for live-fire training, the impact on other training at an urban training site, and the too-often exceptional preparation times taken for unit members to fire a low number of rounds, it is important to carefully consider the benefit to be gained from such training before forgoing similar training using less-than-lethal rounds. While we recognize the need for live-fire exercises in some instances, we do not consider this capability to be a necessary characteristic of large urban facilities designated to support Level 4 JUO training.

Targetry

There have been some notable innovations in urban training targets, often suggested by individual range personnel, whose ideas spread rapidly through a dedicated and open-minded community. Silhouettes now find complements in animated, remote-controlled, three-dimensional dummies that can change character or behavior from iteration to iteration, replicating an enemy at one time and an innocent civilian at another. Bullet traps and moving target arrays, portable target sets, and other innovations similarly enhance realism for trainees. Many of these advances have been accomplished at little expense; the resolution they provide is sufficient for quality training despite their low cost.

Such innovation should be encouraged, as should the more formal developments in targetry that proceed apace with it. Targets and target arrays for pilots could particularly benefit from further development. To better prepare pilots for the actualities of urban operations, moving targets need to be intermixed with arrays of innocent civilians and private vehicles; reproduction of dust, light, electronic, and other interference during engagements; and a general increase in the complexity of the targeting process.

Closing Thoughts

A joint urban training strategy is a starting point for more work. It provides guidance, and it suggests a framework for understanding. But most of all, it imparts a responsibility to develop programs, plans, and guidance that address the many details needed to implement the strategy. It advises how those implementing should write doctrine (itself another form of guidance), spend funds, design instruction, and modify organizations in support of the objectives that initially motivated development of the strategy.

Urban environments are increasingly the norm in military operations. The purview of the soldier, marine, sailor, or airman—looking right or left, ashore from offshore, or beneath from an aircraft—is very likely to be characterized by dense population and manmade structures. Fighting in which villages, towns, and cities do not impact operations is now the exception. Providing support in exclusively rural environments and conducting stability operations not influenced by urban areas are rarities. Training should reflect this new reality. No combatant command major exercise, no senior leaders conference, no theater-level training event should lack a significant urban component. Most training should demonstrate that urban considerations pose major, and likely the primary, challenges those at any echelon will confront. "Train as you fight" is a maxim too frequently ignored when it comes to urban environments. Rare indeed is the training event that steps beyond fighting to encompass a broader spectrum of conflict encompassed in the concept of understand, shape, engage, consolidate, and transition. Rarer yet is that which does so when the focus is the operational or strategic level of war.

In short, this study is but a starting point for much more effort. It is an opportunity for many to participate in the refinement, augmentation, and constant maintenance of the JUO training strategy so that it will become and remain an effective tool in making the lack of preparedness that stimulated *America's First Battles* an ever more distant memory.

Joint Training Definitions

Doctrinal Definitions

joint. Connotes activities, operations, organizations, etc., in which elements of two or more Military Departments participate (JP 1-02, *Department of Defense Dictionary of Military and Associated Terms*, April 12, 2001, as amended through May 23, 2003, p. 275).

joint force. A general term applied to a force composed of significant elements, assigned or attached, of two or more Military Departments operating under a single joint force commander (JP 1-02, *Department of Defense Dictionary of Military and Associated Terms*, April 12, 2001, as amended through May 23, 2003, p. 279).

joint training. Training, including mission rehearsals, of individuals, units, and staffs using joint doctrine or joint tactics, techniques, and procedures to prepare joint forces or joint staffs to respond to strategic, operational, or tactical requirements considered necessary by the Combatant Commanders to execute their assigned or anticipated missions (Department of Defense Directive Number 1322.18, Subject: Military Training, September 3, 2004, p. 10).

joint urban operations. All joint operations planned and conducted across the range of military operations on or against objectives on a topographical complex and its adjacent natural terrain where manmade construction and the density of noncombatants are the dominant features (JP 1-02, *Department of Defense Dictionary of Military and Associated Terms*, April 12, 2001, as amended through May 23, 2003, p. 291).

Definitions Developed in This Study

joint urban training event. A joint training event conducted across the range of military operations on or against objectives on a topographical complex and its adjacent natural terrain where manmade construction and the density of noncombatants are the dominant features.

levels of joint urban operations training (Note: For the purposes of this study, SOF is considered a separate Service).

Level	Description
0	Single-Service urban training event with no participation by other services.
1	Urban training event in which two or more separate Service components orchestrate their activities to achieve common objectives at one or more levels of war (strategic, operational, or tactical). A single Service's actions dominate the event. Participation by other Services plays a minor or superficial role in the accomplishment of assigned missions or the attainment of specified objectives in the training scenario. Limited interaction occurs between command echelons other than within Services; significant vertical coordination may take place within Service components. There is no substantial joint headquarters or other joint synchronization-element participation.
2	Urban training event in which two or more separate Service components orchestrate their activities to achieve common objectives at one or more levels of war (strategic, operational, or tactical). While the event involves greater reliance on one Service throughout or during some phases of the training event, frequent and substantive coordination between at least two Services significantly influences the accomplishment of assigned missions or the attainment of specified objectives in the training scenario. Limited interaction occurs between command echelons other than within Services, though significant vertical coordination may take place within Service components. There may or may not be participation by a joint headquarters or other joint synchronization element.
3	Urban training event in which two or more separate Service components orchestrate their activities to achieve common objectives at one or more levels of war (strategic, operational, or tactical). The

event might involve greater reliance on one or more Services than others during some phases. Virtually continuous orchestration of multiple Services' capabilities in a single primary environment (i.e., land, air, sea, or space) within and between phases dominates training and is essential to the accomplishment of assigned missions or attainment of training scenario objectives. Significant interaction occurs between command echelons across Service boundaries as well as vertically between Service and joint echelons. A joint headquarters, joint effects cell (JEC), or other joint synchronization element orchestrates joint organization and inter-Service capabilities.

4   Urban training event in which two or more separate Service components orchestrate their activities to achieve common objectives at two or more levels of war (strategic, operational, or tactical). The event might involve greater reliance on one or more Services than others during some phases, but virtually continuous orchestration of two or more Services' capabilities in multiple primary environments (i.e., land, air, sea, and space) within and between phases is essential to the accomplishment of assigned missions or the attainment of specified objectives in the training scenario. Significant interaction occurs between command echelons across Service boundaries, as well as vertically between Service and joint echelons. A joint headquarters, joint effects cell (JEC), or other joint synchronization element orchestrates joint organization and inter-Service capabilities.

measures of module potential. The familiar concepts of "crawl, walk, run" are employed here in evaluating the extent to which a module can meet joint urban training requirements. The approach provides two benefits. First, use of a "building-block" methodology both in designing a joint urban training strategy and in developing training in support of that strategy has proven effective in the past: Establishing a fundamental level of ability serves as a foundation for developing more-complex skills. Second, focusing on the ultimately desired attainment of a "run" level of preparedness helps to provide insights into which modules have the greatest return on investment. Other variables being equal, a module that supports attaining "run" status in meeting multiple requirements is likely to be more attractive than one capable of supporting fewer or lower measures of meeting readiness needs. In assigning a given measure of module potential, it is assumed that the module is employed to its full or near-full potential; e.g., assigning "run" to a given requirement with re-spect to Module 2 (facility capable of supporting a unit of company or smaller size with supporting elements) would presume the presence of OPFOR, noncombatant role players, and use of blue-tip or other train-ing munitions at such a facility.

Measures of module potential are entered in the requirements versus modules matrix as C (crawl), W (walk), R (run), or S (supporting task).

These measures support a building-block approach to joint urban training (i.e., developing "crawl" skills before advancing to "walk" and "run" in turn) while also reflecting which modules support attaining the "run" level of operational readiness. The following definitions are used in this study:

- **crawl (C).** Attainment of foundation skills necessary as precursors to de-veloping more-advanced skills or combinations of skills. Being able to es-tablish basic air-ground communications in an urban environment, for example, is essential prior to coordinating close air support. A module supporting a "crawl" measure of ability would have to support develop-ment of base-level skills translatable to application under actual opera-tional conditions in the field.

- **walk (W).** Achievement of greater sophistication in task accomplishment and the ability to coordinate several "crawl"-level or other "walk"-level skills in servicing mission accomplishment. Having the skills to commu-nicate ground-to-air, transmit target grid coordinates or successfully pro-vide laser designation, and conduct accurate post-strike battle damage as-sessment (BDA) would together constitute a "walk" measure of preparedness. A module supporting attainment of a "walk" measure would require managing several skills under realistic field conditions se-quentially or simultaneously as demanded by the situation.

- **run (R).** Accomplishment of complete operational preparedness (combat readiness, if an operation might involve combat action). A "run" status implies proficiency in all supporting tasks and the orchestration of those tasks in accomplishing assigned missions. Being able to successfully co-ordinate close air support under any feasible conditions, even when one or more alternative means of doing so are impractical (e.g., talking a pilot onto an urban target given the failure of GPS and laser designation equipment), would constitute a "run" measure of readiness. A module supporting attainment of "run" status would have to provide sufficient challenge to replicate the most adverse operational conditions.

- **support (S).** Meeting a training requirement in a supportive capacity. The training requirement cannot be met via the module alone, but employment of the module adds realism, provides additional challenges, or otherwise enhances one or more other modules in accomplishing a requirement.

The following supplemental discussion of a "crawl," "walk," "run" metric evaluation is provided for readers interested in the design of virtual or constructive training capabilities or in linking one or both with live training:

"crawl" for L,V,C (coordination live and virtual, live and constructive, or virtual and constructive training modules):[1] Simulations are able to present some of the context of the mission and environmental conditions (e.g., terrain, plan, tasks, resources, interdependencies) in graphic form. Simulations can also be used to provide feedback during after-action reviews or other instruction if physical facilities used during training are instrumented. There is no real-time coordination between simulations and live training.

"walk" for L,V,C (coordination live and virtual, live and constructive, or virtual and constructive training modules): Simulations provide additional excursions or expand the situation presented in the live training scenario by replicating higher and adjacent echelon organizations, specialized systems, not-yet-fielded technologies, or other factors that cannot feasibly be integrated in a live event. Simulations require sequential and/or concurrent real-time integration between live, virtual, and/or constructive training. Simulations may be used to address shortfalls experienced during live training events.

---

[1] In some ways, simulations can be seen as supporting the primary physical facilities; in others, they are a closely coupled tool.

"run" for **L,V,C** (coordination live and virtual, live and constructive, or virtual and constructive training modules[2]): Live, virtual, and constructive training are tightly coupled throughout the training event.

Alternatively, there might be five ways to connect simulations and facilities, each of which has different implications:

1. Separate part-task training: The facility and the simulation have common terrain and scenario, but there is no real-time link between them. The physical facility is limited in scope, concentrating on individual tasks or functions. The simulation is similarly specialized (and likely portable), concentrating on those aspects that the facility is not suited for.

2. Linked single-task proficiency training: The simulation and facility are connected. Selected tasks are given real-time supporting context from the simulation/simulator. There may be messages, events, images, and feedback from the simulation supporting the training process. The simulations primarily support the physical facility under such conditions.

3. Linked team training. This might entail a team as small as a fire team or as large as a headquarters unit. The physical site and simulations both contribute to the training experience, and they are closely linked, with real-time communications, GPS updates, fires, and decision aids. The simulations can be geographically distributed. Players join and withdraw from the exercise as it progresses. (Urban Resolve is probably the most complex current example of linked team training.)

4. Simulation-only training. This might take the form of refresher training after soldiers have finished physical site training, or it might be used for a wide variety of specialty training, including database query about cultural information, familiarization with terrain and plans, and UAV or UGV training (for example, pilots now routinely connect their UAV controllers to PCs to train). This may be sufficient for achieving operational proficiency, much like 777 pilots being allowed to train almost entirely on simulators.

---

[2] Except in very specialized missions, this will require major live-training, extensive incidents linked to sophisticated virtual and constructive simulations capabilities such as that employed during JFCOM's Urban Resolve experiments.

5. Facilities-only training. In the near-term, this is probably the only way to get the noise, dust, explosions, physical demands, stress, reactions, and emotion of battle. It may also be the only way to model Phase IV (support and/or stability) operations realistically.

Consolidated Joint Urban Training Requirements

Avoid fratricide

Communicate in the urban environment

Conduct airspace coordination

Synchronize joint rules of engagement

Conduct stability operations in the urban environment

Conduct support operations in the urban environment

Conduct urban HUMINT operations

Conduct urban SIGINT, IMINT, MASINT, COMINT, ELINT, and other intelligence efforts

Conduct urban operations exercises

Integrate urban operations with other relevant environments

Coordinate maneuvers in the urban environment

Coordinate multinational and interagency resources

Govern in the urban environment

Identify critical infrastructure nodes and system relations

Navigate in the urban environment

Plan urban operations

Provide common situational awareness

Provide fire support

Provide security during urban transition operations

Rehearse/war-game urban operations

Conduct urban noncombatant evacuation operations (NEOs)

Conduct U.S. domestic urban operations

Conduct urban combat search and rescue (CSAR)

Conduct urban operations during and after a WMD event

Consolidate success in the urban environment

Disembark, base, protect, and move in urban environments

Engage in the urban environment

Orchestrate resources during urban operations

Shape the urban environment

Sustain urban operations

Transition to civilian control

Understand the urban environment

Achieve simultaneity in meeting requirements

Conduct training across multiple levels of war

Full List of Identified Requirements

Task/Requirement	Source (numbered cites are UJTL numbers)
Training exercises held at urban training sites are by and large more helpful for infantry units than for their air and other supporting elements.	Unpublished RAND work*
Scheduling and accessibility of training facilities is problematic	Miller et al. (2004: 2)
Facility flexibility is critical	Pendleton interview
Bring jointness to the lowest appropriate level	Wood, JUO Integrating Concept brief, 4 June 04, p. 9
"The JFC should: plan for JUO with the full range of joint assets in mind; train interactively from the task force level down to the lowest tactical level with these joint assets."	Handbook for JUO (2000:EX8)
Fight together:	
Control strategically significant land area	ST 1.6.1
Control operationally significant areas	OP 1.5
Conduct raids in the joint operations area (JOA)	OP 1.2.4.5, DRAFT JTT TA 3.2.X
Conduct penetration, direct assault, and turning movements	OP 1.2.4.6
Conduct direct action in the JOA	OP 1.2.4.7
Conduct offensive operations in the JOA	OP 1.2.5
Conduct defensive operations in the JOA	OP 1.2.6
Conduct seizure of key nodes within urban portions of JOA	Suggested in (JUO) Training Facility Study Phase III Final Report, 2001, p. 27
Conduct joint force targeting	OP 3.1
Employ operational firepower	OP 3
Understand and match fire capabilities to targets	Unpublished RAND work*
Conduct "danger close" fire support in the urban environment	Unpublished RAND work*

Task/Requirement	Source (numbered cites are UJTL numbers)
Conduct time-sensitive targeting	DRAFT JTT TA 3.2.X, Unpublished RAND work*
Employ fire-support coordination measures	OP 3.1.7
Coordinate immediate targets for two or more components	OP 3.1.8
Engage the adversary comprehensively	Wood, JUO Integrating Concept brief, 4 June 04, p. 32
Attack operational targets	OP 3.2
Conduct fire support	OP 3.2.6, JTT TA 3.2.1
Conduct close air support (CAS)	JTT TA 3.2.2
"Talk on target" in an urban environment	Unpublished RAND work*
Provide CAS integration for surface forces	OP 3.2.1, Miller et al. (2004: 2)
Conduct interdiction operations	OP 3.2.5, JTT TA 3.2.3
Provide operational counter-mobility	OP 1.4
Conduct mine and countermine operations	JTT TA 1.4, JTT TA 1.3
Provide counter-psychological and counter-deception operations	OP 6.2.12 and 6.2.11
Coordinate operational information operations (IO)	OP 5.6
Employ PSYOP in the JOA	OP 3.2.2.1
Employ tactical information operations	JTT TA 5.6
Conduct attacks using nonlethal means	OP 3.2.2.4, JTT TA 3.2.6, Unpublished RAND work*
Conduct joint suppression of enemy air defenses (J-SEAD)	OP 3.2.4, JTT TA 3.2.4
Maneuver together:	
Conduct operational movement, maneuver, and force positioning	OP 1, OP 1.2
Precision navigation	GAO/NSIAD-00-63NI, p. 30; Miller et al. (2004: 2)
Conduct activities in ad hoc or "dynamic teaming" formations	Wood, JUO Integrating Concept brief, 4 June 04, p. 19, Unpublished RAND work*
Conduct airlift in JOA	OP 1.1.2.1, JTT TA 1.1.1
Secure LZ/DZs	JTT "supporting task"
Select and mark LZs in urban training	Unpublished RAND work*
Conduct forcible entry: airborne, amphibious, and air assault	OP 1.2.4.3
Conduct air assault operations with another Service	JTT TA 1.2.1
Conduct airborne operations; perform tactical-unit-level airborne operations with another Service.	JTT TA 1.2.2

Task/Requirement	Source (numbered cites are UJTL numbers)
Conduct amphibious assault and raid operations	JTT TA 1.2.3
Coordinate battlespace maneuvers and integrate with firepower	JTT TA 3.3
Conduct integrated armor/mechanized and infantry operations in an urban environment	CAMTF 1999 Initial Draft, p. D-152; Unpublished RAND work*
Conduct aviation operations in urban canyons	Miller et al. (2004: 20)
Conduct aviation route planning and navigation in/over large urban areas	USSOCOM Comments in: J-8 (JUO) Training Facility Study Phase III Final Report, 2001, p. 14
Manage unit adjacencies and transitions:	
Conduct joint reception, staging, onward movement, and integration (JRSOI) in the JOA	OP 1.1.3, OP 1.2.4.4, Draft JTT TA 4.3.x
Coordinate the transition of joint forces to and from tactical battle formations	OP 1.2.1
Coordinate rotation planning	OP 4.4.2.1
Conduct passage of lines	JTT TA 1.2
Conduct force link-up, relief in place, and passage of lines	JTT TA 5.5.1
Manage mission transitions	Derived from author analysis
Transition from wartime operations to military operations other than war (MOOTW)	JTT supporting task
Conduct shaping activities and humanitarian relief concurrently with combat operations.	Unpublished RAND work*
Repeatedly transition from combat to stability operations and support operations and back again	Unpublished RAND work,* letter of instruction for MCWL support for the 1st Marine Division Stability and Support Operations 1-04.
Avoid fratricide:	
Provide positive identification of friendly forces within the JOA	OP 5.1.11; ST 5.1.9
Provide for combat identification	JTT TA 6.5
Coordinate activities of conventional and Special Operations forces in the urban area	Unpublished RAND work*
Understand weapon effects in urbanized terrain	Derived from author analysis
Treat, evacuate, and transport casualties:	
Conduct patient evacuation (CASEVAC)	OP 1.6; JTT supporting task; CAMTF 1999 Initial Draft, p. D-152
Provide for health services in the JOA	OP 4.4.3

Task/Requirement	Source (numbered cites are UJTL numbers)
Conduct search and rescue in an urban environment	OP 6.2.9; GAO/NSIAD-00-63NI, p. 29; Unpublished RAND work*
Base and protect the force:	
Provide protection for operational forces, means, and noncombatants	OP 6, OP 6.2, ST 6, TA 6
Conduct defensive countermeasures against threat precision engagement	OP 6.6.2
Conduct antiterrorism operations	UJTL TA 6.1 (Deleted from UJTL)
Conduct rear-area security—security operations of designated rear-area units that contribute to the security of the entire joint force.	JTT TA 6.3
NBC	JTT supporting task
Conduct deliberate, hasty, and snap VCPs	Letter of instruction for MCWL support for the 1st Marine Division Stability and Support Operations 1-04; Unpublished RAND work*
Protect and assist the urban population	Wood, JUO Integrating Concept brief, 4 June 04, p. 7
Detect/disarm booby traps	GAO/NSIAD-00-63NI, p. 28
Detect sniper	GAO/NSIAD-00-63NI, p. 28
Integrate C4ISR:	
Determine the enemy's theater strategic capabilities and intentions	ST 2.4.1.2
Provide operational intelligence, surveillance, and reconnaissance	OP 2
Identify operational issues and threats	OP 2.4.1.1
Threat detection	GAO/NSIAD-00-63NI, p. 30
Determine the enemy's operational capabilities, course of action, and intentions	OP 2.4.1.2
Identify friendly/enemy/neutral centers of gravity	OP 2.4.1.3
Establish joint force targeting guidance	OP 3.1.1
Assess battle damage on operational targets	OP 3.1.6.1
Provide airspace control and deconfliction	OP 6.1.3, Miller et al. (2004: 20)
Provide rules of engagement	OP 5.4.3
Develop intelligence	UJTL TA 2
Train every soldier to be a HUMINT collector	Unpublished RAND work*
Integrate intelligence personnel and activities with conventional forces	Unpublished RAND work*
Disseminate tactical warning information and attack assessment	JTT TA 2.4

Task/Requirement	Source (numbered cites are UJTL numbers)
Establish, operate, and maintain baseline information exchange	JTT TA 5.2.1
Communications	GAO/NSIAD-00-63NI, p. 30; JTT supporting task; Unpublished RAND work*
Communicate using non–radio-centric signal plans.	7th Marine TTP observations, May 2004
Urban situational awareness	Unpublished RAND work,* JTT supporting task
Establish a joint task force single common picture	GAO/NSIAD-00-63NI, p. 30
Coordinate interactions with the civilian population	Derived from author analysis
Conduct theater psychological activities	ST 3.2.2.1, CAMTF 1999 Initial Draft, p. D-151
Coordinate/conduct civil-military engineering/civil-military operations in-theater	ST 4.4.2, OP 4.7.2, UJTL TA 4.6 (Deleted from UJTL), Draft JTT 4.5
Coordinate/provide law enforcement and prisoner control	ST 4.4.3, OP 4.6.4
Plan and conduct community relations program	ST 5.6.3, OP 5.8.3
Obtain support for U.S. forces and interests	ST 8.3
Assist host nation in populace and resource control	OP 1.5.5
Identify noncombatant issues	Suggested in (JUO) Training Facility Study Phase III Final Report, 2001, p. 28
Conduct peace operations in the JOA	OP 3.3; ST 8.2.8
Transition to civil administration	OP 4.7.4
Conduct evacuation of noncombatants from the JOA	OP 6.2.6; ST 8.4.3, JTT TA 6.4
Conduct populace and resource control (PRC)	Draft JTT 4.5.1
Conduct emergency Service operations	Draft JTT 4.5.2
Conduct foreign humanitarian assistance	Draft JTT 4.5.3
Public affairs	JTT supporting task
Communicate with locals and interface with indigenous authorities	Unpublished RAND work*
Conduct stability operations and support operations	Unpublished RAND work*
Operate in a culturally aware fashion	Unpublished RAND work*
Be prepared for the presence of noncombatants and a wide range of noncombatant behaviors	Unpublished RAND work*
Conduct initial governance and take the first steps toward nation-building	Unpublished RAND work*

Task/Requirement	Source (numbered cites are UJTL numbers)
Positively influence (shape) the perceptions of the indigenous population	Unpublished RAND work*
Conduct shaping operations and combat or force-protection activities contemporaneously	Unpublished RAND work*
Engage in public relations activities during stability operations and support operations	Unpublished RAND work*
Integrate CA and PSYOPs into conventional force operations	Unpublished RAND work*
Conduct CA at the company level	7th Marine TTP observations, May 2004
Conduct policing and civil governance during stability operations and support operations	Unpublished RAND work*
Conduct satellite patrolling	Letter of instruction for MCWL support for the 1st Marine Division Stability and Support Operations 1-04
Conduct population control	GAO/NSIAD-00-63NI, p. 30
Coordinate with government agencies, NGOs, PVOs, other governments, etc.	Derived from author analysis
Provide support to allies, regional governments, and international organizations or groups	ST 8.2
Coordinate civil affairs in-theater	ST 8.2.2
Coordinate foreign humanitarian assistance	ST 8.2.3
Coordinate humanitarian and civic assistance programs	ST 8.2.4
Assist in restoration of order	ST 8.2.7
Coordinate multinational operations within the theater	ST 8.2.10
Cooperate with and support NGOs in the theater	ST 8.2.11
Cooperate with and support PVOs	ST 8.2.12
Provide theater support to other DoD and government agencies	ST 8.4
Assist in combating terrorism	ST 8.4.2
Coordinate and integrate regional interagency activities	ST 8.5
Provide politico-military support to other nations, groups, and government agencies	OP 4.7
Provide security assistance in the JOA	OP 4.7.1
Coordinate and integrate joint/multinational and interagency support	OP 5.7
Integrate host-nation security forces and means	OP 6.5.5

Task/Requirement	Source (numbered cites are UJTL numbers)
Practice/rehearse interagency cooperation in training	Unpublished RAND work*
Logistics:	
Distribute supplies and provide transport services	JTT TA 4.2
Conduct CSS in the urban environment	OP 4, CAMTF 1999 Initial Draft, p. D-154, GAO/NSIAD-00-63NI, p. 30
Conduct convoy operations in an urban environment	Unpublished RAND work*; Letter of instruction for MCWL support for the 1st Marine Division Stability and Support Operations 1-04
Training needs, not elsewhere classified:	
Identified training gap (1999): lack of understanding of the 3D aspects of the urban environment	CAMTF 1999 Initial Draft, p. D-151
Identified training gap (1999): lack of understanding of other important characteristics—historical, ethnic and religious, medical threats/hazards, population density (especially with respect to intelligence)	CAMTF 1999 Initial Draft, p. D-151
Identified training gap (1999): heavy units rarely conduct combined arms training specific to MOUT	CAMTF 1999 Initial Draft, p. D-152
Civilian-dressed combatants should be incorporated in urban training; they should employ tactics not in keeping with international agreements (e.g., using civilians as shields) and attack friendly forces from the rear after they are bypassed	Unpublished RAND work*
Examples of enemy adaptation during urban operations in Iraq should be incorporated in doctrine and training	Unpublished RAND work*
Half of urban skills are lost in six months, virtually all in a year's time; units trained in urban operations should undergo frequent refresher courses, including in-theater training, to maintain their skill sets	Unpublished RAND work*
Urban operations demand that personnel be trained to deal with the unexpected: (1) teach soldiers how to think, (2) provide training exercises against a "thinking" and "adaptive" enemy, (3) inoculate against stress	Unpublished RAND work*

Task/Requirement	Source (numbered cites are UJTL numbers)
Training for languages should be scenario based, i.e., "VCP phrases" or "personnel search phrases"	7th Marine TTP observations, May 2004
During training, add in time to build complacency; instead of conducting lane training and live-fire training that marine units cycle through quickly, build lanes and ranges where marines can sit for hours before engagements occur	7th Marine TTP observations, May 2004
Train to build/design VCPs, not just man them	7th Marine TTP observations, May 2004
Provide 5.03 training for large-unit, combined arms, joint, and operational training in urban environments	4-15-04 JUO Selection and Training (S&T) programs report
Understand the complex urban environment—understand the terrain, the people, the culture, and the infrastructure in totality	Wood, JUO Integrating Concept brief, 4 June 04, p. 23
Face an expert OPFOR	Miller et al. (2004: 2)
Command and control elements could benefit from joint operations training and mission rehearsals	Unpublished RAND work*
Training operations for regular infantry forces should incorporate the wide array of tactical scenarios currently practiced in Afghanistan and Iraq, including sudden changes from close combat to the "soft knock"	Unpublished RAND work*
Facility requirements:	
Need training simulations capable of modeling WMD attacks in urban areas and urban personnel recovery	USJFCOM Comments in J-8 (JUO) Training Facility Study Phase III Final Report, 2001, p. 12
Increase realism of training by having more varied and operationally likely urban layouts and OPFORs that create contemporary situations (i.e., heavy noncombatant presence, suicide attacks, etc.)	Unpublished RAND work*
Cleanness and limits to allowable damage restrict realism of training; Flaka Island (off Kuwait) and other abandoned towns have more realistic rubble and hazard, and can be shot up a little	Unpublished RAND work*
Need for better facilities at home stations for urban operations basics	Unpublished RAND work*

Task/Requirement	Source (numbered cites are UJTL numbers)
Enhance realism: buildings or houses at MOUT facilities have no furniture inside them or obstacles that limit movement inside; facilities with furniture and obstacles will increase confidence and proficiency in building-clearing	Unpublished RAND work*
Even the Joint Readiness Training Center urban "box" is quite small relative to the size of urban areas often confronted during actual operations	Unpublished RAND work*
Make training for stability and support operations as innovative, unpredictable, and difficult as that for warfare in urban areas	Unpublished RAND work*
Range design and training need to incorporate the many requirements for more realistic live-fire events, to include M1 and M2 tables, quick-fire/close-quarter combat ranges, and engagements from convoy vehicles	Unpublished RAND work*
Urban training facilities should accurately simulate anticipated operational environments; the design of street layouts should mimic the twisted and turned nature of those in Iraq and many other countries; opposing forces should deliberately attempt to incite incidents of fratricide	Unpublished RAND work*
Training should include not only operations in the urban area proper, but the transition from rural to urban environments and the reverse	Unpublished RAND work*
Future training programs for rotary-wing aircraft crews should include the provision of live-fire training areas, practice of night operations over urban areas with realistic heat, light, population, and other signatures, the inclusion of crew effects in simulator exercises, and larger urban training sites (either purpose-built or via training adapted to other locations)	Unpublished RAND work*
Pop-up targets in MOUT training centers lack heat signatures, limiting their training value to UH-60s or other aircraft with forward-looking infrared radar (FLIR)	Unpublished RAND work*
Training for helicopter crews should be conducted over actual or simulated urban environments in which building height can interfere with targeting and flight profiles (for both the aircraft and munitions); targets should also be placed in areas where noncombatants and infrastructure are modeled so that the avoidance of collateral damage can be practiced	Unpublished RAND work*

Task/Requirement	Source (numbered cites are UJTL numbers)
Be cautious not to learn the wrong lessons from purpose-built urban training facilities; train in and over actual cities when possible, and include "terrain walks" as part of exercises	Unpublished RAND work*
Need for a very large urban training facility in which a training unit could become "lost" and which OPFOR knows much better than the training unit	Miller et al. (2003:95)
Night training/night operation—many facilities are only open (or staffed) during daylight hours	United States Special Operations Command, Global Special Operations Forces Range Study (2003:I-5)
Need for better-integrated call for fire/full-spectrum air-to-ground training	United States Special Operations Command, Global Special Operations Forces Range Study (2003:V-1)
"The AC-130U simulator at Hurlburt Field is state-of-the-art; the EW station on the simulator, however, is highly unrealistic and is basically just a position for a crewmember to sit while the rest of the crew works a mission"	United States Special Operations Command, Global Special Operations Forces Range Study (2003:V-11)
Army's definition of "overhead fire" prevents live fire from AC-130s during CAS training on Army ranges	United States Special Operations Command, Global Special Operations Forces Range Study (2003:V-11&12)
Since nothing can be dropped from aircraft on or near live troops, simulated ground bursts on the ground, connected with flight simulations, are a possibility	Miller et al. (2004: 2)
Easily reconfigurable city (for air and ground navigation issues/familiarity)	Miller et al. (2004: 2)
Facility inherent OPFOR, traveling OPFOR, specific culture OPFOR	Miller et al. (2004: 27)
Specific physical requirements for aviation training	NTI study, 2001 (not releasable to the general public)
Place to land a helicopter	Derived from author analysis
Roof suitable for helicopter landing with targets on roof that can be fired on before landing	Derived from author analysis
Simulated muzzle flashes and pop-up live-fire targets (vehicle and other)	Derived from author analysis
Rail yard or port facility with nearby POL tanks.	Derived from author analysis
Bridge targets	Derived from author analysis
Moving targets	Derived from author analysis
Firepits and spotlights to distract IR	Derived from author analysis

Task/Requirement	Source (numbered cites are UJTL numbers)
Need to simulation air defense radar signatures on selected ranges	NTI study, 2001 (not for public release)
Tarpaper buildings	Pendleton interview
"High-end" need for large and complex urban terrain capable of integrating CAS and brigade-size ground elements	JWFC, JFCOM J-7 Capabilities Group, undated brief
"Small-end" need for realistic action with role players, force on force, etc.	JWFC, JFCOM J-7 Capabilities Group, undated brief
Props, including role-player costumes, OPFOR weapons, culturally appropriate village/city signs, furniture, materials to use as "goods" in village marketplaces, role-player vehicles	Letter of instruction for MCWL support for the 1st Marine Division Stability and Support Operations 1-04
Urban facilities need to reflect short line of sight, interiors and subterranean structures, clutter, target movement, identification (friend, foe, or neutral)	4-15-04 JUO S&T programs report
Diverse man-made structures	4-15-04 JUO S&T programs report
Complexity in the urban environment (which training facilities must implicitly mirror)	Wood, JUO Integrating Concept brief, 4 June 04, p. 10
Physical terrain—urban canyons, vertical terrain, subsurface maneuver space	Derived from author analysis
Systems and patterns of activity—political, cultural, economic, legal, informational, infrastructure	Derived from author analysis
Population density and diversity	Derived from author analysis
Cell phones, radios for OPFOR and role players	Derived from author analysis
Planning for JUO must take into account physical, infrastructure, commercial, residential, and socioeconomic factors (so, by implication, training facilities must make them real factors)	Handbook for JUO (2000:EX4)
Challenges in urban areas (implicitly, to train to them, they must be simulated in training)	Handbook for JUO (2000:I-8)
Presence of noncombatants	Derived from author analysis
Presence of civil government institutions	Derived from author analysis
Presence of NGOs	Derived from author analysis
Presence of local and international media	Derived from author analysis
Potential sources of host-nation support (labor, construction material, and medical supplies)	Derived from author analysis
Complex social, cultural, and governmental interaction that supports urban habitation	Derived from author analysis
Location of key transportation hubs	Derived from author analysis
"Cultural"/noncombatant OPFOR	Miller et al. (2004: 4)
Goats and/or dogs as part of cultural OPFOR	Derived from author analysis

Task/Requirement	Source (numbered cites are UJTL numbers)
Service-level requirements, either new or unsatisfied:	
Urban navigation training should emphasize terrain association and use of GPS systems	Unpublished RAND work*
Urban patrolling needs to be trained better	Unpublished RAND work*
Combat mount/dismount from heavy trucks	Unpublished RAND work*
Navigation—train on sketch maps and on scale of maps forces will actually fight on	Unpublished RAND work*
Mortar positioning in urban areas	Unpublished RAND work*
Urban target detection by aircraft	Unpublished RAND work*
Improve training and practice with grenades; include more-challenging urban training on grenade ranges	Unpublished RAND work*
Rotary-wing: pilots need to constantly visually scan their surroundings and avoid becoming fixed on a single ground-based target; this should be adapted into training operations	Unpublished RAND work*
Training for landing rotary-wing aircraft in cities, which is difficult and insufficiently practiced	Unpublished RAND work*
Up-to-date electronic-warfare (EW) training	United States Special Operations Command, Global Special Operations Forces Range Study (2003:V-12)
Environment	
Weather	JTT supporting tasks
Training needs to consider urban environmental characteristics unique to a given AOR (i.e., the whole world does not have European-style urban centers and infrastructure)	CENTCOM Comments in J-8 (JUO) Training Facility Study Phase III Final Report, 2001, p. 11
Training and doctrine should include coverage of the risks associated with hazardous materials, gas, other fuels, and chemicals; likely locations of these substances (e.g., jewelry shops) should be identified in written and training guidance	Unpublished RAND work*

* Russell W. Glenn, Christopher Paul, and Todd C. Helmus, *Men Make the City: Joint Urban Operations Observations and Insights from Afghanistan and Iraq*, Santa Monica, CA: RAND Corporation, forthcoming in 2006; Russell W. Glenn and Todd C. Helmus, *Men Make the City 2: More Joint Urban Operations Observations and Insights from Afghanistan and Iraq*, Santa Monica, CA: RAND Corporation, forthcoming in 2006.

RAND Urban Training Facility Survey

Instructions for RAND Urban Training Facility Survey

Thank you for taking the time to complete the RAND urban training facility survey. Find the survey on the second worksheet in this Excel file (access by clicking on the "Facility Survey" tab at the bottom of this page).

This survey is part of a RAND effort in support of OSD P&R, J-9 JFCOM and J-7 JFCOM. POCs in the sponsors' offices are available on request.

RAND has been asked to develop a detailed strategy for joint urban operations training to support funding decisions and improved force-wide urban operations readiness in the period 2005–2011. This effort will have direct application to preparing military personnel from all four Services.

This survey will provide us with detailed information about the capabilities and training activities at key urban training facilities. Yours is one such facility.

This page contains further explanations to assist in answering the questions on the facility survey worksheet. Please refer to this section when what we are asking in the survey is in any way unclear. If you have any questions regarding this survey, please contact Christopher Paul (cpaul@rand.org) or the project principal investigator, Russell Glenn (Russell_Glenn@rand.org).

Thank you in advance for your support of our research effort.

Complete one column of data for each urban training complex/MOUT at your installation. If you think of multiple "ranges" as part of one complex, treat them in a single column.

Guiding principle:
We want as much information as possible. The cells in the worksheet will hold a large amount of information. Please don't just put "yes" or "no" if the actual facts are more extensive.
A whole sentence or even whole paragraph will fit in a spreadsheet cell, and can be very valuable to us as we attempt to understand your facility.

If you cannot complete part of the survey because you don't know the answer, that is OK. Please complete what you can and return it to us not later than December 1, 2004. We would much rather have a mostly complete survey on time than never receive a fully completed survey. Please provide contact information if you can suggest an individual who might be able to provide the omitted information. If a missing datum proves essential to us, we can follow up with that POC.

Row	Question
Basic information	
3	Installation or project name—how do you refer to the range? What do the users call the range? What is the facility/complex's official name? Any and all of these bits of information are welcome (especially nicknames/abbreviations).
4	State/country—State in which the facility is located, or country, if OCONUS.
5	Major Command—the major command in charge or the range or at the facility's location.
6	Location—the base at which the facility is located or other location information as appropriate.
7	Service—the Service with primary responsibility for the facility. If joint or multiple, indicate (and provide any details)
8	POC—A point of contact regarding the facility; either an individual or generic office to whom one could address further questions regarding the facility.
9	Email for that POC.
10	Phone for that POC.
11	Primary user or projected primary user—who is (or is projected to be) the primary user of the facility?
12	Brief description—what is the facility? What kind of range or ranges does it include?
13	Construction cost—how much did (or will) the facility cost to build?
14	What kind of appropriation paid (will pay) for the construction of the facility?
15	As of the date of survey, is the facility complete?
16	When (what year) was (will be) the facility completed?
17	When was construction begun (will begin) on the facility? If you don't know, don't sweat it.
18	Are there pending and approved upgrade to the facility planned? If so, for what FY? Any additional details welcome!
Availability and load	
20	Facility utilization rate—how much training goes on at the facility? Express in units meaningful to you (and provide units). Events per year? Total personnel trained? Man-hours of training per month?

21	Average length of training cycle/event—when a user uses the facility, how long do they use it for? What is the length of the average exercise conducted at the facility? Feel free to provide more information, like: MAC range use averages 3–6 hours; MOUT complex use averages 3 days.
22	How much of the time a unit spends at the facility is actual training time, and how much is set up and preparation, or clean up and departure, or simply idle non-training time?
23	Recycle time—how much time/effort is required to get the facility ready for the "next" unit after one unit completes an event/exercise? Please note if that is time spent by the unit completing training (they can't leave until they clean up) or by facility/range personnel.
24	How many personnel are required to prepare/maintain the facility for exercises? How many of each of officers, enlisted, DoD civilians, contractors?
25	Are there formal procedures for users to book or reserve the facility, or is use ad hoc or as needed? Is the facility wholly dedicated to one unit's training, so no reservation is required? Any important details regarding the scheduling system not covered in the questions that follow welcome here.
26	Does any unit/Service have scheduling priority? What is the hierarchy? What benefit does having priority entail?
27	Does any user regularly get "bumped" by a higher priority user? Who and why? How late in the day can a priority user bump a lower priority user?
28	Is the facility full booked, or is there excess training capacity? Does it sit idle often, or only if there is a scheduling mix-up? When the facility is in use, on average, how much of it is in use?
29	Is there a waiting list of facility use? How long is it (in terms of days/weeks of wait)? How does that work?
30	How far in advance do users need to reserve the facility? Does that vary based on user priority? How?
31	Is the facility available year round, or closed in certain seasons, or for regular long maintenance periods?
32	If the facility is seasonally closed, when, and why?
33	Is the facility available to services other than the one(s) listed in row 7, above?
34	Is the facility actually used by other services? Which? How often? To what extent (how big a formation)?
35	Is the facility available to civilian agencies (either federal agencies or federal/state/local law enforcement)?
36	Is the facility sufficient to support a JTF exercise? Has one ever been held there? When/how often?

Physical plant	
39	How many acres does the facility occupy?
40	How much square footage is enclosed in buildings?
41	Is there another relevant size measure to describe the facility? What are the units of measure? How big is the facility on that scale?
42	Is there room to expand the facility? How much acreage is available for expansion before running into impassible terrain, other ranges, non-base land, other structures, residential areas, environmentally protected areas, etc.?
43	What is the nature of the terrain immediately surrounding the facility? How could this impact expansion or the types of exercises that are/could be run at the facility?
44	Are combined arms exercises possible at the facility? Which, and to what extent? Example, can rotary A/C land at the facility? How many? Can tank-infantry integration be practiced in the facility?
45	Can the facility accommodate armored/mechanized vehicles? Which ones? To a limited extent (i.e., "as long as they don't tear up the curbs")?
46	How many tanks/armor could the facility accommodate? Could a section of armor actually train there, or would it be infantry getting practice interacting with a small number of tanks?
47	How many trucks/wheeled vehicles would be reasonable to have operating in the facility? (Not asking how many could fit if facility used as a parking lot).
48	Does the facility allow ground maneuvers? Can ground troops operate in the facility?
49	What is the training capacity of the facility? What size formations can it handle of different types? (Example, brigade of infantry, company of tanks, 2 helicopters). How many different units can train simultaneously? To what extent can the facility be compartmentalized for use by different users?
50	Is there a rotary wing aircraft LZ at the facility?
51	Is that rotary LZ in the urban area of the facility, or adjacent to it (at the "edge" of the complex/facility)?
52	Is there a fixed wing LZ at the facility/range/complex?
53	Is there a Drop Zone at or adjacent to the facility?
54	Is the facility amphibious accessible? Can the facility be used to practice amphibious assault of an urbanized area?
55	How far is the facility from the point at which amphibious forces exit the water?
56	Does the facility have AAR (after action report) capability? What kind (video, O/C debrief, etc.)?

57	Is the facility (or does it contain) an official MAC (MOUT assault course) or UAC (urban assault course)?
58	Is the facility (or does it contain) an official MOUT collective training facility?
Facility infrastructure	
60	How many buildings are part of the training complex?
61	What types of buildings are represented? Residences? Commercial? Municipal? School? Hospital? Mix?
62	How many of the buildings are two story?
63	How many of the buildings are three or more story?
64	What are the buildings made of? Cinderblock? Wood? 10 Wood and 3 poured concrete?
65	How many of the buildings have furniture? How would you characterize the furnishings? Rudimentary? Fully furnished?
66	Are any of the buildings reconfigurable? Can they be moved, or can the locations of doors or windows or interior walls be changed? If so, how?
67	How "dense" is the urban terrain? What is the average distance between buildings? Is it "rural/sparse," "American suburb standard," "small town main street," "3rd world jumble"? How wide are the streets?
68+	Rows 69–80 are yes/no checklists for the listed elements.
Training assets	
82	Does the facility have its own on-site opposition force (OPFOR)? Where are they drawn from?
83	How big is the OPFOR? (Numbers of personnel and/or size of formations). Number enlisted, officer, DoD civilian, contractor.
84	How long is an OPFOR member tour? Where are they drawn from? Where do they return to at end of tour?
85	Does the facility have its own observer/controllers (O/Cs)?
86	How many O/Cs? Number enlisted, officer, DoD civilian, contractor.
87	Are there other role-play assets? What are they? Example—role-play civilian volunteers, role-player vehicles, role-player props (such as goats, or market goods, or costumes), etc.?
88	Can the presence of crowds be simulated? If so, is that with regularly available role-play assets or with special assets brought in for specific exercises? Explain!
89	Is the facility (or part of the facility) DIS (Distributive Interactive Simulation) capable? To what extent? Has DIS been used there before?

Live fire capabilities	
91	Does the facility allow direct fire? What kinds? (Small arms, machine gun, tank, Air to Surface fires, etc.)
92	Does the facility allow indirect fire? What kinds? (Mortar, tube or rocket artillery, other?)
93	What is the upper limit on size/caliber/weight of allowable live fire? Over what proportion of the facility is live fire allowed?
94	Are there battle effects simulators in the facility? What kinds (smoke generators, etc.)?
95	Is there targetry in the facility? What kinds (fixed, radio controlled, computer controlled, etc.)?
96	Are there reconfigurable targets? (i.e., can they be moved or changed to neutral/friendly from hostile, etc.)?
97	Are there thermal signature targets? How many?
98	Does targetry include (or include the option for) friendly/neutral targets?
99	Are there remote control targets (either radio or computer)? How many?
100	Is there a grenade house/grenade range?
101	Is there a structure/range for practicing demolitions and/or door or wall breaching? Briefly describe!
102	Is there a demolition effects simulator? Briefly describe?
103	Is there a way to include close air support or simulated close air support? Briefly describe!
104	Can you use Simunitions on the facility? Are Simunitions easily available to training units? What kinds?
105	Can MILES gear be used at the facility? Is it available at the facility (or larger base) or do training units need to find/bring their own?
106	Is there a way to simulate the use of non-lethal munitions at the facility? Briefly describe!
Connectivity/instrumentation	
108	How many of the buildings are instrumented in some way?
109	Describe the instrumentation. Cameras? How many per building? Motion sensors, etc? Microphones? What else?
110	Is the instrumentation night capable? All of it?
111	Are there cameras? Both internal and external cameras? Over what proportion of the facility?
112	Are instruments/cameras digitally linked (or linkable)? To what extent? Describe!
113	Are instruments connected to computers?
114	Are instruments connected in such a way that the facility could broadcast or receive information to integrate to an exercise elsewhere? To what extent?

Simulators	
116	Are there simulators at the facility? What kind? (Walk-through simulation, cockpit simulation, etc.)
117	How many simulators are there? What is maximum simulation load, in terms of individuals, aircrews, etc.?
118	How much did the simulators cost? If known, break out cost of hardware vs. software vs. supporting equipment.
119	How much of that is non-recurring (fixed/one-time cost), and how was (will be) that broken down over years of development?
120	What are the costs associated with networking the simulators? Explain the nature of these costs.
121	What does it cost to maintain the simulators? (Include licensing, memberships, etc.)
122	Are there other recurring costs? How much? For what?
123	How many staff (officer, enlisted, DoD civilian, contractor) are used to run the simulators?
124	How often are simulators used? How often are they a part of exercises/rotations at the facilities? Are they only used by a certain type or types of users?
125	Does the use of simulators/simulation have any special requirements? Does it require special personnel? What type? Are these personnel a regular part of facility staff?
126	Do the simulators have post-processing needs? Are facilities to fulfill such needs available? Describe.
127	Are materials presented in an After Action Report format? How much effort is required to produce the AARs?
128	How much prep/setup/shakedown time is required before initial use of simulators? How much effort is required to shakedown between uses?
Scenarios supported	
	Which of the following scenarios can be supported at the facility? Which of them are "ready to go" and which would require additional materials be brought in by the training unit?
130	Force on force (FoF)
131	Force on targetry (FoT)
132	HA—humanitarian assistance
133	NEO—noncombatant evacuation operation
134	Peacekeeping/riot control
135	SOF special scenarios
136	NBC (nuclear/biological/chemical) weapons
137	Medical capabilities

Throughput	
139	What is the total annual throughput of persons trained at the facility?
140	What is the largest unit type/formation that can train collectively at the facility?
141	How many different units use the facility? Can you estimate the total units supported in BDE equivalents?
142	How frequent is training activity? Either average exercises per week, or average idle days per month, or % of days facility is available that it is in use. Whatever measure of activity is meaningful to you.
143	Other measure of throughput? How do you think about/measure the amount of training that goes on at your facility?
Budget	
	If you are able to answer these questions about facility budget, please do. If they are outside your portfolio, please suggest a POC who might track/have this sort of information.
145	What are the annual costs to operate and use the facility? To the extent possible, provide details about these costs. Major categories could include labor (military and civilian to operate to include aggressor force, observers/controllers, and other role play assets), consumables, utilities, and contracts.
146	DoD Facility Category comes from the *DoD Facilities Cost Factors Handbook*. The Engineer or Installation Manager should have these codes.
147	Each service has a more detailed code called a service category code.
148	What are the initial facility construction costs, i.e., the initial investment in special purpose urban training structures, facilities, ranges? These would be historical costs for existing facilities and program or budget costs for planned facilities. Please provide separate costs for particular components of the facility if funded separately.
149	What are the other non-recurring investment costs for such items as initial procurement of vehicles, machinery, equipment, targets, etc.? What had to be purchased to open the facility?
150	Were all non-recurring costs in the first year or spread out over several years? Also, if non-recurring costs are cyclical (e.g., every third year) or annual, let us know.
151	How long will the facility or its components be expected to be useful? In other words, how often will it need to be replaced?

152	What are the annual costs to sustain the facility? To the extent possible, provide details about these costs. Major categories could include labor (military and civilian to maintain) and maintenance (repair/replace/periodic normal upkeep, and replacement for destruction during training), lease/rent, and contracts.
153	What are the incremental costs for scheduling and using the facility? For example, if a unit must provide personnel on the day of an exercise (as opposed to permanently assigned personnel), include those costs here.
154	What are the specific elements of that incremental cost for using the facility? Who pays them?
155	Are there costs associated with cleaning/preparing the facility for re-use after an exercise?
156	If there are costs not included in row 145 to maintain an empty or non-used facility, please provide.
157	Are there any other costs not included elsewhere?
158	What budget or appropriation pays for major upgrades to the facility?
159	How many major upgrades have occurred?
Existing limitations	
161	Are there limitations to the airspace use over the facility? If so, what kinds, and how limiting are they (altitude restrictions, civilian use airspace, type of A/C restricted, etc.)?
162	Is environmental encroachment a factor at the facility? Can certain kinds of training not be done due to environmental protection issues? What are the limitations?
163	Is operational security an issue for training at the facility—is the facility easily observed from public access lands/areas that might result in compromise/observation of protected TTP?
164	Is unit liability a restriction, either liability for property damage, or in some other way? Explain if so.
165	Is public acceptance an issue for the facility? Is the facility close enough to civilian areas that noise or night-time activities are restricted? To what extent?
166	Are there requirements training units must fulfill before using? (Like having their own trained Range Safety Officer, etc)?
167	Are there any other limitations to the kinds of training that can be done at the facility, or factors that would prevent joint training at the facility that aren't clearly enumerated above?

Row	Characteristic/Feature	Complex 1
	Basic Information	
3	Installation or project name	
4	State/country	
5	Major command	
6	Location	
7	Service	
8	POC	
9	Email	
10	Phone	
11	Primary user (projected)	
12	Brief description	
13	Construction cost (in millions)	
14	Type: 1) RDT&E; 2) procurement; 3) MILCON; 4) O&M; 5) MILPERS	
15	Facility complete?	
16	Year completed/to be completed	
17	Year construction to begin/began	
18	Programmed development/upgrades (FY)	
	Availability and load	
20	Facility utilization rate	
21	Average length of training cycle/event?	
22	% of time that a unit is on site devoted to actual training	
23	Recycle time (time required to repair/reset to go again?)	
24	# of full time personnel to maintain and prepare facility for exercises	
25	Formal scheduling procedure for units to "book" the range/facility?	
26	Who has scheduling priority?	
27	Who gets "bumped" and how often?	
28	Fully booked/scheduled or excess capacity?	
29	Waiting list?	
30	Length of time users need to book in advance?	
31	Available year-round?	
32	When is not available and why?	
33	Available to other services?	

Row	Characteristic/Feature	Complex 1
34	Used by other services? Which? How often?	
35	Available to civilian agencies?	
36	JTF exercise capable?	
	Physical plant	
38	Size	
39	Acreage	
40	Sq footage enclosed	
41	Other size measure? (units?)	
42	Can the site be expanded if necessary? By how much acreage?	
43	What is the nature of the terrain immediately surrounding the facility?	
44	Combined arms possible (which, and to what extent?)	
45	Accommodate armor/mechanized?	
46	Maximum number of armor vehicles	
47	Maximum practical number of trucks/wheeled vehicles:	
48	Allow ground maneuvers?	
49	Capacity (size of troop formation, number of vehicles, number of units)	
50	ACFT LZ (rotary)	
51	LZ adjacent to urban or "in" urban area?	
52	ACFT LZ (fixed)	
53	Drop zone	
54	Amphibious accessible?	
55	Distance from amphib landing site to facility?	
56	AAR site capability	
57	Official MOUT assault course (MAC)?	
58	Official MOUT collective training facility?	
	Facility infrastructure	
60	Number of buildings	
61	Type of buildings	
62	Number of two-story buildings	
63	Number of 3 or more story buildings	
64	Type of construction of buildings (wood, concrete, cinderblock, etc.)	

Row	Characteristic/Feature	Complex 1
65	Number of buildings with furniture	
66	Buildings reconfigurable?	
67	How close are buildings to each other, on average?	
68	*Other structural elements*	
69	Subterranean elements	
70	3rd world shanty town	
71	UN check point capabilities	
72	Embassy complex	
73	Drug labs	
74	Port facilities	
75	Concurrent sites	
76	H2O potable/non-potable	
77	Fastrope capability	
78	Electricity	
79	Kickable/breachable doors?	
80	Notable other?	
	Training assets	
82	Dedicated OPFOR available?	
83	Size of dedicated OPFOR	
84	Length of tour of OPFOR personnel	
85	Observer/controllers available?	
86	Number of O/Cs (Off, Enl, DoD Civ, Contr)	
87	Other role-play assets	
88	Can the presence of civilian crowds be simulated?	
89	Site DIS (distributed interactive simulation) capable?	
	Live fire capabilities	
91	Direct fire (type)	
92	Indirect fire (type)	
93	Maximum caliber/weight/size of allowable live fire	
94	Battle effects simulator	
95	Targetry (mechanical, electric, computer controlled)	

Row	Characteristic/Feature	Complex 1
96	Reconfigurable targets	
97	Thermal signature targets	
98	Targetry both hostile and neutral/friendly	
99	Remote controlled targets	
100	Grenade/grenadier structure	
101	Demolition/breach	
102	Demo effects simulator	
103	SIMCAS/CAS	
104	Simulations	
105	MILES lasers	
106	Can nonlethal weapon use be simulated at the facility?	
107	*Connectivity/instrumentation*	
108	Number of buildings instrumented	
109	Description of instrumentation	
110	Instrumentation night capable	
111	Cameras	
112	Digital linkage	
113	Computer connectivity	
114	Prepared for connection to other site simulation?	
	Simulators	
116	Type(s) of simulators	
117	Number of simulators/number of personnel able to use simulators	
118	Purchase costs of simulations and simulators (computers, software, stations)	
119	Non-recurring (investments) and timing ($ year1, year2, etc.)	
120	Network costs (linkages, telecomm)	
121	Maintenance costs of sims (including license fees, membership, etc.)	
122	Other recurring costs	
123	Staffing req. for running, updating, programming (Off, Enl, DoD Civ, Contr)	
124	Frequency of usage during rotations	
125	Special reqs. (observer/controllers, OPFOR, SAFOR specialists, etc.) (Off, Enl, DoD Civ, Contr)	

Row	Characteristic/Feature	Complex 1
126	Post-processing needs (e.g. Starwars center at NTC)	
127	Materials preparation for AARs	
128	Staff and time required to shake down sims prior to initial use	
	Scenarios supported	
130	Force on force (FoF)	
131	Force on targetry (FoT)	
132	HA	
133	NEO	
134	Peacekeeping/riot control	
135	SOF special scenarios	
136	NBC weapons	
137	Medical capabilities	
	Throughput	
139	Annual personnel throughput	
140	What is the largest unit type that train collectively at the facility	
141	Units supported in BDE equivalents	
142	Frequency of training activity	
143	Other measure of throughput?	
	Budget	
145	Annual operating cost	
146	DoD Facility Analysis category	
147	Service category code	
148	Construction cost ($M)	
149	Other non-recurring investment costs ($M)	
150	Timing of construction/non-recurring costs ($year1, year2,etc)	
151	Economic/useful life (years)	
152	Annual sustainment cost ($M)	
153	Cost to run an exercise	
154	Breaks down how (elements of exercise cost)	
155	Cost to recycle	
156	Cost to sit empty	
157	Other costs	

158	Source of funding for upgrades	
159	Major upgrades in the last 5 years	
	Existing limitations	
161	Airspace use over facility?	
162	Environmental factors	
163	Operational security	
164	Unit liability	
165	Public acceptance	
166	Requirements to use (e.g., RSO training)	
167	Other	
168	DATE FORM COMPLETED:	

Facility Summary

The full list of sources is given in the Bibliography. It includes Dominant Maneuver Division, J8, Joint Chiefs of Staff, *Joint Urban Operations (JUO) Training Facility Study Phase III Final Report*, Washington, DC, 2001; Ann Miller, Matt Grund, Deborah Jonas, *Joint National Training Capability: Functional Area Training Resource Requirements*, Alexandria, VA: Center for Naval Analyses, Research Memorandum CRM D0008243.A2/FINAL, August 2003; and email correspondence with sponsor, interviews, site visits, and site surveys.

Service	State/Country	Installation or Project Name	Brief Description	Completeness of Data[a]	Module[b]	Programmed Development/Upgrades[c]	Planned Year of Upgrade	Major Command[d]
Army Ranges								
Army (ARNG)	AK	Ft Greely	MAC	•	<, x2			USARPAC NGB
Army	AK	Ft Wainwright (Yukon)	MAC, shoot house	•	^, 9			USARPAC
Army	AK	Ft Chaffee	MOUT	•	3			
Army	AK	Ft Richardson	Shoot house	•	9			USARPAC
Army	AK	USARAK DTA	CACTF, modular MOUT	•	3	3		USARAK
Army	AL	Ft McClellan	CTF small	•	3			
Army	AR	Camp Robinson	MAC	•	<			
Army (USAR)	CA	Camp Parks	Shoot house	•		9	FY06	USARC
Army (ARNG)	CA	Camp Roberts	UAC, CACTF, shoot house	•		^, 3, 9	FY07–FY11	NGB
Army (ARNG)	CA	Camp San Luis Obispo	MAC, CTF small	••	^, 3 x2			NGB
Army	CA	Ft Hunter Liggett	Shoot house	•	9	9	FY06	USARC

a Indicates extent of data collected by RAND: • little information beyond name and location, least confidence in data; •• some detailed information, but less than 50 percent of survey fields; ••• almost complete information (visit, complete survey, other highly detailed data).

b Best guess at closest "module" (see Chapter Five) represented by existing training facilities; "completeness of data" provides an index of confidence in that assessment. See Appendix G for complete key; common entries: ^ = MAC or UAC, 3 = platoon purpose-built facility; 9 = shoot house.

c Primary source is not publicly available; it is not clear whether these proposed actions will be fully implemented or whether the notations are comprehensive.

d Command is noted where known; otherwise, cell is left blank.

Service	State/ Country	Installation or Project Name	Brief Description	Complete- ness of Data[a]	Module[b]	Programmed Develop- ment/ Upgrades[c]	Planned Year of Upgrade	Major Command[d]
Army	CA	Ft Irwin	NTC, CACTF, CTF large	•	3, 4, 8	3, 3+	FY05, FY06, FY07	FORSCOM
Army	CO	Ft Carson	MAC, CACTF, shoot house, CTF large	•••	3	3, 9	FY06	FORSCOM
Army (ARNG)	IA	Camp Dodge	MAC/UAC, MOUT	••	^, 3			NGB
Army (ARNG)	ID	Orchard Range Gowan	MAC, CACTF, shoot house	•		^, 3, 9	FY08, FY10, FY11	NGB
Army (ARNG)	IN	Camp Atterbury	UAC, CACTF	••		^, 3, 9	FY07–FY11	NGB
Army	GA	Ft Benning	MAC, CACTF, shoot house	•••	^ x 2, 9	3	FY08, FY09	TRADOC
Army	GA	Ft Stewart	MAC, UAC, CACTF, shoot house	•	^, 3	^, 3, 9	FY07, FY08	FORSCOM USASOC
Army	HI	Schofield Barracks	MAC, CACTF, live fire MOUT	•	^, 3	3 x2	FY06	USARPAC
Army	KS	Ft Riley	UAC, CACTF, shoot house	•••		^, 3, 9	FY05, FY07	FORSCOM
Army	KY	Ft Campbell	UAC, CACTF, shoot house	•	^ x4, 3, 9 x2	^ x2	FY05, FY06	FORSCOM, USASOC
Army	KY	Ft Knox	UAC, shoot house, CTF large	••	3	^, 9	FY07, FY09	TRADOC

[a] Indicates extent of data collected by RAND: • little information beyond name and location, least confidence in data; •• some detailed information, but less than 50 percent of survey fields; ••• almost complete information (visit, complete survey, other highly detailed data).

[b] Best guess at closest "module" (see Chapter Five) represented by existing training facilities; "completeness of data" provides an index of confidence in that assessment. See Appendix G for complete key; common entries: ^ = MAC or UAC, 3 = platoon purpose-built facility; 9 = shoot house.

[c] Primary source is not publicly available; it is not clear whether these proposed actions will be fully implemented or whether the notations are comprehensive.

[d] Command is noted where known; otherwise, cell is left blank.

Service	State/Country	Installation or Project Name	Brief Description	Completeness of Data[a]	Module[b]	Programmed Development/Upgrades[c]	Planned Year of Upgrade	Major Command[d]
Army	LA	Ft Polk	JRTC, live fire village, shoot house, CTF large	•••	3 x2, 3+, 4, 7, 8, 9	3	FY05	FORSCOM
Army (ARNG)	MA	Devens RFTA	UAC	••		^	FY08	USARC NGB
Army	MD	Aberdeen Proving Grounds	Joint Warfighting Range Complex Reconfigurable MOUT	••				
Army (ARNG)	ME	BOG Brook/Riley	CTF small	•	3 x2			NGB
Army	MI	Ft Custer Training Center	MAC, CTF small	•	^, 3			
Army (ARNG)	MN	Camp Ripley	MOUT small, shoot house	•	3, 9			NGB
Army	MO	Ft Leonard Wood	UAC, CTF small	••	3	^	FY06	TRADOC
Army (ARNG)	MS	Camp Shelby	UAC, CACTF, shoot house	•		^, 3, 9	FY08	NGB
Army (ARNG)	MT	Ft William Henry Harrison	MAC	••	^			NGB

[a] Indicates extent of data collected by RAND: • little information beyond name and location, least confidence in data; •• some detailed information, but less than 50 percent of survey fields; ••• almost complete information (visit, complete survey, other highly detailed data).
[b] Best guess at closest "module" (see Chapter Five) represented by existing training facilities; "completeness of data" provides an index of confidence in that assessment. See Appendix G for complete key; common entries: ^ = MAC or UAC, 3 = platoon purpose-built facility; 9 = shoot house.
[c] Primary source is not publicly available; it is not clear whether these proposed actions will be fully implemented or whether the notations are comprehensive.
[d] Command is noted where known; otherwise, cell is left blank.

Service	State/ Country	Installation or Project Name	Brief Description	Complete- ness of Data[a]	Module[b]	Programmed Develop- ment/ Upgrades[c]	Planned Year of Upgrade	Major Command[d]
Army	NC	Ft Bragg	MAC, UAC, CACTF, shoot house, CTF large	••	^ x3, 3 x2	^, 9	FY05, FY06	FORSCOM, TRADOC, USASOC
Army (ARNG)	NE	Greenlief Training Site	CTF small	••	3			NGB
Army (ARNG) (USAR)	NJ	Ft Dix	UAC, CACTF, shoot house, CTF small	•	^, 3 x2	3, 9	FY10, FY11	USARC NGB
Army	NJ	Ft Monmouth	CTF small	•	3 x2			AMC
Army	NY	Ft Drum	MAC, UAC, CACTF, CTF large	••	3	^, 3 x2	FY05	FORSCOM
Army (ARNG)	OK	Camp Gruber	UAC, CACTF	•••	^, 3			NGB
Army	OK	Ft Sill	UAC	•	^ x2			TRADOC
Army (ARNG)	PA	Ft Indiantown Gap	MAC, UAC	•	^, 9	^, 9	FY06	NGB
Army	SC	Ft Jackson	UAC, CTF small	•	3	^	FY07	TRADOC
Army	TX	Camp Bullis	UAC, CACTF	•		^, 3	FY05, FY08	MEDCOM
Army	TX	Ft Bliss	MOUT	••		3, 9	FY06	TRADOC
Army	TX	Ft Hood	MAC, UAC, CACTF, shoot house, CTF small, CTF large	••	^, 3 x2	^, 3, 9	FY05, FY07, FY09	FORSCOM

a Indicates extent of data collected by RAND: • little information beyond name and location, least confidence in data; •• some detailed information, but less than 50 percent of survey fields; ••• almost complete information (visit, complete survey, other highly detailed data).

b Best guess at closest "module" (see Chapter Five) represented by existing training facilities; "completeness of data" provides an index of confidence in that assessment. See Appendix G for complete key; common entries: ^ = MAC or UAC, 3 = platoon purpose-built facility; 9 = shoot house.

c Primary source is not publicly available; it is not clear whether these proposed actions will be fully implemented or whether the notations are compre- hensive.

d Command is noted where known; otherwise, cell is left blank.

Service	State/Country	Installation or Project Name	Brief Description	Completeness of Data[a]	Module[b]	Programmed Development/Upgrades[c]	Planned Year of Upgrade	Major Command[d]
Army (ARNG)	UT	Camp Williams	MAC, CTF small	••	^, 3			NGB
Army	UT	Dugway Proving Ground	MAC	•	^			ATEC
Army	VA	Ft A.P. Hill	UAC, CACTF, shoot house	•		^, 3, 9	FY05, FY07	MDW
Army (ARNG)	VA	Ft Pickett	MAC, UAC	••	^, 3	^, 3, 9	FY05, FY10, FY11	NGB
Army	WA	Ft Lewis	MAC, shoot house, CTF	••	^ x7, 3, 9			FORSCOM USASOC
Army	WA	Ft Lewis (Yakima)	UAC, CACTF, shoot house	•	9	2	FY07, FY09	FORSCOM
Army (USAR)	WI	Ft McCoy	MAC, shoot house, CTF small	••	^ x2, 3	3, 9	FY06, FY10	USARC
Army (ARNG)	WV	Camp Dawson	MAC	•	^			NGB
Army	Afghanistan	Bagram	Afghani village	•				
Army	Germany	Baumholder LTA	CTF large	••	3			USAREUR
Army	Germany	Boeblingen LTA	MAC	••	^			USAREUR

a Indicates extent of data collected by RAND: • little information beyond name and location, least confidence in data; •• some detailed information, but less than 50 percent of survey fields; ••• almost complete information (visit, complete survey, other highly detailed data).
b Best guess at closest "module" (see Chapter Five) represented by existing training facilities; "completeness of data" provides an index of confidence in that assessment. See Appendix G for complete key; common entries: ^ = MAC or UAC, 3 = platoon purpose-built facility; 9 = shoot house.
c Primary source is not publicly available; it is not clear whether these proposed actions will be fully implemented or whether the notations are comprehensive.
d Command is noted where known; otherwise, cell is left blank.

Service	State/Country	Installation or Project Name	Brief Description	Completeness of Data[a]	Module[b]	Programmed Development/Upgrades[c]	Planned Year of Upgrade	Major Command[d]
Army	Germany	Grafenwoehr Training Area	UAC, CACTF, Shoot house, CTF small	••	3	^, 3, 9	FY06	USAREUR
Army	Germany	Hohenfels	CMTC CTF small, CTF large	••	3, 3+			USAREUR
Army	Germany	Klosterforst	CTF small	•	3			USAREUR
Army	Germany	Mainz-Layenhof	CTF small	•	3			USAREUR
Army	Germany	Schweinheim-Aschaffenburg LTA	CTF small	•	3			USAREUR
Army	Germany	Stuttgart	UAC	•		^		
Army	Italy	Vicenza	UO training facility	•	^			
Army	Japan	Okinawa	MOUT	•	3	^	FY07	USASOC
Army	Korea	Rodriguez	UAC, CACTF, shoot house, CTF large	•••	3	^ x2, 9	FY06, FY07	EUSA
Army	Kuwait	Camp Udari	Mobile MOUT	•				

a Indicates extent of data collected by RAND: • little information beyond name and location, least confidence in data; •• some detailed information, but less than 50 percent of survey fields; ••• almost complete information (visit, complete survey, other highly detailed data).

b Best guess at closest "module" (see Chapter Five) represented by existing training facilities; "completeness of data" provides an index of confidence in that assessment. See Appendix G for complete key; common entries: ^ = MAC or UAC, 3 = platoon purpose-built facility; 9 = shoot house.

c Primary source is not publicly available; it is not clear whether these proposed actions will be fully implemented or whether the notations are comprehensive.

d Command is noted where known; otherwise, cell is left blank.

Service	State/Country	Installation or Project Name	Brief Description	Completeness of Data a	Module b	Programmed Development/Upgrades c	Planned Year of Upgrade	Major Command d
NAVY RANGES								
Navy	AZ	Yuma Training Range Complex		•				
Navy	CA	San Clemente Island	Maritime MOUT	••		3	FY05	
Navy	NV	Fallon NAS	MOUT	••		3	FY08	
Navy	VA	Ft. Story	MOUT and tactical breacher	••		3	FY09	NSWG-2 on Army installation
Navy	Albania	Sazan Island		•				
Navy		Battle Fleet Tactical Trainer	Special trainer	•	^			
MARINE RANGES								
Marines	AZ	Urban Target Facility (UTC) or "Yodaville"	Urban CAS	•••	8 x2			
Marines	CA	Camp Pendleton	Combat town, unsophisticated city, sophisticated city	•••	3 x4	3	FY08	
Marines	CA	29 Palms Range	MAC, Mega MOUT	•••	3, 4	1, 8?	Proposed	
Marines	NC	Camp LeJeune	CTF	••	3 x2	3	FY06	

a Indicates extent of data collected by RAND: • little information beyond name and location, least confidence in data; •• some detailed information, but less than 50 percent of survey fields; ••• almost complete information (visit, complete survey, other highly detailed data).

b Best guess at closest "module" (see Chapter Five) represented by existing training facilities; "completeness of data" provides an index of confidence in that assessment. See Appendix G for complete key; common entries: ^ = MAC or UAC, 3 = platoon purpose-built facility; 9 = shoot house.

c Primary source is not publicly available; it is not clear whether these proposed actions will be fully implemented or whether the notations are comprehensive.

d Command is noted where known; otherwise, cell is left blank.

Service	State/Country	Installation or Project Name	Brief Description	Completeness of Data[a]	Module[b]	Programmed Development/Upgrades[c]	Planned Year of Upgrade	Major Command[d]
Marines	VA	Naval Support Activity Northwest	CQB facility, Simunitions, range	•	^, 3 x2			
Marines	VA	Quantico	MOUT II, combat town	•	3			
Marines	Guam	Anderson AFB South		••				
Marines	Japan	Camp Hansen		•	3			
AIR FORCE RANGES								
Air Force	CA	George AFB (BRAC)		•	18			
Air Force	FL	Avon Park	MOUT	••	3 x2			
Air Force	FL	Eglin/Hurlburt AFB	JUO training center	••		3	FY06	
Air Force	GA	Moody AFB Grand Bay Range	MOUT	••	3	3	unknown	
Air Force	ID	Mountain Home AFB	Juniper Butte bombing range, Saylor Creek range	•	8			
Air Force	LA	Barksdale AFB	Air Warrior II, airfield used with Ft. Polk JRTC	•				
Air Force	NJ	Ft Dix/Eagle Flag	Live training, air base ops, convoy ops	•				
Air Force	NV	Nellis Test & Training Range	MOUT, urban target complex high technology TTR	••	3 x5, 4, 8 x5	3, 8	FY05	

[a] Indicates extent of data collected by RAND: • little information beyond name and location, least confidence in data; •• some detailed information, but less than 50 percent of survey fields; ••• almost complete information (visit, complete survey, other highly detailed data).

[b] Best guess at closest "module" (see Chapter Five) represented by existing training facilities; "completeness of data" provides an index of confidence in that assessment. See Appendix G for complete key; common entries: ^ = MAC or UAC, 3 = platoon purpose-built facility; 9 = shoot house.

[c] Primary source is not publicly available; it is not clear whether these proposed actions will be fully implemented or whether the notations are comprehensive.

[d] Command is noted where known; otherwise, cell is left blank.

Service	State/Country	Installation or Project Name	Brief Description	Completeness of Data [a]	Module [b]	Programmed Development/Upgrades [c]	Planned Year of Upgrade	Major Command [d]
Air Force	SC	Shaw AFB	Poinsett electronic combat range target	•				
Air Force	TX	Camp Bullis	MOUT	••	3			
Air Force	TX	Ft Bliss/Holloman AFB Centennial Bombing and Gunnery Range	Urban target complex	•	3	3, 8	FY05 FY06	
Air Force	TX	Lackland AFB	MOUT	•	3			
Air Force	UT	Utah Test & Training Range	UOT	••	3			
PRIVATE, INTERNATIONAL, or "OTHER" RANGES								
Other	IN	Muscatatuck	Joint ops urban training	•••	17			ING
Other	MA	NERTC	Joint combined ops in urban areas	••	3 x2			MANG
Other		Mobile MOUT	Afghanistan, Iraq, misc. CONUS	••				
Other	NC	Blackwater	Training complex	••	4			
Other	CA	Segall Studios	Training complex	•				
Other	NV	Former Nuclear Test Site	Counterterrorism center	•			Proposed	
Other	ND	Various	Abandoned town	•			Proposed	

a Indicates extent of data collected by RAND: • little information beyond name and location, least confidence in data; •• some detailed information, but less than 50 percent of survey fields; ••• almost complete information (visit, complete survey, other highly detailed data).
b Best guess at closest "module" (see Chapter Five) represented by existing training facilities; "completeness of data" provides an index of confidence in that assessment. See Appendix G for complete key; common entries: ^ = MAC or UAC, 3 = MAC or UAC, 3 = platoon purpose-built facility; 9 = shoot house.
c Primary source is not publicly available; it is not clear whether these proposed actions will be fully implemented or whether the notations are comprehensive.
d Command is noted where known; otherwise, cell is left blank.

Service	State/Country	Installation or Project Name	Brief Description	Completeness of Data[a]	Module[b]	Programmed Development/Upgrades[c]	Planned Year of Upgrade	Major Command[d]
Other	NM	Playas	Abandoned town	•••	17			
Other	VA	FBI Hogan's Alley	Law enforcement training facility	•				
Other	Germany	Hamelburg	MOUT	•				
Other	Germany	Erlangen	MOUT	•				
Other	Germany	Bergenhoen	MOUT	•				
Other	Israel	Camp Lahisch	Warfare center	•				
Other	Netherlands	Oostdorp	Urban training facility	••				
Other	Netherlands	Marnehuizen	Urban training facility	••				
Other	Morocco	Tan Tan Training Area	Ft Aoreora, ERC building area	•				
Other	Kent, UK	Lydd and Hythe	Used by Army	•	3			
Other	Wiltshire, UK	Salisbury Plain	Used by Army	•	3			
Other	Tain, UK	Castlemartin	Field training center	•				

a Indicates extent of data collected by RAND: • little information beyond name and location, least confidence in data; •• some detailed information, but less than 50 percent of survey fields; ••• almost complete information (visit, complete survey, other highly detailed data).

b Best guess at closest "module" (see Chapter Five) represented by existing training facilities; "completeness of data" provides an index of confidence in that assessment. See Appendix G for complete key; common entries: ^ = MAC or UAC, 3 = platoon purpose-built facility; 9 = shoot house.

c Primary source is not publicly available; it is not clear whether these proposed actions will be fully implemented or whether the notations are comprehensive.

d Command is noted where known; otherwise, cell is left blank.

Service	State/ Country	Installation or Project Name	Brief Description	Complete- ness of Data[a]	Module[b]	Programmed Develop- ment/ Upgrades[c]	Planned Year of Upgrade	Major Command[d]
Other	Salis- bury Plains, UK	Copehill Down vil- lage	Urban training center	••				
Other	Shorn- cliffe, UK	Ricksborough bar- racks		••				

[a] Indicates extent of data collected by RAND: • little information beyond name and location, least confidence in data; •• some detailed information, but less than 50 percent of survey fields; ••• almost complete information (visit, complete survey, other highly detailed data).

[b] Best guess at closest "module" (see Chapter Five) represented by existing training facilities; "completeness of data" provides an index of confidence in that assessment. See Appendix G for complete key; common entries: ^ = MAC or UAC, 3 = platoon purpose-built facility; 9 = shoot house.

[c] Primary source is not publicly available; it is not clear whether these proposed actions will be fully implemented or whether the notations are compre- hensive.

[d] Command is noted where known; otherwise, cell is left blank.

Training Retention

Knowledge and skill retention are of particular interest in this study because retention affects the demand for training throughput, influencing training frequency requirements and thereby potentially contributing to gaps between requirements and capabilities. Unfortunately, it is very difficult to accurately pinpoint rates of skill retention. Just as different skills require different teaching methods, skills decay at varying rates.

Factors Affecting Retention

Many factors influence retention. These are interrelated and highly context-driven, affected by the characteristics of the learner and the skill. Some important considerations in retention/skill decay are

- Similarity of the new, "target" skill to existing skills possessed by the learner;
- Level of automaticity of both the existing and new skills (skill performance is fluid and individualized; the learner easily adapts behaviors to new settings);[1]

---

[1] Kurt Kraiger, Kevin J. Ford, and Eduardo Salas, "Application of Cognitive, Skill-Based and Affective Theories of Learning Outcomes to New Methods of Training Evaluation," *Journal of Applied Psychology, Monograph,* Vol. 2, No. 2, 1993, pp. 311–328.

- Whether the practice time for the new skill was "massed" (done all at once in a short time period) or "distributed" (done over a longer time, with time between practice sessions);
- Length of time between last use and current use (delay);
- Complexity of the skill;
- Whether the skill is "motor-" or cognition-centric;
- Amount of automaticity desired in the performance of the target skill.[2]

Not only will different learners have varying skill retention rates, but one learner will differentially retain different types of skills. Relearning skills includes the concept of "savings," that is, the amount of time it takes to regain proficiency. Different kinds of knowledge (motor, cognitive) are retained differently and are more or less easily accessed. Riding a bicycle is a skill with a great deal of "savings" in that most people can get back on a bicycle after years of delay and ride fairly well. Practice is needed to perform more-complicated maneuvers, but the basic skills remain intact and easily accessible. On the other hand, it may take a long time to relearn how to program in a computer language the individual last used long ago.[3]

Despite significant interest in this area, there are no clear formulas to determine how often a given individual or team needs to practice a certain task. For instance, it would be very difficult to determine how frequently a unit should conduct training on room-clearing in order to maintain combat proficiency, or how often it should practice low-threat interrogation skills for use at vehicle check points.

There have, however, been some important attempts to model skill decay. The U.S. Army Research Institute for the Behavioral and

---

[2] Barbara Raymond and Matthew W. Lewis, email correspondence, January 11, 2005.

[3] Barbara Raymond and Matthew W. Lewis, correspondence.

Social Sciences (ARI) has identified a set of 10 criteria for predicting retention:[4]

1. Presence of job aids
2. Quality of job aids
3. Number of steps
4. Sequence requirements
5. Feedback
6. Time requirements
7. Mental requirements
8. Number of facts
9. Difficulty of facts
10. Motor-control requirements

ARI applied this model to 16 common individual-level tasks. Subject-matter experts evaluated each task against the above criteria, and researchers then used the data to generate a procedural complexity score. This score was converted to a predicted retention rate, defined as the time at which 70 percent of the subjects would receive a "go" (passing score).[5] For example, the task of "maintaining your assigned protective mask" was predicted to be retained without recurrent training for more than two years, while "moving under tactical fire" required training every 10 weeks; and it was predicted that "evaluating a casualty" would need training every two weeks to maintain the 70 percent standard.

While the ARI model provides a useful foundation for beginning to assess individuals' skill retention, it does not readily accommodate collective retention in general or joint urban training skills in particular. Collective skills, especially those of a large, distributed joint force team, would be well beyond the bounds of this model.

---

[4] U.S. Army Research Institute for the Behavioral and Social Sciences, "The Computer Backgrounds of Soldiers," *ARI Newsletter*, Vol. 12, No. 2, Winter 2002, pp. 11–12, available online at http://www.hqda.army.mil/ari/pdf/winter2002_vol12_no2.pdf (accessed January 14, 2005).

[5] Ibid., p. 11.

Additionally, the criteria used here are not suited to the assessment of retention of higher-order cognitive processing capacity or affective outcomes. Finally, and very importantly, trainee characteristics such as general cognitive ability, motivation to learn, opportunities to perform trained skills, and variables such as age, rank, education, and experience are not accounted for in the model.

Other organizations and industries are also interested in learning retention. While results are not consistently clear, and more needs to be understood before results from other disciplines can be generalized, it is sometimes possible to draw inferences from other studies. A study of flight-skill decay supported the bicycle-riding example: Motor or control-oriented skills are less prone to decay with lack of use than are cognitive/procedural skills. The study therefore encouraged the use of recurrent cognitive (knowledge) training methods.[6] Much joint force training involves skills that make demands on cognitive and procedural abilities. Virtually all skills at higher echelons do so, thus likely necessitating fairly frequent retraining.

Computer Simulations May Aid Retention

The regular use of skills is one of the best ways to prevent skill decay. As discussed above, the trick is to determine the frequency that is sufficient for any given skill and learner. The appropriate form of refresher training is also unclear. To illustrate this challenge, paramedics tested on pediatric resuscitation were randomized into four groups and given a combination of knowledge examinations and/or mock resuscitation practice at six-month intervals. Twelve months after the initial training, all of the scores had returned to pre-education base-

---

[6] J. M. Childs and W. D. Spears, "Flight-Skill Decay and Recurrent Training," *Perceptual and Motor Skills*, Vol. 62, No. 1, February 1986, pp. 235–242, available online at http://www.ncbi.nlm.nih.gov/entrez/query.fcgi?CMD=Display&DB=pubmed (accessed January 14, 2005).

line levels regardless of the intervention.[7] The researchers recommended seeking novel ways to increase retention and ensure readiness.

One possible way to maintain skills is through computer-assisted learning. A study in the use of automatic external defibrillators found that computer-assisted learning could help maintain skills when interspersed with traditional training methods. The study concluded that computer learning may offer a cost-effective way to maintain skills,[8] since it has the flexibility to be conducted in different places and without instructors.

There are many computer-assisted training tools available to the military. Computer games, in particular, with sophisticated virtual warfare, are promising and cost-effective training tools. While they cannot replicate the real-world experience of war, they do provide a way to approximate critical tasks. For instance, simulating an exercise for a company commander can save the expense of assembling the entire company for a field training exercise. Further, computers offer an excellent opportunity to collect measurement data—a task that is time- and resource-intensive in live environments. However, whether skills developed in an online environment are easily transferred to the battlefield remains an open question. Anecdotal reports of sports video-game usage indicate that good live football players make good virtual football players, but whether the inverse is true has not been established.[9]

---

[7] Eustacia Su, Terri A. Schmidt, N. Clay Mann, and Andrew D. Zechnich, "A Randomized Controlled Trial to Assess Decay in Acquired Knowledge Among Paramedics Completing a Pediatric Resuscitation Course," *Academic Emergency Medicine,* Vol. 7, 2000, pp. 779–786, available online at http://www.aemj.org/cgi/content/abstract/7/7/779 (accessed January 14, 2005).

[8] Ibid.

[9] "Playing to Win," Technology Quarterly section, *The Economist,* Vol. 373, No. 8404, December 4, 2004, pp. 24–25.

Concluding Comments

The scale and complexity of joint urban training activities underscore the need for training planners to know more about learning retention. A better understanding of knowledge and skill retention, both individual and collective, would provide trainers with means to improve training strategies and dramatically improve training-material retention and scheduling of facilities. The result would be an even more effective U.S. joint force.

Matrix of Modules vs. Requirements

The symbols used in the matrix presented in this appendix are defined as follows:

C = crawl (necessary for basic building blocks in this joint training requirement area)

W = walk (necessary for intermediate building blocks in this joint training requirement area)

R = run (necessary for advanced/complex levels of mastery of this task/requirement)

S = support (adds realism or otherwise enhances training in support of a requirement)

Module	Avoid fratricide	Communicate in the urban environment	Conduct airspace coordination	Synchronize joint rules of engagement	Conduct stability operations in urban environment	Conduct support operations in urban environment	Conduct urban HUMINT operations	Conduct other urban intelligence efforts	Conduct urban operations exercises	Integrate urban ops with other relevant environments	Coordinate maneuver in the urban environment	Coordinate multinational and interagency resources	Govern in the urban environment	Navigate in the urban environment	Plan urban operations	Provide common situational awareness	Provide fire support	Provide security during urban transition operations	Rehearse/war-game urban operations	Conduct urban noncombatant evacuations (NEO)	Conduct U.S. domestic urban operations	Conduct urban combat search and rescue (CSAR)	Conduct urban ops during and after a WMD event	Consolidate success in the urban environment	Disembark, base, protect, move in urban environment	Engage in the urban environment	Orchestrate resources during urban operations	Shape the urban environment	Sustain urban operations	Transition to civilian control	Understand the urban environment	Achieve simultaneity in meeting requirements	Identify critical infrastructure nodes & system relations	Conduct training across multiple levels of war
Purpose-built facilities																																		
1. Battalion and larger purpose-built facility	R	R	R	R	R	R	W	R	R	W	R	R	W	W	R	R	W	W	R	W	W	W	W	W	W	R	R	W	W	W	W	W	R	W
2. Company purpose-built facility	W	W	C	W	C	C	C	W	W	C	W	C	C	C	R	W	C	C	R	C	C	R	C	C	C	C	C	C	W	W	W	W	W	W
3. Platoon purpose-built facility	W	W	C	C	C	C	C	C	C	C	C	C	C	C	R	C	C	C	C	R	C	C	C	C	C	C	C	W	C	C	C	C	W	W
7. Hybrid facility	R	R	W	R	W	W	W	W	R	W	R	R	W	C	R	R	W	W	R	W	W	W	W	W	R	R	R	W	R	W	W	R	W	W
8. Air-ground facility	W	C	R	C	C	C			W		W				C	W	W	W			C			C				C	W	W			C	C
Use of populated urban areas																																		
10. Terrain walks	W	C	C	C	C		C	C	W	C	C	C	R	C	C		C	W	C	W	C	C	C	C	C	C		C			W	C	R	W

Module	Avoid fratricide	Communicate in the urban environment	Conduct airspace coordination	Synchronize joint rules of engagement	Conduct stability operations in urban environment	Conduct support operations in urban environment	Conduct urban HUMINT operations	Conduct other urban intelligence efforts	Conduct urban operations exercises	Integrate urban ops with other relevant environments	Coordinate maneuver in the urban environment	Coordinate multinational and interagency resources	Govern in the urban environment	Navigate in the urban environment	Plan urban operations	Provide common situational awareness	Provide fire support	Provide security during urban transition operations	Rehearse/war-game urban operations	Conduct urban noncombatant evacuations (NEO)	Conduct U.S. domestic urban operations	Conduct urban combat search and rescue (CSAR)	Conduct urban ops during and after a WMD event	Consolidate success in the urban environment	Disembark, base, protect, move in urban environment	Engage in the urban environment	Orchestrate resources during urban operations	Shape the urban environment	Sustain urban operations	Transition to civilian control	Understand the urban environment	Achieve simultaneity in meeting requirements	Identify critical infrastructure nodes & system relations	Conduct training across multiple levels of war
15. Use of buildings scheduled for demolition	W	W	S	C	C	C	C	C	W	C	W			C	C	C	W	C	C	C	C	W	R	C	W	W	C	C	C	C	C	C	C	C
16. Use of public facilities during hours of closure	C	W	S	C	W	W	C	C	C	C	W	R		C	C	C	C	W	W	W	R	W	W	W	C	W	W	C	W	C	C	W	W	W
Alternative/other training concepts																																		
17. Use of abandoned domestic urban areas	W	W	R	R	R	W	R	R	R	R	W	R	R	R	W	R	R	W	R	R	W	W	W	W	R	R	R	W	R	W	W	R	R	R
18. BRAC'd military installations	R	W	R	R	R	W	W	R	W	R	W	R	R	W	R	R	R	R	W	W	R	R	R	W	R	W	W	R	R	R	W	R	R	R
21. Abandoned factories	W	C	W	C	C	C	C	C	C	C	C	C	C	C	C	C	C	C	C	C	C	C	C	C	C	C	C	W	W	C	C	C	W	W

Module	Avoid fratricide	Communicate in the urban environment	Conduct airspace coordination	Synchronize joint rules of engagement	Conduct stability operations in urban environment	Conduct support operations in urban environment	Conduct urban HUMINT operations	Conduct other urban intelligence efforts	Conduct urban operations exercises	Integrate urban ops with other relevant environments	Coordinate maneuver in the urban environment	Coordinate multinational and interagency resources	Govern in the urban environment	Navigate in the urban environment	Plan urban operations	Provide common situational awareness	Provide fire support	Provide security during urban transition operations	Rehearse/war-game urban operations	Conduct urban noncombatant evacuations (NEO)	Conduct U.S. domestic urban operations	Conduct urban combat search and rescue (CSAR)	Conduct urban ops during and after a WMD event	Consolidate success in the urban environment	Disembark, base, protect, move in urban environment	Engage in the urban environment	Orchestrate resources during urban operations	Shape the urban environment	Sustain urban operations	Transition to civilian control	Understand the urban environment	Achieve simultaneity in meeting requirements	Identify critical infrastructure nodes & system relations	Conduct training across multiple levels of war
22. Abandoned or constructed overseas urban areas	W	W	W	W	W	W	W	W	W	R	R	R	R	W	R	R	R	W	R	R	R	C	R	W	W	R	R	R	C	R	W	W	R	R
24. Classroom instruction	C	C	C	C	C	W	C	C	W	C	W	W	C	W	C	C	C	C	C	C	W	C	C	C	C	C	W	C	C	C	W	W	R	W
25. Conduct of combatant command or JTF headquarters, large-scale schools, or multi-echelon/interagency exercises	W	W	C	R	R	W	W	R	R	W	R	R			R	W	C	C	W	W	W	C	W	W	W	W	R	W	W	W	R	R	R	R

Module	Avoid fratricide	Communicate in the urban environment	Conduct airspace coordination	Synchronize joint rules of engagement	Conduct stability operations in urban environment	Conduct support operations in urban environment	Conduct urban HUMINT operations	Conduct other urban intelligence efforts	Conduct urban operations exercises	Integrate urban ops with other relevant environments	Coordinate maneuver in the urban environment	Coordinate multinational and interagency resources	Govern in the urban environment	Navigate in the urban environment	Plan urban operations	Provide common situational awareness	Provide fire support	Provide security during urban transition operations	Rehearse/war-game urban operations	Conduct urban noncombatant evacuations (NEO)	Conduct U.S. domestic urban operations	Conduct urban combat search and rescue (CSAR)	Conduct urban ops during and after a WMD event	Consolidate success in the urban environment	Disembark, base, protect, move in urban environment	Engage in the urban environment	Orchestrate resources during urban operations	Shape the urban environment	Sustain urban operations	Transition to civilian control	Understand the urban environment	Achieve simultaneity in meeting requirements	Identify critical infrastructure nodes & system relations	Conduct training across multiple levels of war
Simulation capabilities																																		
26. Tactical behaviors in and around structures	W	S	C	C	W	C	C	C	C	S	W	S	S	C	W	C	W	C	W	S	S	C	C	W	W	W	S	C	W	S	W	W	W	S
27. Higher-echelon planning and coordination	C	S	W	W	W	C		C	C	C	C	W	S	C	W	C	W	C	W	S	S	S	S	W	W	C	C	W	W	W	W	W	R	C
28. Joint, multinational, and interagency operations	C	S	W	W	W	C	C	C	C	C	C	W	S	W	C	C	C	W	C	S	S	S	W	C	C	C	C	W	W	W	W	W	R	C
29. Specialized-technology simulation	W	C	C	S	C	C	C	S	C		S	S		W		W	W	C			S	S	S		S	C	S	C			C	S	S	

Module	Avoid fratricide	Communicate in the urban environment	Conduct airspace coordination	Synchronize joint rules of engagement	Conduct stability operations in urban environment	Conduct support operations in urban environment	Conduct urban HUMINT operations	Conduct other urban intelligence efforts	Conduct urban operations exercises	Integrate urban ops with other relevant environments	Coordinate maneuver in the urban environment	Coordinate multinational and interagency resources	Govern in the urban environment	Navigate in the urban environment	Plan urban operations	Provide common situational awareness	Provide fire support	Provide security during urban transition operations	Rehearse/war-game urban operations	Conduct urban noncombatant evacuations (NEO)	Conduct U.S. domestic urban operations	Conduct urban combat search and rescue (CSAR)	Conduct urban ops during and after a WMD event	Consolidate success in the urban environment	Disembark, base, protect, move in urban environment	Engage in the urban environment	Orchestrate resources during urban operations	Shape the urban environment	Sustain urban operations	Transition to civilian control	Understand the urban environment	Achieve simultaneity in meeting requirements	Identify critical infrastructure nodes & system relations	Conduct training across multiple levels of war
30. Scenario-variant generation	C	S	S	S	S	S	S	S	S	S	S	S	S	S	S	S	S	S	S	S	S	S	S	S	S	S	S	S	S	S	S	S	S	S
32. Geographically distributed joint simulation		W	S	S			S	S	S	S						S	S	C	S	S	S	S	S	S	S	S	S	S	S	S	S	C	S	S
33. Environmental degradation and urban biorhythm	S			S	S	S	S	S	S	S	S	S	S			S	S	S	S	S	S	S	S	S	S	S	S	S	S	C		S		

Module	Avoid fratricide	Communicate in the urban environment	Conduct airspace coordination	Synchronize joint rules of engagement	Conduct stability operations in urban environment	Conduct support operations in urban environment	Conduct urban HUMINT operations	Conduct other urban intelligence efforts	Conduct urban operations exercises	Integrate urban ops with other relevant environments	Coordinate maneuver in the urban environment	Coordinate multinational and interagency resources	Govern in the urban environment	Navigate in the urban environment	Plan urban operations	Provide common situational awareness	Provide fire support	Provide security during urban transition operations	Rehearse/war-game urban operations	Conduct urban noncombatant evacuations (NEO)	Conduct U.S. domestic urban operations	Conduct urban combat search and rescue (CSAR)	Conduct urban ops during and after a WMD event	Consolidate success in the urban environment	Disembark, base, protect, move in urban environment	Engage in the urban environment	Orchestrate resources during urban operations	Shape the urban environment	Sustain urban operations	Transition to civilian control	Understand the urban environment	Achieve simultaneity in meeting requirements	Identify critical infrastructure nodes & system relations	Conduct training across multiple levels of war
Training support elements																																		
34. Infra-structural trappings	S		S	S	S		S	S	S	S		S	S		S	S	S		S	S	S	S	C	S	S	S	S	S	S	S	C	S	S	
35. OPFOR	S	C	S	S	W	S	C	S	S	S	S	S	S	S	S	S	S	S	S	S	S	S	S	S	S	S	S	W	S	S	S	W	S	W	S
36. Non-combatant role-players	S	W	S	S	W	W	C	C	S	S	S	S	S	S	S	S	S	S	S	S	W	S	S	W	S	S	C	W	C	W	W	S	W	S
37. Targets	S		S	S			S	S					S				S			S	S	S		S	C	S	S	S					S	
38. Instrumentation/connectivity	S	W	S	S	S	S			S		S	S		S		S	S	S		S	S	S	S	S	S	S		S			C	S	S	
39. Joint force headquarters	W	W	W	W	W	W	C	C	S	C	W	C	S	S	S	S	S	S	S	S	C	C	S	S	C	S	C	S	C	S	C	W	C	S	S	C

Bibliography

Books

Adkin, Mark, *Operation Urgent Fury: The Battle for Grenada*, London: Leo Cooper, 1989.

Collins, Larry, and Dominique Lapierre, *O Jerusalem!* NY: Simon & Schuster, 1972, pp. 50, 104.

Glenn, Russell W., *Reading Athena's Dance Card: Men Against Fire in Vietnam*, Annapolis, MD: Naval Institute Press, 2000.

Huntington, Samuel P., *The Soldier and the State: The Theory and Politics of Civil-Military Relations*, New York: Vintage, 1957.

Articles

Barnes, William F., "An Overview of Internal Training: Maintaining Individual Proficiency Within the SJFHQ (CE)," *Joint Center for Lessons Learned Quarterly Bulletin,* No. 6, June 2004, pp. 26–30.

Childs, J. M., and W. D. Spears, "Flight-Skill Decay and Recurrent Training," *Perceptual and Motor Skills,* Vol. 62, No. 1, February 1986, pp. 235–242, online at http://www.ncbi.nlm.nih.gov/entrez/query.fcgi?CMD=Display&DB=pubmed (accessed January 14, 2005).

Colinmore, Ed, "Iraqis' Roles Aid Training at Fort Dix," *Philadelphia Inquirer,* December 30, 2004, online at http://ebird.afis.osd.mil.ebfiles/e20041230343620.html (accessed January 5, 2005).

Cosenza, Charles W., "Standing Joint Force Headquarters (Core Element): Its Origin, Implementation and Prospects for the Future," *Joint Center for Lessons Learned Quarterly Bulletin*, No. 6, June 2004, pp. 3–8.

Crane, David, "Valhalla Training Center, LLC: The Future of Tactical Training Schools?" *Defense Review*, online at http://www.defensereview. com/modules.php?name=News&file=article&sid=545, posted July 8, 2004.

Deutsch, Kevin, "Blasts Will Trigger Terrorism Drill," *The Miami Herald*, May 15, 2004.

"Dutch Army Orders Urban Training Equipment," *Military Technology*, Vol. 27, May 2003, p. 46.

Erwin, Sandra I., "Dangerous Convoy Duties Prompt Expanded Training for Truck Crews," *National Defense*, December 2004, online at http:// www.nationaldefensemagazine.org/issues/2004/Dec/DangerousConvoy Duties.htm (accessed October 26, 2005).

Erwin, Sandra I., "On-the-Move, Combined Arms Training Available to Soldiers," *National Defense,* November 2000, online at http:// www.nationaldefensemagazine.org/issues/2000/Nov/On-the-Move.htm (accessed October 26, 2005).

"Exercise Program," *Inside the Pentagon*, May 27, 2004, p. 5.

"Facing Urban Inevitabilities," *Jane's International Defense Review*, August 2001, pp. 39ff.

Fawcett, John, "Distributed Mission Operations and Distributed Mission Training," *Military Technology*, Vol. 28, April 2004, pp. 24–30.

Garreau, Joel, "Reboot Cap: As War Looms, the Marines Test New Networks of Comrades," *The Washington Post*, March 24, 2004, p. C1.

Glasser, Susan B., "Urban Combat Training Gets Real," *The Washington Post Foreign Service,* February 4, 2003, p. A17, online at www. washingtonpost.com/ac2/wp-dyn/A20810-2003Feb3 (accessed February 3, 2003).

Gourley, Scott, "Training for the Ambush," *Military Training Technology*, October 27, 2004, online at http://www.military-training-technology. com/article.cfm?DocID=663 (accessed October 26, 2005).

Grossman, Elaine M., "No Time to Spare Before Iraq Deployment? Trainers Make House Calls," *Inside the Army*, January 17, 2005.

Harris, David, *Support to the Warfighter: Fort Lewis Gets Major Urban Warfare Site*, online at http://www.hq.usace.army.mil/cepa/pubs/jan03/story16.htm.

Hodge, Nathan, "More Training Facilities for Urban Combat on Order," *Defense Today*, June 8, 2004, p. 1.

Kauka, Dick, "Antiterrorism Center Planes: Muscatatuck Set to Be Training Site," *The Courier-Journal*, Louisville, KY, September 2, 2004.

Kneebone, Roger, and David ApSimon, "Surgical Skills Training: Simulation and Multimedia Combined," *Medical Education*, Vol. 35, Issue 9, September 2001, p. 909, online at http://www.blackwell-synergy.com/links/doi/10.1046/j.1365-2923.2001.00997.x/abs/ (accessed January 14, 2005).

Kraiger, Kurt, Kevin J. Ford, and Eduardo Salas, "Application of Cognitive, Skill-Based and Affective Theories of Learning Outcomes to New Methods of Training Evaluation," *Journal of Applied Psychology Monograph*, Vol. 2, No. 2, 1993, pp. 311–328.

Krulak, Charles C., "The Strategic Corporal: Leadership in the Three Block War," *Marines Magazine*, January 1999, online at http://www.au.af.mil/au/awc/awcgate/usmc/strategic_corporal.htm (accessed January 12, 2005).

Louwage, Pam, "Marines Storm N.O. for Combat Training," *New Orleans Times-Picayune*, April 1, 1998, p. B1.

Naylor, Sean D., "The Coming Brigade Shuffle: How Adding New Combat Units Will Radically Alter the Army's U.S. Footprint," *Army Times*, January 31, 2005.

Naylor, Sean D., "Playing to Win," *The Economist*, Vol. 373, No. 8404, December 4, 2004, Technology Quarterly section, pp. 24–25.

"Predicting the Retention of Proficiency at 16 Common Tasks," *ARI Newsletter*, Vol. 12, No. 2 (undated), pp. 10–12.

Pugliese, David, "Mock City May Be Built for Antiterrorism Training," *Army Times*, March 8, 1999, p. 27.

Rivera, Jorge, and Joseph A. Giunta, Jr., "10th Mountain Division's Own Urban Battleground," *Training and Simulation*, Summer 2001.

Robar, Jason, "Multiplayer Technology: A Primer," *Modeling, Simulation, and Training*, Issue 6, 2004, pp. 26–29.

Robson, Seth, "Soldiers Learn Art of Clearing Rooms in Shoot-House," *Pacific Stars and Stripes*, July 28, 2004.

Sheridan, Greg, "Down Under Base Not on US Agenda," *The Australian*, June 18, 2004.

Slear, Tom, "JNTC: An Efficient Solution to a Vexing Problem," *Modeling, Simulation, and Training*, Issue 6, 2004, pp. 14–19.

Stot, Jim, "AFSOC Range Initiatives," Air Force Special Operations Command, Briefing, May 17, 2004.

Su, Eustacia, Terri A. Schmidt, N. Clay Mann, and Andrew D. Zechnich, "A Randomized Controlled Trial to Assess Decay in Acquired Knowledge Among Paramedics Completing a Pediatric Resuscitation Course," *Academic Emergency Medicine*, Vol. 7, 2000, pp. 779–786, online at http://www.aemj.org/cgi/content/abstract/7/7/779 (accessed January 14, 2005).

"Terrorism Drill Tests Cruise Ship Crisis at Port Canaveral," *Florida Today*, May 19, 2004, online at http://www.floridatoday.com/topstories/051904terrordrill.htm (accessed May 19, 2004).

Tiron, Roxana, "Army Training Fails to Address 'Enormity of the Urban Problem,'" *I/ITSEC Showdaily*, December 8, 2004, pp. 1, 3.

Tiron, Roxana, "Pilots Spurring Training, Tactics Revolution," *National Defense*, June 2004, online at http://www.nationaldefensemagazine.org/issues/2004/Jun/Pilots.htm (accessed January 11, 2005).

Tiron, Roxana, "SOCOM a Trailblazer for Joint Training," *National Defense*, February 2004, online at http://www.nationaldefensemagazine.org/issues2004/Feb/SOCOM_a_Trailblazer.htm (accessed October 26, 2005).

United States Marine Corps, *Concept Report: Large Scale MOUT Feasibility Study*, Marine Air Ground Task Force Training Command (MAGTFTC) and Marine Corps Air Ground Combat Center (MCAGCC), Twentynine Palms, CA, April 2003.

U.S. Army Research Institute for the Behavioral and Social Sciences, "The Computer Backgrounds of Soldiers," *ARI Newsletter*, Vol. 12, No. 2, January 10, 2005, online at http://www.hqda.army.mil/ari/pdf/winter2002_vol12_no2.pdf (accessed January 14, 2005).

Wallace, William S., et al., "A Joint Context for Training at the Combat Training Centers," *Military Review*, Vol. 84, September–October 2004, pp. 4–11, online at http://www.leavenworth.army.mil/milrev/English/SepOct04/indxso04.htm (accessed January 9, 2005).

Young, R. Michael, et al., "An Architecture for Integrating Plan-Based Behavior Generation with Interactive Game Environments," *Journal of Game Development*, March 2004, online at liquidnarrative.csc.ncsu.edu/pubs/jogd.pdf (accessed October 26, 2005).

Manuals, Emails, and Miscellaneous Sources

2005 Guide to Military Installations Worldwide, Springfield, VA: Military Times Media Group, November 2004.

"A/G Ranges and Units," Washington, DC: Department of the Air Force, briefing, undated.

Air Force Doctrine Center, *Air Warfare, Air Force Doctrine Document 2-1*, Washington, DC, Department of the Air Force, January 22, 2000.

Air Force Doctrine Center, *Counterland, Air Force Doctrine Document 1-1.2*, Washington, DC, Department of the Air Force, August 27, 1999.

AMEC, CGI, EDAW, SAIC, and SHCA, *Northeast Regional Training Center Homeland Defense/Homeland Security*, Feasibility Study Technical Report prepared for the Massachusetts National Guard, March 8, 2004.

Analysis of Alternatives for Providing Joint Urban Warfare Training Capabilities, Alexandria, VA: Center for Naval Analyses, February, 2004.

"Army Co-Sponsorship of the Joint Urban Fires and Effects Joint Test and Evaluation Nomination," U.S. Army Combined Arms Center and Fort Leavenworth, KY, Memorandum for Deputy Chief of Staff, G-8, United States Army, Washington, DC, and Director of Training, Office of the Deputy Chief of Staff, G-3, United States Army, Washington, DC, August 27, 2004.

"BRAC 2005," Washington, DC: Headquarters U.S. Air Force SAF/IEB, briefing, 2003.

Brocato, Becky, *Joint Experimental Range Complex Phasing Plan*, Yuma, AZ: U.S. Army Yuma Proving Ground, February 2, 2004.

Capabilities Group, JWFC, JFCOM J-7, "Joint Tactical Tasks Development," briefing, March 25, 2004.

Capabilities Group, JWFC, JFCOM J-7, "Joint Urban Operations Training Focus Study," briefing, n.d.

Cape Cod Commission (in conjunction with the Community Working Group), *Executive Summary of the Massachusetts Military Reservation Master Plan Final Report*, Barnstable County, MA: Cape Cod Commission, online at http://www.capecodcommission.org/MMR/text.html (accessed December 15, 2004).

Cares, Jeffrey R., "The Use of Agent-Based Models in Military Concept Development," *Proceedings of the 2002 Winter Simulation Conference*, online at http://www.informs-cs.org/wsc02papers/123.pdf (accessed February 4, 2005).

Ceranowicq, Andy, and Mark Torpey, "Adapting to Urban Warfare," *Interservice/Industry Training, Simulation, and Education Conference (I/ITSEC) Proceedings*, Orlando, FL, 2004, online at www.alionscience.com/pdf/Adapting_to_Urban_Warfare.pdf (accessed October 26, 2005).

Chairman of the Joint Chiefs of Staff, *Joint Training Manual for the Armed Forces of the United States*, Chairman of the Joint Chiefs of Staff Manual CJCSM 3500.03A, Washington, DC, September 1, 2002.

Chairman of the Joint Chiefs of Staff, *Joint Training Master Plan 2002 for the Armed Forces of the United States*, Chairman of the Joint Chiefs of Staff Instruction CJCSI 3500.02C, Washington, DC, August 14, 2000, online at http://www.dtic.mil/cjcs_directives/cdata/unlimit/m350004.pdf (accessed April 19, 2004).

Chairman of the Joint Chiefs of Staff, *Joint Training Policy for the Armed Forces of the United States*, Chairman of the Joint Chiefs of Staff Instruction CJCSI 3500.01B, Washington, DC, December 31, 1999.

Chairman of the Joint Chiefs of Staff, *Universal Joint Task List (UJTL)*, Chairman of the Joint Chiefs of Staff Manual CJCSM 3500.04C, Washington, DC, July 1, 2002.

Chisholm, Patrick, "Tutoring for Future Combat," *Military Training Technology*, Online Edition, September 8, 2003, online at www.mt2-kmi.com/archive.cfm?DocID=219 (accessed February 4, 2004).

Combined Arms MOUT Task Force Study Group (CAMTF), "Combined Arms (MOUT) Resource Requirements and Training Strategy Initial Draft," Ft. Benning, GA, December 17, 1999.

Combined Arms MOUT Task Force Study Group (CAMTF), "Urban Operations (UO) Resource Requirements and Combined Arms Training Strategy," Volume V, Final Report, Ft. Benning, GA, September 30, 2001.

Commanding General, Marine Corps Warfighting Laboratory (MCWL), "Letter of Instruction for MCWL Support for the 1st Marine Division Stability and Support Operations 1-04," Quantico, VA, December 24, 2003.

Concept Paper: NTC Urban Operations Training Requirements, National Training Center G3, October 2, 2000.

Concept Report, Large Scale MOUT Feasibility Study, Appendix A, prepared by Ecology and Environment in association with The Onyx Group, M2 Technologies, Fenner Associates, and Hunter Pacific Group, Twenty-nine Palms: MAGTFTC MCAGCC, 2004.

Course schedule for July 6–July 16, 2004, USAF Air-Ground School, Nellis AFB, NV, file name AirGndOpsUrb8_04.xls, undated.

"Cross Service Range Use Standardization Working Group Aviation Range Safety Sub-Group Charter," Office of the Secretary of Defense Personnel & Readiness, Washington, DC, draft, undated.

DA Form 1391, submitted to Congress in FY2003 and FY2005 Department of the Army Budget Estimates for Military Construction, Washington, DC: U.S. Army.

Department of the Army, Military Operations on Urbanized Terrain (MOUT) FM 90-10, Washington, DC, August 15, 1979.

Department of the Army Cost Analysis Manual, Washington, DC: U.S. Army Cost and Economic Analysis Center, May 2002.

Department of the Army Economic Analysis Manual, Washington, DC: U.S. Army Cost and Economic Analysis Center, February 2001.

Department of the Army Headquarters, *The Army Universal Task List*, FM 7-15, Washington, DC, August 2003.

Department of Defense, *Department of Defense Dictionary of Military and Associated Terms*, JP 1-02, April 12, 2001, as amended through May 23, 2003.

Department of Defense Directive Number 1322.18, Subject: Military Training, Office of the Secretary of Defense Personnel & Readiness, September 3, 2004.

Department of Defense United States Joint Urban Operations, "Integrating Concept," Version .95, working draft, June 2, 2004.

Department of the Navy, *OPNAV Instruction 3500.38A/USCG Commandant Instruction M3500.1A Universal Navy Task List (UNTL)*, Washington, DC, May 1, 2001.

DoD Facilities Cost Factors Handbook, Version 2.0, Washington, DC: Department of Defense, April 2000.

DoD Tiger Team, "Department of Defense Plan for Integrating National Guard and Reserve Component Support for Response to Attacks Using Weapons of Mass Destruction, January 1998," online at http://www.defenselink.mil/pubs/wmdresponse/ (accessed February 4, 2005).

Dominant Maneuver (DM) Division, J-8, Joint Chiefs of Staff, *Joint Urban Operations (JUO) Training Facility Study Phase III Final Report*, Washington, DC, 2001.

"Facilities Unit Costs—Military Construction," *PAX* (Programming Administration and Execution System), *Newsletter*, No. 3.2.2, January 9, 2004, online at http://www.hq.usace.army.mil/cempt/e/ec/pax/paxtoc.htm (accessed January 10, 2005).

General Accounting Office (GAO), *Military Capabilities: Focused Attention Needed to Prepare U.S. Forces for Combat in Urban Area*, GAO/NSIAD-00-63N1, Washington, DC, February 2000.

General Accounting Office (GAO), *Military Training: DOD Report on Training Ranges Does Not Fully Address Congressional Reporting Requirements*, GAO-04-608, Washington, DC, June 2004.

"George AFB Becoming Southern California International Airport (Changed to S. Cal. Logistics Airport)," online at http://www.eltoroairport.org/issues/george-102898.html (accessed February 8, 2005).

"George Air Force Base Being Developed as Southern California Logistics Airport," online at http://www.eltoroairport.org/issues/george-afb.html (accessed February 8, 2005).

Glenn, Russell W., Christopher Paul, and Todd C. Helmus, *Men Make the City: Joint Urban Operations Observations and Insights from Afghanistan and Iraq*, Santa Monica, CA: RAND Corporation (forthcoming).

"Global Access Southern California Logisitics Airport—George AFB, Victorville, CA, BRAC 88," online at http://www.afrpa.hq.af.mil/ols/george.htm (accessed February 8, 2005).

Hall, Lt Col Richard D., Future Plans Officer, Operations & Training Directorate, MAGTF Training Command, Marine Air Ground Combat Center, Twentynine Palms, CA, "MOUT Initiatives," Santa Monica, CA: RAND Corporation, briefing, August 2, 2004.

J8 DM Joint Urban Training Facility Study, Phase III Final Report, prepared by Booz, Allen, Hamilton, Washington, DC: Department of the Air Force, n.d.

J9, Joint Experimental Analysis Division, Joint Urban Warrior 2004, Joint Urban Operations, Final Report, Washington, DC: Department of the Air Force, June 2004.

Johnson, Mark, Rich Guilli, and Scott Oberg, *Integration of JCATS and CCTT in an LVC Exercise*, online at www.jtepforguard.com/pubs/04E-SIW-066.pdf (accessed February 4, 2005).

Joint Chiefs of Staff, *Doctrine for Joint Urban Operations*, Joint Publication 3-06, Washington, D.C, September 16, 2002.

Joint Chiefs of Staff, *Handbook for Joint Urban Operations*, Washington, DC, May 17, 2000.

Joint Defense Capabilities Study Team, *Joint Defense Capabilities Study: Improving DOD Strategic Planning, Resourcing, and Execution to Satisfy Joint Capabilities, Final Report*, January 2004.

"Joint Training Requirements Analysis Team (JTRAT) Working Group Outbrief," U.S. Joint Forces Command Joint Warfighting Center, Washington, DC, briefing, April 14, 2004.

"JTT Construct," U.S. Joint Forces Command, Washington, DC, briefing, May 10, 2004.

Lockwood, David E., *Military Base Closures: Implementing the 2005 Round*, Congressional Research Service, Report for Congress, January 4, 2005.

Lopez, Daniel H. "Playas, New Mexico . . . Imagine the Possibilities," New Mexico Institute of Mining and Technology, Socorro, NM, briefing, November 2, 2004.

"Marine Corps Task List 'Road Ahead,'" United States Marine Corps Development Command, briefing, December 3, 2004.

Matsumura, John, et al., *Exploring Advanced Technologies for the Future Combat Systems Program*, Santa Monica, CA: RAND Corporation, MR-1332-A, 2002, online at http://www.rand.org/publications/MR/MR1332/ (accessed February 4, 2004).

"McKenna Urban Experimental Complex," compact disk provided by Soldier Battle Lab, U.S. Army Infantry Center, ATTN: ATZB-WC, Ft. Benning, GA, undated.

McLaughlin, CTR David M., handouts from RAND JUO training facilities brief 6/30/04, email message to Maj Clinton J. Chlebowski, July 23, 2004.

Memorandum, Command Training Guidance (CTG) for Training Years 2004–2006, Ft. McPherson, GA: Headquarters, United States Army Reserve Command, December 12, 2003.

Memorandum, Command Training Guidance (CTG) for Training Years 2005–2007, Ft. McPherson, GA: Headquarters, United States Army Reserve Command, October 2, 2004.

Memorandum, Command Training Guidance (CTG) for Training Years 2005–2007. Ft. Jackson, SC: United States Army Reserve Readiness Command, November 4, 2004.

Memorandum, Internal Control Plan for Management of the Department of the Navy 2005 Base Realignment and Closure (BRAC) Process Policy Advisory Two, Washington, DC: Department of the Navy, June 27, 2003.

Memorandum, Operation IRAQI FREEDOM (OIF) Major Combat Operations (MCO) Lessons Learned (LL) for Joint Urban Operations (JUO) DOTMLPF Change Recommendation Package (DCR), Washington, DC: The Joint Staff, January 31, 2005.

Memorandum, Transformation Through Base Realignment and Closure, Washington, DC: Secretary of the Army, December 12, 2002.

Memorandum, Transformation Through Base Realignment and Closure, Washington, DC: Secretary of Defense, November 15, 2002.

Memorandum for Director, Combined Arms and Tactics Directorate, U.S. Army Infantry School, Trip Report—Fast Train VI – Wargaming Division, Marine Corps Warfighting Laboratory (MCWL), Quantico, VA, 10 Sep 2004, September 13, 2004.

Memorandum for Infrastructure Executive Council Members, Infrastructure Steering Groups Members, Joint Cross-Service Group Chairman, 2005 Base Closure and Realignment Selection Criteria, Washington, DC: The Under Secretary of Defence Acquisition, Technology and Logistics, January 4, 2005.

Memorandum for Record, USJFCOM Joint Training Requirements Analysis Team (JTRAT) Working Group 13–14 April 2004, April 15, 2004.

Memorandum of Agreement on Overarching Range Cooperative Guidelines Between the Department of the Air Force and the Department of Navy, signed by General (USAF) John P. Jumper (June 16, 2003), Admiral V. E. Clark (June 22, 2003), and General (USMC) M. W. Hagee (June 3, 2003).

"Military Capabilities: Focused Attention Needed to Prepare U.S. Forces for Combat in Urban Areas," Washington, DC: U.S. General Accounting Office, February 2000.

"Military Operations on Urbanized Terrain Facility," online at http://www. jrtc-polk.army.mil/JRTCExercise/MOUT.HTM, posted July 16, 2001.

"Military Training: DOD Report on Training Ranges Does Not Fully Address Congressional Reporting Requirements," Washington, DC: U.S. General Accounting Office, June 2004.

Miller, Ann, Robert Book, and Peter Kusek, *Analysis of Alternatives for Providing Joint Urban Warfare Training Capabilities*, Alexandria, VA: Center for Naval Analyses, Research Memorandum CRM D0009201.A2/ FINAL, February 2004.

Miller, Ann, Matt Grund, and Deborah Jonas, *Joint National Training Capability: Functional Area Training Resource Requirements*, Alexandria, VA:

Center for Naval Analyses, Research Memorandum CRM D0008243.A2/FINAL, August 2003.

Miller, Dale D., Annette C. Janett, and Melissa E. Nakanishi, *An Environmental Data Model for the OneSAF Objective System*, online at www.wood.army.mil/TPIO-TD/02F-SIW-082-FINAL.doc (accessed February 4, 2005).

"MOUT Survey Sites," DMSO MOUT Training Sites.ppt, source unknown, undated.

Muscatatuck Survey Response, provided by LTC Ken McCallister, INARNG, Deputy J5/7, October 2004.

"Muscatatuck Urban Training Center," Indiana National Guard, briefing, October 13, 2004.

Northeast Regional Training Center Homeland Defense/Homeland Security, *Feasibility Study Technical Report*, prepared for Massachusetts National Guard, Milford, MA, March 8, 2004.

Office of the Secretary of Defense, "FY06-11 OSD Program Review Service LVC Training Systems Modernization," Washington, DC, briefing, September 15–16, 2004.

Office of the Under Secretary of Defense (Personnel and Readiness), *Report to Congress, Implementation of the Department of Defense Training Range Comprehensive Plan*, Washington, DC, February 2004.

"Options for Changing the Army's Overseas Basing," Washington, DC: Congressional Budget Office, May 2004.

Programming Cost Estimates for Military Construction, Training Manual 5-800-4, Washington, DC: Headquarters, Department of the Army, May 1994.

"Project Kingpin: Overarching Instrumentation Policy and Standards Development to Support Joint Force Training and Joint Force Testing: Scope/Framework/Process Definition," SRI project draft sponsored by Office of the Secretary of Defense Personnel & Readiness, January 21, 2005.

Raymond, Barbara, "Skills and Perishability," Santa Monica, CA: RAND Corporation, email discussion with Matthew W. Lewis, January 11, 2005.

"Report to Congress: Implementation of the Department of Defense Training Range Comprehensive Plan—Ensuring Training Ranges Support Training Requirements," Office of the Secretary of Defense Personnel & Readiness, Washington, DC, February 2004.

Robinson, LTC Bill, *Modeling and Simulation Support to Homeland Security and Defense,* U.S. Joint Forces Command Joint Warfighting Center, briefing, October 2003, online at http://www.mel.nist.gov/div826/msid/sima/simconf/proc/ftp/robinson.pdf (accessed February 4, 2005).

"ROE for 30 JTT Review," U.S. Joint Forces Command, briefing, May 10, 2004.

Schank, John, Harry Thie, Clifford Graf, Joseph Beel, and Jerry M. Sollinger, *Finding the Right Balance: Simulator and Live Training for Navy Units,* Santa Monica, CA: RAND Corporation, MR-1441-NAVY, 2002.

Short, Paul B., compiled comments from Johnny W. Brooks and LtCol William P. McLaughlin on RAND 6/30/04 handouts, attachment to email to Christopher Paul, July 26, 2004.

"Southern California Logistics Airport (SCLA), George Air Force Base (GAFB)," online at GlobalSecurity.org, http://www.globalsecurity.org/military/facility/george.htm (accessed February 8, 2005).

"Title 10. Armed Forces, Subtitle A. General Military Law, Part III. Training and Education," United States Code Service, Washington, DC, approved December 23, 2004.

Townsend, Major John A., "Urban Operation Training Strategy Review," Briefing Combat Arms Center-Training/Collective Training Directorate (CAC-T/CTD), Ft. Leavenworth, KA, briefing, summer 2004.

"Training for Military Operations on Urbanized Terrain," Army Training Circular 90-1, Washington, DC: Department of the Army, April 1, 2002.

Training Ranges, Training Circular 25-8. Washington, DC: Headquarters, Department of the Army, April 5, 2004.

"UK Urban Operations Advisor's Course, 1/04, 27 September–1 October 2004," Land Warfare Centre (British Army), 2004.

United States Air Force, *Air Force Task List (AFTL)*, Washington, DC, Department of the Air Force, Air Force Doctrine Document 1-1, August 12, 1998.

United States Air Force, *Leadership and Force Development*, Air Force Doctrine Document 1-1, February 18, 2004, online at https://www.doctrine.af.mil/Library/Doctrine/afdd1-1.pdf (accessed April 19, 2004).

United States Department of Defense, "Department of Defense Directive," Number 1322.18, Washington, DC, September 3, 2004.

United States Department of Defense, *Implementation of the Department of Defense Training Range Comprehensive Plan: Ensuring Training Ranges Support Training Requirements*, Report to the Congress, February 2004.

United States Marine Corps, *Concept Report: Large Scale MOUT Feasibility Study*, Marine Air Ground Task Force Training Command (MAGTFTC) and Marine Corps Air Ground Combat Center (MCAGCC), Twentynine Palms, CA, April 2003.

"Urban Patrolling 1," student handout, United States Marine Corps Basic Officer Course, The Basic School, Marine Corps Development Command, Quantico, VA, undated.

"US Air Force Cost and Planning Factors," Air Force Instruction 65-503, Washington, DC: Secretary of the Air Force, February 4, 1994.

U.S. Army Training and Doctrine Command, *Military Operations on Urbanized Terrain Advanced Concept Technology Demonstration (MOUT ACTD) Site Analysis Study*, Fort Monroe, VA, October 1995.

"USJFCOM Joint Training Requirements Analysis Team Working Group," U.S. Joint Forces Command, briefing, April 13–14, 2004.

U.S. Joint Forces Command, "Joint Lessons Learned: Operation IRAQI FREEDOM Major Combat Operations," coordinating draft, March 1, 2004.

U.S. Joint Forces Command, "USJFCOM Joint Training Requirements Analysis Team (JTRAT) Working Group 13–14 April 2004," briefing, May 10, 2004.

U.S. Joint Forces Command, "Western Range Complex Joint National Training Capability Horizontal Training Exercise," briefing, January 2004.

U.S. Special Forces Command, *Tiger Team Report: Global Special Operations Forces Range Study,* McDill Air Force Base, FL, January 27, 2003.

Vick, Alan, et al., *Aerospace Operations in Urban Environments: Exploring New Concepts,* Santa Monica, CA: RAND Corporation, 2000.

Waller Todd & Sadler Architects, *Urban Target Complex,* Air Combat Command, November 5, 2002.

Weith, Gordon, "Joint Urban Operations Training Facilities Study (RAND Study)," email message, sent July 26, 2004.

"Western Range Complex Joint National Training Capability Horizontal Training Exercise, January 2004," U.S. Joint Forces Command, briefing, January 2004.

Wood, Robert, "Joint Urban Operations Integrating Concept," Joint Urban Operations Office, briefing, June 4, 2004.

Wood, Scott, et al., "An Intelligent-Agent Framework for Supervisory Command and Control," *2004 Command and Control Research and Technology Symposium Proceedings,* San Diego, CA.

Interviews

Anderson, Spencer, with Russell W. Glenn, Nellis AFB, NV, August 11, 2004.

Axelberg, Mark, Lt Col, U.S. Army; Maj Everett Baber, U.S. Army; Sgt Maj Henry Legge, U.S. Army; Marty Martinson, and Capt Sven Myrberg, U.S. Army, with Brian Nichiporuk, Christopher Paul, and Barbara Raymond, Shughart-Gordon Urban Training Complex, Ft. Polk, LA, October 26, 2004.

Bagordia, Rajiv, Scalable Network Technologies, at Culver City, CA, with Randall Steeb, April 21, 2004.

Bailey, Robert, Systems Consultants Services Ltd., Henley-on-Thames, UK, with Russell W. Glenn, December 21, 2004.

Briant, Mike, Captain, British Army, SO3 Artillery Urban Operations Wing, Land Warfare Centre, Copehill Down Village, UK, with Russell W. Glenn, December 16, 2004.

Cawthorn, Richard, Systems Consultants Services Ltd., Henley-on-Thames, UK, with Russell W. Glenn, December 21, 2004.

CJSOTF, Afghanistan Commander, Deputy Commander, and Command Sergeant Major, Bagram, Afghanistan, with Russell W. Glenn and Todd C. Helmus, February 15, 2004.

Clifton, Gary, Director of Training, Blackwater Training Center, Moyock, NC, with Russell W. Glenn, September 9, 2004.

Condon, Mike, Capt, U.S. Army, Training Support Division, National Training Center, Ft. Irwin, CA, with Russell W. Glenn, August 10, 2004.

Courage, Miles, SCS (Systems Consultants Services Ltd.) Mothership Ltd., Henley-on-Thames, UK, with Russell W. Glenn, December 21, 2004.

Cousens, Richard, Centre for Defence and International Security Studies, Henley-on-Thames, UK, with Russell W. Glenn, December 21, 2004.

Davis, Eddie, Vice-Deputy Director, U.S. Army Soldier Battle Laboratory, Ft. Benning, GA, with Russell W. Glenn, November 1, 2004.

Dutton, Paul, Major, British Army, Operational Training and Advisory Group (OPTAG), Folkestone, UK, with Russell W. Glenn, December 17, 2004.

Ferguson, Dean, Captain, British Army, Staff Officer 3 (SO3) Coord Urban Operations Wing, Land Warfare Centre, Copehill Down Village, UK, with Russell W. Glenn, December 16, 2004.

Finch, Terry, Bill Ash, Andy Chatelin (Range Management), Maj Bill Russel and Staff Sergeants Baker and McCarty, U.S. Marine Corps, Camp Pendleton, CA, with Dr. Christopher Paul and Barbara Raymond, May 27, 2004.

Gumz, Bob, G3 Training Support Division, National Training Center, Ft. Irwin, CA, with Russell W. Glenn, August 10, 2004.

Hain, David Bruce, Col, U.S. Army, Task Force Mojave, National Training Center, Ft. Irwin, CA, with Russell W. Glenn, August 10, 2004.

Harward, Robert S., Capt, U.S. Navy, reviewer comments to Russell W. Glenn during telephone conversation, January 31, 2005.

Howsden, William A., MOUT Complex Manager, Bagram, Afghanistan, with Russell W. Glenn, February 16, 2004.

Hutchinson, T.M.O., Major, British Army, Officer Commanding (OC) Urban Operations Wing, Land Warfare Centre, Copehill Down Village, UK, with Russell W. Glenn, December 16, 2004.

Indiana National Guard, "Muscatatuck Urban Training Center," IN, briefing, July 29, 2004.

Jobe, Robert, Maj, U.S. Air Force, 6 Combat Training Squadron Assistant Director of Operations, Nellis AFB, NV, with Russell W. Glenn, August 12, 2004.

"JOUST Input to JUO Masterplan," briefing, updated August 2003 (via P. Craig).

Le Moyne, John, Commandant, U.S. Army Infantry School, "U.S. Army Urban Operations Training Strategy," briefing, February 2001.

Lyde, John, Systems Consultants Services Ltd., Henley-on-Thames, UK, with Russell W. Glenn, December 21, 2004.

Marine Corps Security Force Training Company representatives, Northwest Naval Support Activity, Chesapeake, VA, with Russell W. Glenn, September 9, 2004.

Miller, Ron, Hurlburt Field, FL, with Russell W. Glenn, October 29, 2004.

Moore, Scott E., Hurlburt Field, FL, with Russell W. Glenn, October 29, 2004.

Naval Special Warfare Group Two representatives, Norfolk, VA, with Russell W. Glenn, September 10, 2004.

Oerlemans, H. J. R., Lieutenant Colonel, Dutch Army, Oostdorp and Marnehuizen, The Netherlands, with Russell W. Glenn, December 7–8, 2004.

Parker, Don, G3 Training Support Division, National Training Center, Ft. Irwin, CA, with Russell W. Glenn, August 10, 2004.

Pearce, Lee, Maj, U.S. Army, U.S. Army Infantry School, Ft. Benning, GA, with Russell W. Glenn, November 1, 2004.

Pickard, Sandy L., DSP Plans, Ft. Irwin, CA, with Russell W. Glenn, August 10, 2004.

Reischl, Tim, Deputy G3, National Training Center, Ft. Irwin, CA, with Russell W. Glenn, August 10, 2004.

Ringgold, Lloyd, Lt Col, U.S. Air Force, 98th OSS Director of Operations, Nellis AFB, NV, with Russell W. Glenn, August 11, 2004.

Roland-Price, Alan, Systems Consultants Services Ltd., Henley-on-Thames, UK, with Russell W. Glenn, December 21, 2004.

Schwalm, H. E., Tampa, FL, with Russell W. Glenn, October 28, 2004.

Shaw, Timothy S., Head, Synthetic Environment Applications Laboratory (SEA Lab), Penn State University, with Russell W. Glenn, November 15, 2004.

Sierawski, Jim, Vice President for Training, Blackwater Training Center, Moyock, NC, with Russell W. Glenn, September 9, 2004.

Stone, Dr. Douglas S., G2, National Training Center, Ft. Irwin, CA, with Russell W. Glenn, August 10, 2004.

Taylor, B. J., Hurlburt Field, FL, with Russell W. Glenn, October 29, 2004.

Twitty, Stephen (LTC, U.S. Army), Commander, 3-15 Infantry Battalion, 3rd Infantry Division Baghdad, Iraq, with Arthur Durante, undated.

Van Houten, Johan, Lieutenant Colonel, Dutch Army, Oostdorp and Marnehuizen, The Netherlands, with Russell W. Glenn, December 7–8, 2004.

Watt, Gregory, LTC, U.S. Army, G3, National Training Center, Ft. Irwin, CA, with Russell W. Glenn, August 10, 2004.

Watt, J. N., Lieutenant Colonel, British Army, Staff Officer 1 (SO1) Training/Chief Instructor, Operational Training and Advisory Group (OPTAG), Folkestone, UK, with Russell W. Glenn, December 17, 2004.

Wilson, Donald, Systems Consultants Services Ltd., Henley-on-Thames, UK, with Russell W. Glenn, December 21, 2004.

Woolfolk, Ned W., Director, Combined Arms and Tactics Directorate, U.S. Army Infantry School, Ft. Benning, GA, with Russell W. Glenn, November 1, 2004.